UNIV 10 0258243 5
WITHDRAWN
D0420900
FROM THE LIBRARY

Basic Cell Culture

Second Edition

The Practical Approach Series

Related **Practical Approach** Series Titles

Human cytogenetics malignancy and acquired abnormalities

Human cytogenetics: constitutional analysis

Macrophages

Cytokine cell biology

Animal cell culture 3/e

Flow cytometry 3/e

Lymphocytes 2/e

Apoptosis

Cell growth, differentiation and senescence

Virus culture

Light microscopy

Growth factors and receptors

Cell separation

Subcellular fractionation

Platelets

Epithelial cell culture

Neural cell culture

Please see the **Practical Approach** series website at

http://www.oup.com/pas

for full contents lists of all Practical Approach titles.

No. 254

Basic Cell Culture

Second Edition

A Practical Approach

Edited by

J. M. Davis

Research & Development Department,
Bio Products Laboratory, Dagger Lane, Elstree,
Hertfordshire WD6 3BX, U.K.

OXFORD
UNIVERSITY PRESS

Great Clarendon Street, Oxford OX2 6DP

Oxford University Press is a department of the University of Oxford.
It furthers the University's objective of excellence in research, scholarship, and education by publishing worldwide in

Oxford New York

Auckland Bangkok Buenos Aires Cape Town Chennai Dar es Salaam Delhi Hong Kong Istanbul Karachi Kolkata Kuala Lumpur Madrid Melbourne Mexico City Mumbai Nairobi São Paulo Shanghai Singapore Taipei Tokyo Toronto

with associated companies in Berlin

Oxford is a registered trade mark of Oxford University Press
in the UK and in certain other countries

Published in the United States
by Oxford University Press Inc., New York

© Oxford University Press, 2002

The moral rights of the author have been asserted

Database right Oxford University Press (maker)

First edition published 1994
Second edition published 2002

All rights reserved. No part of this publication may be reproduced, stored in a retrieval system, or transmitted, in any form or by any means, without the prior permission in writing of Oxford University Press, or as expressly permitted by law, or under terms agreed with the appropriate reprographics rights organization. Enquiries concerning reproduction outside the scope of the above should be sent to the Rights Department, Oxford University Press, at the address above

You must not circulate this book in any other binding or cover and you must impose this same condition on any acquirer

1002582435

British Library Cataloguing in Publication Data
Data available

Library of Congress Cataloging in Publication Data
Basic cell culture : a practical approach / J.M. Davis-2nd ed.
(Practical approach series ; 254)
Includes bibliographical references.
1. Cell culture-Laboratory manuals I. Davis, J. M. (John M.) II. Series.

QH585.2.B37 2001 571.6′38-dc21 2001036910
1 3 5 7 9 10 8 6 4 2

ISBN 0 19 963854 3 (Hbk)
ISBN 0 19 963853 5 (Pbk)

Typeset in Swift by Footnote Graphics, Warminster, Wilts
Printed in Great Britain on acid-free paper
by The Bath Press, Avon

Preface to Second Edition

In editing the second edition of *Basic Cell Culture* my concern has been to retain the successful concepts from the first edition, whilst bringing the content up to date to reflect the considerable advances made in certain areas, and making the coverage of essential topics more comprehensive. This has entailed not only the complete revision of all those chapters which appeared in the first edition, but also the addition of a chapter on microscopy and an authored appendix devoted to Internet resources (an area which has burgeoned beyond all measure since publication of the first edition).

None of this would have been possible without the willingness of all the authors to give unstintingly of their time and wealth of knowledge of their subjects. My unreserved thanks go to them for the exceptional standard of all their contributions.

Elstree J.D.
November 2001

Preface to First Edition

In the last 20 years cell culture has developed from being a tool used only in certain specialized areas of research, to being the cornerstone of probably the world's fastest-growing industry, biotechnology. New applications are appearing all the time, not only in academic research but in fields such as *in vitro* fertilization, artificial organs, and the replacement of animals in medical and toxicological experiments. Thus newcomers to cell culture can easily find themselves performing work which would have been impossible or undreamt of only a few years ago.

The primary aim of this book is to guide the newcomer progressively through all those areas which nowadays are basic to the performance of animal (including human) cell culture. As far as is possible, the chapters have been arranged in a logical order, progressing from the most basic requirements—a place to work and equipment to use—through sterilization, media, and a solid core of essentail techniques, to slightly more advanced aspects—deriving one's own cell lines, growing specialized cells in defined conditions, cloning, and maintaining and controlling the quality of cell stocks. The final chapter is on good laboratory practice (GLP), and is one which I would encourage every cell culture worker to read. This chapter has been included for a number of reasons: in certain types of cell culture laboratory, a knowledge of GLP is as basic and essential as the ability to passage a cell line; alternatively, the reader, having progressed thus far, may find himself in a position where he can envisage wishing to commercialize either his expertise or cell lines, and a knowledge of GLP at an early stage can save much heart-ache (not to mention money and effort) later; ans all the cell culture workers can benefit from considering and adopting at least some of the principles under-lying GLP, although most will be heartily thankful if it does not need implementing in full in *their* laboratory!

Each reader will vary in what he expects of a 'basic' book, and what is a basic requirement for one worker may well seem advanced to another. I have tried to cover 'the basics' relevant to industrial laboratories as well as those relevant to academic institutions, as judged from 20 years' experience across the two spheres. Thus there will unquestionably be more information in this text than any one individual worker will require when beginning to work with cells in culture, so my advice is—hold on to your copy; much of that extra information will become useful as your work progresses. It follows that this text will not only be of use to newcomers to cell culture but also to more experienced workers, for example when expanding their work into new areas, setting up a new laboratory, or simply realizing, as we all do from time to time, that they have forgotten things they once knew!

The other aim of this book is to underpin the somewhat more advanced texts on cell culturewhich are already available in the *Practical Approach* series, in particular *Animal Cell Culture* edited by R. I. Freshney, *Mammalian Cell Biotechnology* edited by M. Butler, and *Cell Growth and Division* edited by R. Baserga. I have attempted to minimize the overlap between the present text and these others, although in order that this book forms a coherent unit a limited amount of overlap has been unavoidable. It is my hope and belief that the newcomer will be better able to benefit from these other texts once he has read this one.

I would finally like to take this opportunity to thank all the authors who have contributed to this volume. Their willingness to share their considerable expertise with a wider audience made the compilation of this text possible, and their unfailing cooperation made editing it a far more pleasurable experience than it might otherwise have been.

Elstree J.D.
March 1994

Contents

Protocol list

Table 1 The early years of cell and tissue culture

Late 19th century	Methods established for the cryopreservation of semen for the selective breeding of livestock for the farming industry
1907	Ross Harrison (1) published experiments showing frog embryo nerve fibre growth *in vitro*
1912	Alexis Carrel (2) cultured connective tissue cells for extended periods and showed heart muscle tissue contractility over two to three months
1948	Katherine Sanford *et al.* (3) were the first to clone — from L cells
1952	George Gey *et al.* (4) established HeLa from a cervical carcinoma — the first human cell line
1954	Abercrombie and Heaysman (5) observed contact inhibition between fibroblasts — the beginnings of quantitative cell culture experimentation
1955	Harry Eagle (6) and later, others, developed defined media and described attachment factors and feeder layers
1961	Hayflick and Moorhead (7) described the finite lifespan of normal human diploid cells
1962	Buonassisi *et al.* (8) published methods for maintaining differentiated cells (of tumour origin)
1968	David Yaffe (9) studied the differentiation of normal myoblasts *in vitro*

Over the next decade, tissue culture became, as well as an experimental research field in its own right, a technology used by many biologists and biotechnologists in the newly established field of molecular biology. Hybridoma technology allowed the mass production of monoclonal antibodies from cell culture medium and recombinant DNA technology made use of cultured cells, either as a source of messenger RNA or gene sequence or as an expression vector for recombinant DNA. In many areas of biomedical research, tissue culture methods were developed which replaced much of the former need for animal experimentation. For instance, *in vitro* pharmacotoxicology became an established discipline for the investigation of drug activities and the design of new forms of therapy.

A recent development in the cell culture field which points the way for future progress, has been the introduction of methods for histotypic culture, using filter well inserts in culture vessels (see Chapter 5, Section 5.2.2). This allows for the co-culture of two or more purified and defined populations of cells, separated but in the same medium so that interactions mediated by soluble factors can occur. For example, the expression of differentiated characteristics not demonstrated by a particular cell type in isolation may be induced in histotypic culture. It is hoped that such technology will further our understanding of the complex cellular interactions that occur within tissues *in vivo,* not only during development, but also those which regulate tissue homeostasis and normal functioning throughout life. Many disease processes are thought to involve a perturbation of cellular interactions (cancer is an obvious example) and the new methods now becoming available for the study of cellular interactions *in vitro* will undoubtedly lead to progress in research into those cellular defects which cause or contribute to disease processes.

In the biomedical field, there are hopes that tissue culture technology will contribute more and more to clinical treatments involving transplants of various kinds. Cultured skin cells are already in use as grafts or even as superior wound dressings in burns and skin ulcer patients; cultured urothelium has been used to correct congenital defects in the urogenital system; transplanted cells carrying genetically engineered DNA sequences are already being trialled or may be used in the future to correct defects in diabetic patients, patients with cystic fibrosis, haemophilia, perhaps some of the rare enzyme deficiency syndromes, and for tumour therapy (see Chapter 5, Section 6.3). There are likely to be a number of situations where it is preferable to correct a defective function via implanted cells rather than by conventional drug therapy. Degenerative disorders such as the myelodysplastic syndromes and Parkinson's disease may respond to implants of haemopoietic or neural stem cells, respectively, to replace lost cell populations. Even so, for most situations, therapy will still involve administering drugs conventionally. In the case of complex protein or hormone therapy, these drugs will increasingly be produced from large-scale cell cultures, of either the appropriate cell type or of cells expressing recombinant genes, in order to avoid some of the drawbacks of microbial expression vectors. Biotechnology is still in its infancy and will undoubtedly develop rapidly in the years leading up to the centenary, in 2007, of tissue culture's first appearance as a new methodology.

2 The laboratory

2.1 Design concepts and layout

Tissue culture laboratories differ from general purpose biomedical research laboratories in that their most important function is to allow the sterile handling of cultured cells. It is practically impossible to work in a totally germ-free environment, so that provision must be made to ensure that cultures and culture media are maintained free from contaminating micro-organisms. These will flourish in the cell culture environment and will rapidly overgrow, and in many cases kill the animal cells by release of toxins, depletion of the nutrient medium, or depressing the medium pH.

In almost all modern laboratories, sterile handling of cell cultures is carried out in laminar flow cabinets or 'hoods', where only the operator's arms enter the sterile work area. All other areas in the laboratory must be kept scrupulously clean, with surfaces clear from unnecessary clutter and swabbed down with an antiseptic cleansing solution at regular intervals. In addition to the sterile handling facility, all tissue culture laboratories must incorporate areas for a variety of other activities; these may be in one room or arranged as a suite of rooms according to the scale of the operation, available space and, of course, budget.

The sterile handling area itself should be positioned so that there is minimal movement of people past or through the clean area. There should be adequate separation of the clean area for sterile handling of cells from the area for 'dirty' operations, including disposal of used culture vessels and a wash-up facility for reusable glassware and equipment. In between, at increasing distance from the sterile handling area to produce a 'sterility gradient' within the laboratory, should be the other facilities essential to cell culture work. These are, in order:

(a) Laminar flow cabinets.

(b) Incubators (CO_2 and, if required, dry, ungassed).

(c) Storage space for sterile equipment and solutions.

(d) Instrument and equipment benches.

(e) Media fridges (sterile working solutions only).

(f) Freezers for culture reagents requiring storage at −20 °C and, if necessary, −70 °C.

(g) Preparation area for media and other solutions.

(h) General cold storage facility for chemicals and non-sterile reagents.

Where possible, in a separate room or rooms:

(i) Liquid nitrogen freezers for frozen cell stocks.

(j) General preparation bench.

(k) Storage area for unopened plasticware.

(l) Sterilization oven and autoclave.

(m) Drying oven.

(n) Water purification system.

(o) Wash-up area with sinks, soak tanks, pipette washer (if needed), and automatic washing machine.

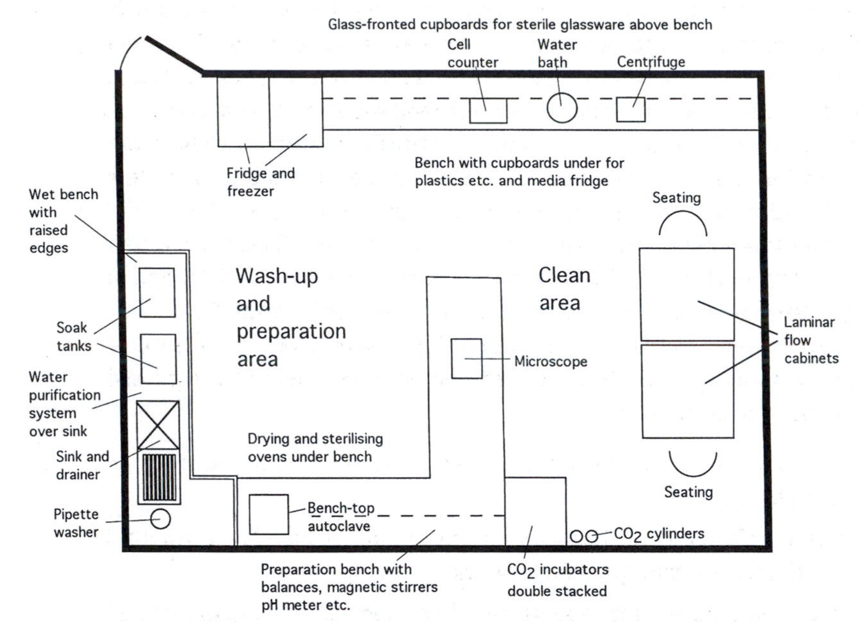

Figure 1 Self-contained tissue culture laboratory, suitable for two to three people.

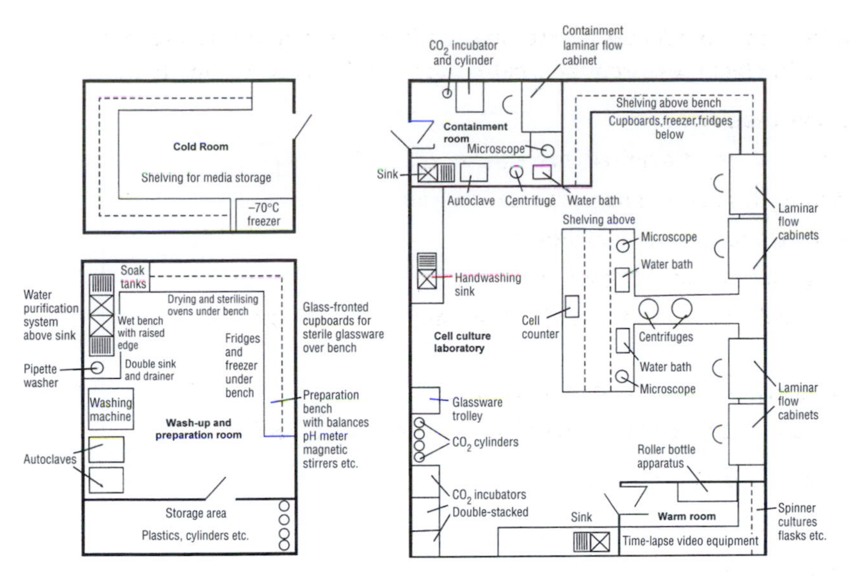

Figure 2 Suite of laboratories comprising a large tissue culture facility, with separate wash-up, preparation, and storage areas, suitable for up to ten people.

Plans of two possible arrangements are given in *Figures 1* and *2*. The first is a single, self-contained tissue culture laboratory whilst the second is for a bigger group, with a clean culture laboratory and a separate, adjacent, wash-up and sterilization facility.

When planning the layout of the tissue culture laboratory, care should be taken to avoid placing the laminar flow cabinets where there is likely to be a disturbance to airflow around the cabinet. Some examples of recommended cabinet location in relation to door openings and structural features such as walls and columns (pillars) are shown in *Figure 3*. Recommendations and restrictions on cabinet locations in relation to the likely routes of traffic through the laboratory are indicated in *Figure 4*. Movement of personnel through the sterile handling area of the laboratory should be minimal and restricted as shown by careful positioning of laminar flow cabinets relative to other structural features. *Figures 3* and *4* and their key are adapted, with permission, from British Standard document BS 5726, and relate particularly to microbiological safety cabinets (MSCs).

In addition to the essential facilities outlined above, several desirable features of a cell culture laboratory suite should be considered, according to the type and scale of the work to be carried out:

(a) A warm room at 37 °C for large-scale cultures in, e.g. roller bottles, time-lapse video equipment, etc.

(b) A laminar flow cabinet away from the sterile handling area for the initial dissection of tissues for primary culture work.

(c) An electronic cell counter for laboratories handling large numbers of cell lines or for those involved in growth kinetics research.

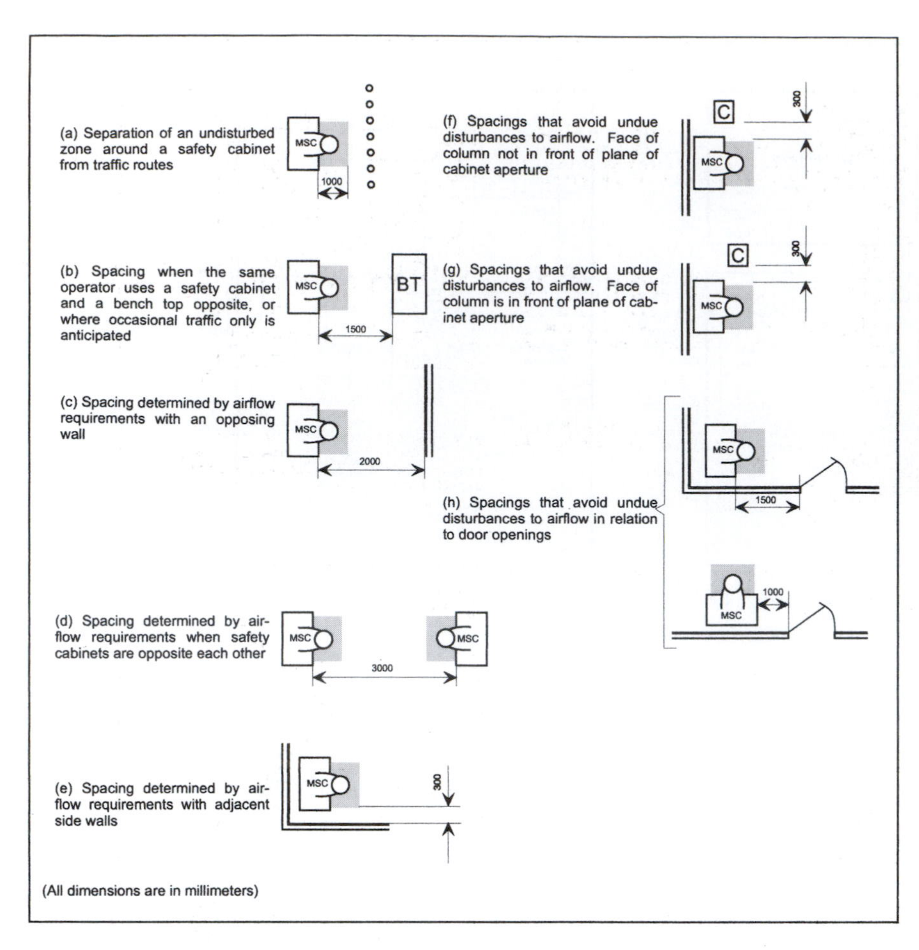

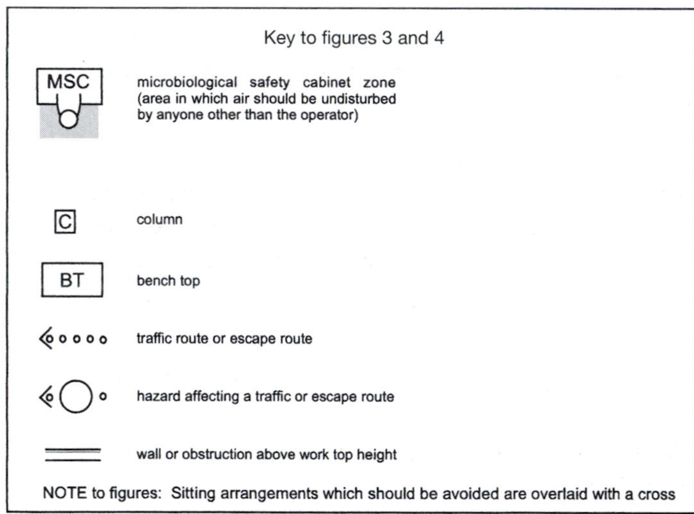

Figure 3 Recommendation for minimum distances for avoiding disturbance to the microbiological safety cabinet and its operator.

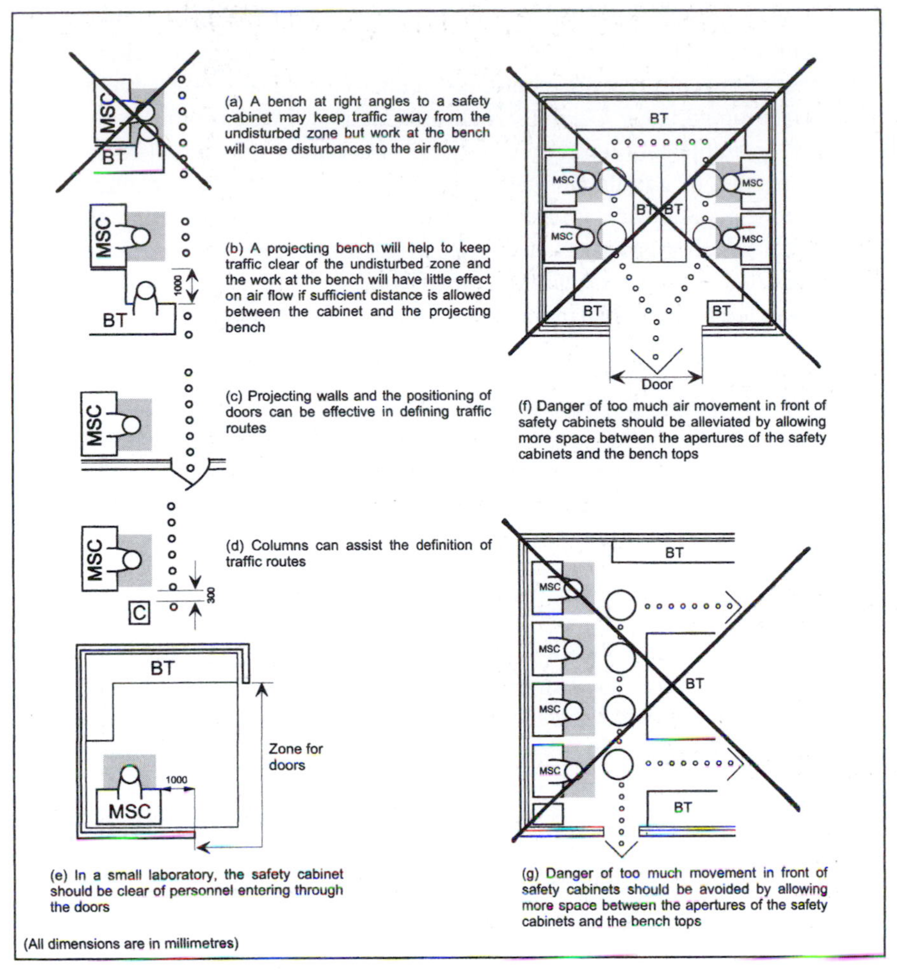

Figure 4 Avoiding disturbances to the microbiological safety cabinet and operator due to other personnel. See key in Figure 3.

(d) Air-conditioning. This will be necessary for laboratories in warm climates and where many heat-generating items of equipment, e.g. laminar flow cabinets, are concentrated into a small space. Incubators may overheat if the ambient temperature is too close to their operating temperature.

(e) A cold room at 4 °C for media storage.

(f) Some form of containment facility for incoming cell lines and potentially hazardous material.

(g) A well-ventilated area for liquid nitrogen freezers with easy access to liquid nitrogen storage tank or delivery bay.

A containment facility is advisable for laboratories which receive biological samples, including cell lines, from elsewhere. Any potentially pathogenic material will also need to be handled here according to regulations for the hazard they pose. A containment facility should comprise a separate room, without direct access to the rest of the culture suite, containing a microbiological safety cabinet of the appropriate category, e.g. Class II, a dedicated incubator which can accommodate sealable, gassable containers to isolate cells in quarantine, and all back-up equipment needed to handle the cells (centrifuge, small refrigerator, etc.). Cell lines new to the laboratory, unless obtained directly from a reputable cell culture collection which guarantees them free from contaminants, should be isolated and handled as potential carriers of mycoplasma, for instance, until shown to be clear. For further treatment of this whole topic, see Chapter 9.

2.2 Services

Tissue culture laboratories have some of their service requirements in common with more general laboratories, for example water supply to sinks, gas taps for Bunsen burners, and above and below bench power sockets. Other services are less widely used outside the culture laboratory.

2.2.1 CO_2

Most incubators for cell culture work depend on a supply of CO_2 to maintain a fixed CO_2 tension in the humidified air within the incubator (see below). Ideally, CO_2 cylinders should be outside the main cell culture laboratory and the gas piped through, to maintain cleanliness in the sterile handling area. Since this may not always be possible, *Figures 1* and *2* show cylinders alongside the incubators. A bank of CO_2 cylinders should be secured to a rack placed near the incubators and the gas fed via a reduction valve on the cylinder head, through pressure tubing to the incubator intake port.

Control of the CO_2 level in gassed incubators is achieved through the inclusion of CO_2 monitors, which increase the initial costs of an incubator but repay this with lower CO_2 usage and better control and reliability. As long as the incubator remains closed, usage is minimal and, in the event of a cylinder running out unexpectedly, the gas phase remains within tolerable limits for quite a long time. It is a useful precaution to have more than one cylinder on line, with an automatic switch-over unit, operated by the fall in pressure as a cylinder empties.

The effect of a leak in the CO_2 system is worth consideration. If this is large enough and is left unchecked over a holiday period, for example, everything connected to that supply will be affected. Thus it may be worth having two completely separate supplies, each to a different bank of incubators. Valuable stock cultures can then be duplicated in independently-supplied incubators to avoid their complete loss in the event of a major leak in one of the CO_2 lines. Prevention is always better than cure, however, and a little care in using the correct tubing for pressurized gasses and securing and testing all connections, should avoid most problems. A wash-bottle filled with soap solution (household washing-up liquid is ideal), squirted around connections, makes a cheap and effective leak detector. In particular, newly connected cylinders should be tested in this way; dirt on either face of the connection can cause significant leakage.

2.2.2 Ultrapure water

Water is used in the culture laboratory for several different purposes which demand a particularly high level of purity. Water is used:

(a) As a solvent or diluent for culture media (when preparing from powdered formulae or media concentrates).

(b) As a solvent for supplements to culture media.

(c) For the final rinsing steps in washing-up reusable glassware which will be used for the preparation, storage, and handling of culture media.

For a small laboratory with limited resources, it may be more cost-effective to buy ready-prepared, sterile, single strength media and supplements from a supplier. It will be advisable also to buy ultrapure water for the occasional preparation of solutions of reagents which cannot be purchased ready to use. In this situation, deionized water from a conventional double distillation apparatus may be adequate for glassware rinsing and other general purposes.

Ideally, a water purification system situated in the wash-up area can serve as a continuous supply of ultrapure water for all purposes. This will also help in the long-term to minimize the costs of media and other sterile tissue culture reagents. Double glass distillation is adequate for the first stage in the purification of tap-water but has now largely been superseded by reverse osmosis (R.O.). This process typically removes about 98% of water contaminants. Tap-water fed to an R.O. unit should first be passed through a conventional water softener cartridge in hard water areas, to protect the R.O. membrane. The output from the R.O. unit can be stored (for a limited time) in an intermediate tank which should be sealed, with hydrophobic sterilization filters on the air inlets.

Water from this tank can be used directly to feed the second-stage, ultra-purification system, which is comprised of a series of cartridges for ion exchange and the removal of organic contaminants by filtration through carbon and other types of filter. Water purity is monitored with a resistivity meter which should reach about 18 Megohm.cm^{-1}. Ultrapurification systems are supplied by companies such as Millipore (Milli-Q system) or Fistreem (NANOpure Infinity system) supplied by Fisher Scientific. Finally, ultrapure water is passed through a micro-pore filter to exclude residual particulate matter, including micro-organisms. Matched R.O. and ultrapurification systems are also available as a complete package, e.g. the Multipure R.O. system from Fisher Scientific.

In the choice of water ultrapurification systems, the rate of pure water generation is important, particularly if large volumes are needed for a washing machine, for instance. You should also be sure that the unit will run off tap-water at high pressure and will not require an intermediate storage facility. The sterilization cycle should be semi-automatic and easy to operate (see below), otherwise its implementation will be 'put off until another day' by busy technicians and membrane damage may result.

Water should be collected and autoclaved or filtered immediately for sterile use. The intermediate storage tank can also be diverted to feed a glassware washing machine for the final, 'distilled water' rinse cycles (see below). Tanks and tubing should be 'light-tight' to prevent algal growth.

Although the cost of the R.O. unit is a substantial initial outlay, provided that care is taken to cleanse and sterilize the unit and its membrane at regular intervals, e.g. monthly, with a proprietary solution containing formaldehyde, the membrane will have a lifespan of several years. There is an automated cleansing cycle on most models which requires minimum effort and cost to run. The initial cost is soon repaid, in much reduced power use and in the saving on replacement of elements etc. and time spent on cleaning and de-scaling a glass still.

3 Equipping the laboratory

3.1 Major items of equipment and instrumentation

3.1.1 Laminar flow hoods

It was common practice in the early days of tissue culture to carry out sterile procedures under a sheet of glass or Perspex, clamped horizontally about 0.3 m above a bench in a still air room. Bench and glass sheet were swabbed down with 70% ethanol and, since most microbial contamination falls from above or is wafted into an open, sterile vessel from a non-sterile source, careful manipulation of sterile material under the plate glass was often successful. Nowadays however, tissue culture laboratories rely on the use of laminar flow hoods for all sterile handling work. The alternative is to supply whole rooms or a suite of individual cubicles with sterile-filtered air. The rooms are kept under slight positive pressure to ensure that non-sterile air is not drawn in. This arrangement may be necessary for some large-scale procedures where cost is not a major consideration.

Laminar flow hoods, then, are widely used in cell culture laboratories. They operate by filter-sterilizing the air taken in, so excluding particles including bacterial and fungal organisms. Air thus sterilized then passes in a vertical downflow on to the work surface. Some other types of cabinet produce a laminar flow of sterile air which is directed horizontally towards the operator, but these do not offer protection to the operator from the potential hazards of some cultured cells.

Laminar flow cabinets are classified according to the degree of safety they offer the user, and a risk assessment must be performed before starting work with any cell line in order to establish which is most appropriate to use (see also Chapter 2). The principles behind the design of the most commonly used laminar flow cabinets are shown in *Figures* 5 and 6. *Figure* 5 represents a standard laminar downflow cabinet, suitable for most cell work and *Figure* 6 shows a Class II microbiological safety cabinet, for handling primate or human material, or cells which may be virally infected or carry other adventitious agents. Both types of cabinet

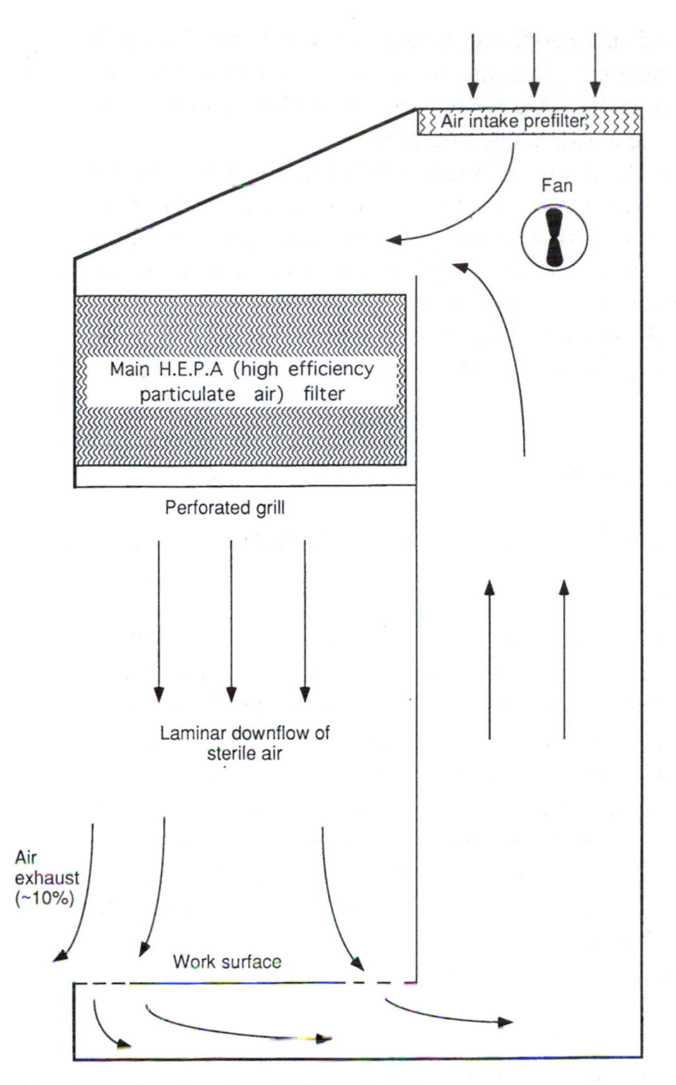

Figure 5 Laminar downflow sterile work cabinet.

provide similar control of airborne microbial contamination. In addition, the Class II cabinet offers greater protection to the operator because:

(a) There is provision of a front window with a lower edge which gives minimum turbulence to air drawn in from outside, whilst allowing adequate access for the operator's arms. The air taken in at the front acts as a safety curtain, preventing aerosols of potentially hazardous material from reaching the operator. In an empty hood, this curtain of air is drawn immediately to exhaust, so does not compromise the sterility of any material inside the working area. However, the operator's arms disturb the airflow and cause eddies. Non-sterile air may then enter the cabinet and it is wise to avoid using the front 10 cm of work surface for sterile handling.

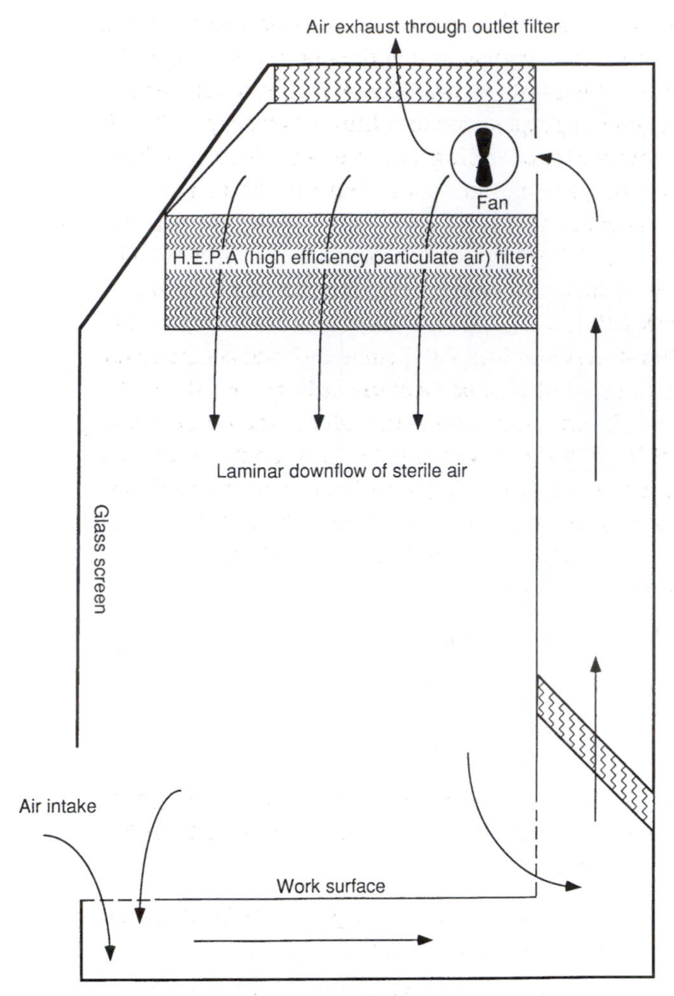

Figure 6 Class II microbiological safety cabinet for sterile work requiring a higher degree of operator protection than provided by an open-fronted sterile work cabinet (see *Figure 5*).

(b) Air flowing across the cultures is exhausted to the laboratory only after passing through an outlet filter; most is recycled within the cabinet as shown in the diagram.

(c) A monitor and alarm system warns when the airflow rate falls below or rises above safe levels.

These and other types of cabinet are discussed further in Chapter 2. All cabinets should be inspected regularly according to current regulations, and maintained carefully; in particular, all spillages should be cleared up immediately and the surfaces wiped down with 70% ethanol (it is useful to keep a spray or wash-bottle of this close at hand). The cabinet should also be cleaned out completely at regular intervals, by removing the detachable work surface.

When choosing a laminar flow cabinet (or hood), a number of factors should be taken into consideration. The first, of course, is that the hood should be of the appropriate class for the work which will be carried out in it (for further discussion, see Chapter 2). For a general purpose tissue culture laboratory, a Class II hood is generally the most appropriate, providing a degree of protection to both worker and cultures, so permitting the handling of all but the most hazardous cell lines. Such hoods are available in a variety of sizes (usually based on nominal width) and can vent either to the laboratory or can be ducted to the outside. Ducting to the outside need only be considered if there is good reason to do so, e.g. regular use of higher risk cell lines or the frequent need to fumigate the hood. Most tissue culture laboratories will find a 120 cm wide hood adequate for most purposes, although certain procedures, particularly at large- or pilot-scale, may be more easily carried out in one that is 180 cm wide. It should be noted that the working height within such hoods varies between manufacturers (an important thing to consider if frequently dealing with large bottles or spinner flasks) and the width is only a very rough guide to the effectively useful working area. The internal dimensions of two Class II hoods, both nominally 120 cm wide but from different manufacturers, are given below.

	Manufacturer A	**Manufacturer B**
Width (cm)	120.5	105.5
Depth (cm)	59.5	50.0
Height (cm)	82.5	68.5

Clearly, the hood from manufacturer A is much the larger, with a working area 36% greater and a working volume 64% greater than that from manufacturer B.

Other factors which may be considered are:

(a) Does the hood require an internal power point and/or gas tap? (These can be fitted on request by most manufacturers if not supplied as standard.)

(b) Is the inside of the hood easy to clean? There should be no difficult, inaccessible spots and the work surface should be easy to remove and replace.

(c) Will it be necessary to open the front of the hood regularly to introduce large vessels? If so, the front glass, once opened, should remain open by itself so that both of the worker's hands are free to handle the vessel.

(d) Would it be useful for the height of the working surface of the hood to be adjustable? If so, some manufacturers can supply height-adjustable stands.

3.1.2 Incubators

Incubators for cell culture work should provide an environment in which the following are controlled:

- temperature
- gas phase
- humidity

The temperature required for most mammalian cells to grow optimally is 37 °C. Incubators should be set half a degree below this for safety and a cut-out operated at 39 °C, above which cells rapidly die. Avian cells prefer a slightly higher temperature, reflecting their higher body temperature, and mammalian skin cells, for instance, a slightly lower temperature. Cells from cold-blooded animals are generally incubated at a constant temperature within their normal exposure range.

Most cell culture work requires that cells are incubated in an atmosphere which contains an elevated CO_2 tension. Culture medium is thus maintained at a physiological pH (7.2–7.4) by the equilibrium established between dissolved CO_2 and the bicarbonate buffer in the medium (see Chapter 3). The gas phase can be adjusted in culture flasks by introducing the correct gas mixture from a cylinder via a plugged pipette and then closing the cap tightly. Alternatively, open vessels can be incubated in a sealable chamber, which can be humidified, gassed, and closed. Several suppliers market such chambers, e.g. ICN Biomedicals, but a good workshop should be able to produce an equally effective version from Perspex, with a removable front fitted with a silicone gasket and butterfly screws to make a tight seal when assembled. Inlet and outlet ports with tubing that can be clamped permit the introduction of a gas mixture. Such chambers then allow the use of less expensive, dry air incubators.

Most tissue culture laboratories, however, use CO_2 incubators as the most accurate and reliable way of providing the right incubation conditions for cells in vented vessels. Indeed, for certain critical, sensitive work, a CO_2 incubator may be essential. Even in tightly-capped flasks, gas phase leaks can occur. Whilst in other incubators this can lead to a (sometimes devastating) rise in pH of the medium due to loss of CO_2, this is not a problem in a CO_2 incubator.

CO_2 incubators can regulate the CO_2 tension in a number of ways. In older-style incubators, a constant flow of a CO_2/air mixture was utilized, with each gas being supplied separately from a different cylinder and mixed according to set proportions by gas burettes. The major drawbacks to this arrangement were the extravagant use of CO_2 and the likelihood that, if the CO_2 ran out unexpectedly, the air supply on its own would rapidly flush out the incubator and allow the pH of media to rise.

The most efficient and widely-used method involves the use of an instrument which measures the CO_2 level in the incubator and activates a valve to draw on pure CO_2 from a cylinder, whenever the gas phase falls below a set value. This is generally 5%, although high bicarbonate media such as Dulbecco's modification of Eagle's minimal essential medium (DMEM), particularly when used for serum-free cultures, should be incubated in a 10% CO_2 atmosphere (see Chapter 3). Simple CO_2 calibration devices, e.g. the Fyrite test kit marketed by Shawcity Ltd., can be purchased to check the monitor readings, but for most purposes the colour of medium in a dish without cells is an accurate guide for the experienced worker.

Humidity is maintained by including a water tray in the bottom of the incubator over which the air is circulated (most incubators have fans to ensure even temperature and humidity throughout). Water trays should be prevented from harbouring micro-organisms by the inclusion of a non-volatile cytostatic reagent such as thimerosal (but not azide, which reacts with metals) or a low concentration (1%) of a disinfectant detergent such as Roccal, supplied by Henry Schein Rexodent. The water level should be topped up regularly (using deionized or R.O. water) and condensation aspirated from other areas of the incubator base. There is, nevertheless, a potential risk of fungal growth in particular, in humidified incubators. The interior should therefore be cleaned out and wiped over with antiseptic at regular intervals; it is important to check when buying a new incubator that the interior can be dismantled to allow adequate cleaning. LEEC Ltd. make simple, reliable, and easily cleaned CO_2 incubators at competitive prices.

When choosing an incubator, it is nevertheless worth considering whether the extra expense of a model with an automatic sterilization cycle (e.g. one of the Heraeus models, marketed by Kendro Laboratory Products), is justified. This may be a distinct advantage if the tissue culture laboratory has a high number of trainees or where there are other, environmental factors which make contamination more likely. If the whole culture laboratory is to be fumigated periodically, care should also be taken that the CO_2 monitor can be removed from the incubator or is of a type that will not be damaged by aldehyde fumigants.

For special applications, CO_2 incubators are available in which the partial pressure of oxygen can also be controlled. As well as CO_2, such incubators require an oxygen and nitrogen supply. CO_2 incubators are also available that can be used in high ambient temperature conditions (e.g. laboratories without air-conditioning or that are constantly exposed to the sun or other heat sources). These use heating or cooling, as necessary, to maintain their set incubation temperature.

3.1.3 Centrifuge

The main use of a centrifuge in a tissue culture laboratory is to spin down cells during tissue disaggregation or in harvesting cell lines for subculture, freezing, analysis, etc. A very simple bench-top centrifuge which will reach 80–100 g, preferably with a variable braking system, is generally all that is needed. Various considerations may indicate the need for a more sophisticated piece of equipment. The volumes of cell suspensions to be handled in a single operation will dictate the bucket size and adaptability needed, for instance the capacity to take four swing-out buckets each holding five 50 ml tubes in one run, followed by a mixture of 15 ml and 50 ml tubes. If primary cultures are anticipated, it may be an advantage to have a cooled centrifuge with a timer for long runs with sensitive cells which have been exposed to proteolytic enzymes. For some laboratories, a bench-top microcentrifuge is a useful addition for high-speed centrifugation of small volumes of reagents which may generate a precipitate, e.g. after thawing from the freezer.

3.1.4 Microscope

An essential instrument for any culture laboratory is a good quality inverted microscope, preferably with phase-contrast optics and a photographic facility.

Cell morphology—the degree of spreading, granularity, membrane blebbing, the proportion of multinucleate cells, vacuolation, and so on—should be monitored regularly for signs of stress in cells. Morphological appearance is a sensitive indicator of problems with culture conditions.

Early signs of microbial contamination can also be detected with a good phase-contrast microscope. Regular checking of cultures under the microscope can help to avoid catastrophic losses of irreplaceable material by ensuring that a problem is identified at an early stage. Also, cells which are used for experiments in a less than healthy state may give variable or erroneous results. When choosing a microscope, select the long or extra-long working distance condenser so that flasks and even roller bottles can be viewed. There is usually no need for objectives above ×20; their depth of field is often too low to obtain a sharp image of all but the very flattest cells. A good low power, wide field objective, e.g. the Nikon ×4, is very useful for scanning cultures for foci or colonies, for instance.

A fuller treatment of the whole topic of the microscopy of living cells is given in Chapter 4.

3.1.5 Fridges and freezers

For most laboratories, domestic larder-type fridges (with no ice box) are adequate. Media storage requires considerable space and it may be more convenient to store unopened bottles in a nearby cold room, if available. The fridge in the tissue culture laboratory can then be reserved for media in current use, with each opened bottle designated for one individual's work with one cell line. Ideally, separate fridges should be used for sterile culture media and for non-sterile solutions, chemical stocks, etc.

Freezers at −20 °C and, ideally, −70 °C, will be needed for storage of sera, solutions, and reagents which are unstable at higher temperatures. At the lower temperature, sera and proteins such as collagenase, which is prone to degradation even at −20 °C, can be stored for extended periods. Some of these lower temperature freezers are available with liquid CO_2 or liquid nitrogen back-up facilities that permit temperature to be maintained even in the event of an electrical supply or compressor failure. Liquid nitrogen freezers for long-term cryopreservation of cells will be considered in Chapter 5.

3.1.6 Miscellaneous small items of equipment

A **water-bath**, set at 36.5 °C, will be needed to warm media etc. before use, to thaw frozen aliquots of reagents or cells, and to carry out enzyme incubations. Care should be taken to change the water regularly (e.g. at weekly intervals) and to add a cytostatic reagent such as thimerosal (but not azide) to prevent microbial growth. A submersible magnetic stirrer is a useful accessory. Some workers prefer not to use water-baths because of a perceived risk of contamination from the water and associated aerosols. In this case a small dry incubator at the same temperature fulfils a similar function but has the disadvantage that heat dispersal is less efficient.

In the area for preparation of reagents for culture use, there will need to be a **balance**, a **pH meter**, and one or more **magnetic stirrers**. The balance should be capable of weighing accurately in the milligram range. Many reagents for culture work are active in the microgram or nanogram range and stocks solutions are usually prepared at 100–1000×. An **osmometer** is a useful, but not essential, piece of equipment, especially if media are prepared from powder or basic ingredients or if a number of additives are used which will affect osmolality. The Fiske One-Ten osmometer supplied by Astell Scientific, for example, measures small volume samples (10 μl) and is quick and easy to use.

Spent medium can be drawn off into an **aspirator jar**, most conveniently situated underneath the laminar flow hood, with a pump to draw a vacuum. *Figure 7* shows such an assembly, which can be adapted to serve two or three laminar flow cabinets by using a two- or three-way connection on the aspirator line and an in-line valve to close off lines not in use. A simple diaphragm pump will pull a vacuum sufficient for one aspirator line but a more powerful pump, such as an oil vane vacuum pump, may be needed if two or more lines are open simultaneously.

The aspirator jar should contain a small amount of detergent such as Decon (Decon Laboratories Ltd.) and a small volume of Chloros (sodium hypochlorite) solution, kept in a wash-bottle standing on a plastic surface (drips will corrode metal), should be drawn through the aspirator line at the end of each work session and between handling different cell lines. This will help to avoid both microbial contamination and cross-contamination between cell lines via the aspirator line. The aspirator jar contents should be disinfected before emptying, preferably into a sink in a separate room. As an alternative to an aspirator jar assembly, if there is a suitable water supply at sufficient pressure nearby, a simple tap syphon is a cheaper and more convenient way of aspirating spent medium, since waste is drawn directly into the mains drainage.

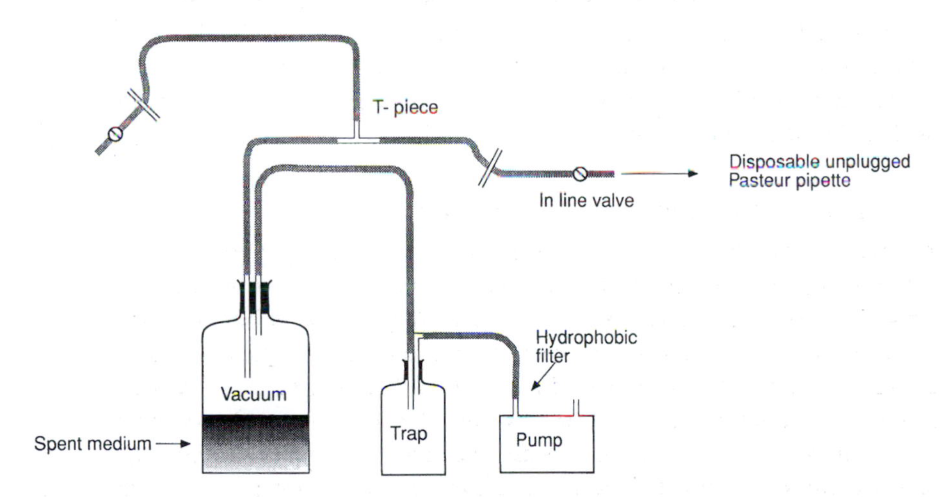

Figure 7 Aspirator jar assembly for withdrawing spent medium.

3.2 Culture plasticware and associated small consumable items

3.2.1 Tissue culture plasticware

This is supplied by a number of specialist companies; the top four or five companies produce very comparable ranges of sterile plastics (usually polystyrene), both in terms of price and quality, but may offer different modifications of a basic design or introduce new, specialist ranges. The most commonly used items of plasticware are given here.

(a) For growing cells ('tissue culture' treated, to produce an electrostatically charged surface for wettability and cell adhesion):
- flat-bottomed flasks
- Petri dishes
- multiwell dishes
- conical flasks (for suspensions)
- roller bottles
- tubes

(b) For handling solutions and cell suspensions:
- volumetric pipettes
- plastic Pasteur pipettes
- micropipette tips
- centrifuge tubes

(c) For storage:
- sample tubes
- Eppendorf tubes
- cryotubes
- larger volume screw-capped bottles

One of the most useful vessels for growing up stocks of cells in culture is the flask, which is widely available with usable surface areas of 25, 75, 175, and 225 cm^2. Flasks have either straight or canted necks and are supplied with a conventional, two position, screw-threaded cap for vented or closed incubations. Alternatively, some suppliers (e.g. Corning Costar, Nunc, Greiner) may offer flasks with a sterile, hydrophobic filter in a perforated cap so that gas exchange can occur without even the slight risk of microbial contamination associated with a loosened cap.

Tissue culture dishes are widely used and are available in 35, 60, 90, and 150 mm diameters with a variety of modifications possible, such as a series of internal wells, grids, etc. Generally, the lids are vented for adequate gas exchange but designed for minimum evaporation. Multiwell dishes are supplied in various sizes, but 96-well (flat- or round-bottomed for different applications) and 24-well are the most widely used. 4- and 6-well plates may have specialist applications, such as the accommodation of filter well inserts (e.g. those supplied by Corning Costar) which allow different types of cells to be grown separately, on one or more membrane supports within the same volume of culture medium.

For large-scale adherent cell cultures, plastic roller bottles can be used, which require a special apparatus within a cabinet or warm room to turn the bottles at a constant rate, thus utilizing a large surface area for growth in a relatively small volume of medium. Increasingly, this type of large-scale culture is achieved in other ways, such as the use of microcarrier beads (e.g. those supplied by Sigma), matrices allowing ingrowth of cells, or meshes of various types to increase the available culture surface area. Cells grown in suspension can be cultured either in spinner flasks, where the medium is constantly circulated by a magnetic stirrer, in conical flasks on a rotating platform (kept in a warm room or dedicated incubator), or in static culture on non-adherent plasticware.

There are a number of other culture vessels for particular uses, such as chamber slides (e.g. Labtek, from Nunc) for growing cells to be used for immunocytochemistry, or triple flasks (Nunc), which provide three stacked surface areas for cell growth within a conventional-sized flask, as a solution to the problem of limited incubator space. Suppliers of tissue culture plastics will advise on new or more specialist items on their lists and most catalogues give detailed information on the specification and uses of their products.

For some cells, the tissue culture surface, even when treated to enhance cell adhesion, does not provide an adequately adhesive surface. Some cells have an absolute requirement, especially in serum-free medium, for an extracellular matrix protein substrate. Consideration of such specific substrates is dealt with in Chapters 5 and 7. It may be sufficient to coat the tissue culture surface with a highly charged polymer such as polylysine or polyornithine at about 10 μg/ml. This is also useful for encouraging attachment of cells which usually grow in suspension, e.g. for staining hybridomas with Hoechst fluorochrome for mycoplasma testing.

3.2.2 Filters for sterilizing tissue culture solutions

Several suppliers of sterilizing filters, such as Millipore or Gelman (Pall Gelman Laboratory), produce disposable, sterile units for attachment to a syringe, which are suitable for filtering small volumes of tissue culture solutions. The pore size should be 0.2 μm or even 0.1 μm in order to be sure of excluding mycoplasma as well as other micro-organisms. For most purposes however, 0.2 μm is the standard pore size for liquid sterilization. It may be necessary to use an assembly with a pre-filter of e.g. 0.8 μm, if solutions are likely to block a low pore size filter readily, e.g. a semi-purified enzyme preparation or a biological fluid.

There are several features which should be considered before choosing the appropriate filter. Some filter membranes, e.g. polysulfone, are low protein binding and essential for proteins where the concentration of the filtered solution is critical, particularly if the molecule is highly charged, as are some of the polypeptide growth factors. Very small volumes of valuable reagents should be filtered through units where the 'dead' space (hold-up volume) is minimal. It is cost-effective, therefore, to have some of these more expensive filters reserved for special purposes. Larger volumes can be sterile filtered either using pre-assembled, disposable units which attach to a vacuum line (see Section 3.1.6) or, more economically, a washable, autoclavable unit in which the filter can be replaced.

Large tissue culture facilities may consider production and sterilization of media 'in-house', in which case a stainless steel pressure vessel, connected to a nitrogen line, may be useful. However, this is outside the requirements of most tissue culture laboratories. Sterile filtration is considered in further detail in Chapter 2.

3.2.3 Pipettes

Those needed for all culture laboratories are of four basic types, with different uses:

- micropipettes with disposable, autoclavable tips
- unplugged Pasteur pipettes
- plugged Pasteur pipettes
- volumetric pipettes

Plastic micropipette tips are supplied by a number of companies to fit the variety of micropipettors on the market (see below). These will be of several types, to suit volumes in the 1–20, 10–200, 100–1000 μl, and up to 10 ml ranges. In addition, extended, fine tips are available for small volumes in narrow diameter tubes. Tips also come with or without a circumferential ridge part way along their length, so that they can be supported in an autoclavable tip carrier and easily attached to the micropipettor without handling. Several companies also market tips with an integral filter for extra safety in sterile handling, but these may have to be purchased pre-sterilized, which adds to their cost. Autoclaving, but not irradiation (which is in any case not feasible for most laboratories), causes the filters to deteriorate.

Disposable plastic Pasteur pipettes (1 ml, with an integral bulb) and volumetric pipettes of all volumes (1–100 ml) can be purchased from most tissue culture plasticware suppliers in pre-sterilized packs. However, this is an avoidable expense if substantial numbers of pipettes are used and there are facilities to wash and sterilize glass pipettes. Generally, glass Pasteur pipettes are supplied non-sterile but ready to use after dry heat sterilization. Both plugged and unplugged glass Pasteurs will be needed and are intended for disposal after a single use. Extra-long unplugged Pasteurs are preferable for reaching to the corners of large flasks for aspiration of spent medium. Standard length, plugged Pasteurs are used for any small-scale, non-volumetric pipetting of tissue culture solutions and for the introduction of gasses into e.g. a flask or sealable chamber, where the cotton wool plug ensures sterility.

Glass volumetric pipettes should be of good quality borosilicate glass, graduated to the tip, preferably with the maximum volume graduation at the top. A relatively wide-bored tip is advisable for fast delivery. A selection of volumes from 5–25 ml is generally required; below this, micropipettes are more accurate and convenient. 5 and 10 ml pipettes are the most frequently used and there should be at least five times the number in daily use, to allow for those out of circulation in soak, wash, or sterilization. Reusable volumetric pipettes will have to be unplugged for washing and replugged before sterilization, care being taken to discard any that are cracked or broken (see below).

3.2.4 Pipetting aids

There are a number of micropipettes available on the market to suit most budgets and needs. Some are autoclavable for additional safety for sterile procedures. For most laboratories, a range of micropipettes covering e.g. 1–20 μl, 20–200 μl, 100–1000 μl, and 1–5 ml will be adequate, perhaps with one or two fixed volume micropipettes, such as a 50 or 100 μl for frequently used volumes. Depending on the number of regular users, sets will need to be replicated. In addition, it is useful to have a multichannel micropipette; this is essential if much work is done with 96-well plates. For many laboratories, an automated aid to pipetting larger volumes is also a high priority. Several companies make lightweight, hand-held, battery- or mains-operated units for this purpose, with loading and dispensing push-button controls and an air filter for sterility. For those on a tight budget, conventional manual pipette bulbs will be adequate and these are available in a range of volumes. **Under no circumstances** should mouth pipetting be considered for tissue culture work.

3.2.5 Glassware for tissue culture use

Laboratories differ in their preferences for glass or plastic for handling and storage of solutions for tissue culture use. There is no doubt that plastic containers are less prone to leaching trace elements which might be deleterious, but this is not usually a problem in practice, provided that high quality borosilicate glass is used and wash-up facilities are good. Glassware has a greater propensity to adsorb substances such as alkaline detergent onto its surface, than plastics such as polypropylene. Whatever the choice, most laboratories will require a selection of volumetric cylinders, flasks, and beakers for preparation and bottles for storage of sterile solutions. Schott bottles (supplied by Merck Ltd.) are particularly suited to this. They are robust and will stand the pressure created by solutions autoclaved with a tightly closed cap (e.g. bicarbonate). The cap is deep and will therefore ensure a good depth of sterile outer screw thread surface and greater safety on the few occasions when pouring cannot be avoided.

3.2.6 Miscellaneous small items

- pipette cans
- tube racks
- haemocytometers
- instrument tins
- pipette bulbs

For the preparation area:

- autoclave tape
- oven tape
- Browne sterilizer control tubes (for oven and autoclave, from Solmedia)
- cotton 'rope' for pipette plugging
- autoclave bags (several sizes)

If reusable volumetric pipettes are used, it is convenient to sterilize them in long, stainless steel, cylindrical cans or aluminium cans with a square cross-section, according to preference. Aluminium cans are less expensive, but will oxidize over a period of time. There should be at least one can per worker for each pipette volume, more if the workload is heavy, but enough so that individual workers can keep a can, once opened, for their exclusive use. Shorter cans are needed for Pasteur pipettes. Oven-sterilizable, tight-lidded flat tins are convenient for sterilizing instruments for dissection, sterile handling of coverslips, etc.

4 Washing reusable tissue culture equipment

All tissue culture equipment which can be washed, sterilized, and recycled should go through the general process given in *Protocol 1*, either manually or through a tissue culture-dedicated automatic washing machine with both acid rinse and distilled water rinse facilities.

Protocol 1

Washing and sterilization of reusable labware

Equipment and reagents

- Rigid plastic soak tanks, cylinders, and beakers (for small items and instruments)
- Detergent, e.g. Micro (International Products Corporation) for manual washing; low-foam for machine, as recommended by manufacturer
- Ultrapure water
- Chloros solution for soaking
- AnalaR (or similar high purity) HCl for manual washing; formic or acetic acid for machines

Method

1. Soak in Chloros solution (except metals), or directly in detergent.
2. Wash with detergent (by soaking or machine).
3. Rinse with tap-water (continuously or with sequential changes).
4. Rinse with dilute acid (except metals).
5. Rinse with tap-water as above.

Protocol 1 continued

6. Rinse with distilled/reverse osmosis/ultrapure water (two or more times).
7. Dry in hot air.
8. Cap or cover for temporary storage, e.g. on preparation bench.
9. Prepare for sterilization (see Chapter 2).
10. Sterilize in autoclave or dry oven.
11. Store for use (in dedicated cupboards or racks).

Problems with inadequately washed equipment are common, but mostly avoidable by attention to points discussed in the following sections.

4.1 Soaking

Many tissue culture solutions contain protein. If glassware is left unrinsed and unwashed for any length of time after use or is allowed to dry out, there are several consequences:

(a) Residual medium will support microbial growth and act as a source of air-borne contaminants in the laboratory.

(b) Dried-on protein will be difficult to remove by normal washing procedures.

(c) Inadequately washed glassware, when sterilized, will leach contaminating material (possibly toxic products of heat denatured proteins, e.g. hydrocarbons) into new solutions.

The following procedures are therefore advisable. Soak tanks should be situated near the wash-up area, one for heavy-duty glassware such as medical flats, bijou bottles, etc. and one for fine glassware such as beakers and cylinders, easily broken. The soak tanks can be filled with tap-water but must contain Chloros or an equivalent solution to limit microbial growth. This should be changed regularly. Presept tablets (Johnson & Johnson Medical Ltd.) are a convenient way of achieving this. If the wash-up area is outside the main laboratory, it is acceptable to collect glassware in e.g. a bucket, kept beside the work station, provided that it is taken to the soak tank as soon as possible at the end of a session. Glassware for soaking should:

(a) Be rinsed under a cold tap.

(b) Have any tape removed.

(c) Have marker pen labelling removed with acetone (a carefully labelled wash-bottle kept by the tank is useful).

(d) Be immersed **completely** in the soak tank.

4.2 Washing

When the tanks contain a load for washing, glassware can be taken straight from the tank. Different laboratory washing machine manufacturers, e.g. Lancer,

Chapter 2
Sterilization

Peter L. Roberts
Research & Development Department, Bio-Products Laboratory, Dagger Lane, Elstree, Hertfordshire WD6 3BX, UK.

1 Introduction

Effective sterilization techniques are essential prerequisites for practising the art of cell culture. Cell cultures provide an ideal environment for the growth of micro-organisms including viruses, and it is only by excluding such agents that cells can be successfully cultured and meaningful experimentation can be carried out. In this chapter the various methods used for sterilization will be reviewed including some theoretical background but concentrating on practical aspects of their use. In addition to cellular micro-organisms, the unique problems posed by viruses and prions will be considered. For further general background information see refs. 1–3.

1.1 What is sterilization?

The methods used for sterilization have a long history. Pasteur first used heat to preserve wine (1864), and, even before this, heat was used in the canning industry (*c.* 1700) although with no understanding of the microbial agents involved. Other sterilization techniques such as filtration were first used in the late 1800s.

Sterilization is defined as a process which removes all living things. In practice we are really concerned with micro-organisms, i.e. protozoa, fungi, bacteria, mycoplasma, viruses, and prions, all of which are invisible to the unaided eye. Although sterility is an absolute term, in practice methods are used that give a high probability that an item is sterile. Sterilization methods vary in their severity and the properties of the item needing to be sterilized also need to be considered. Micro-organisms also vary in sensitivity/suitability for different methods of inactivation or removal.

1.2 The importance of sterility in cell culture

When applied to a cell culture, sterility means the absence of any contaminating organism apart from the cells of interest. Such sterile cultures are essential tools in research and diagnostics, and also for the manufacture of various biological products. The presence of contamination can have a wider range of deleterious effects. Examples include:-

(a) Complete destruction of the cell culture in the worst case.

(b) Problems and artefacts in research or diagnostic studies.

(c) Effects on cell-derived products, e.g. yield, stability.

(d) Compromising the safety of cell-derived therapeutic products.

(e) Contamination of other cultures also being handled in the laboratory.

1.3 The use of antibiotics

When a cell culture is free of contaminating micro-organisms, sterility can be maintained by the use of good aseptic technique as outlined in this and other chapters. This represents the first and best line of defence, but the use of antibiotics may also have a role in certain circumstances (see Chapters 6 and 9).

2 Basic principles

Sterilization techniques are designed to kill or remove a wide range of micro-organisms including, in size order, the protozoa and fungi, bacteria, mycoplasma, and finally viruses. In all cases the agents are composed of single cells, of varying complexity, except for the viruses which are basically composed of only nucleic acid surrounded by a protein coat and in some cases a lipid-containing envelope. In addition to these, agents with an apparently even simpler structure exist, i.e. prions (4), which are believed to be composed of protein alone (see Section 9.2).

The principal target(s) on which a method of sterilization has its effect varies with the technique and with the micro-organism concerned but may include the nucleic acid, proteins, or membranes. Micro-organisms also vary with regard to their sensitivity to the method being used. For instance:

(a) Bacterial endospores are resistant to heat.

(b) Prions are resistant to heat, irradiation, and detergent.

(c) Non-enveloped viruses are resistant to organic solvents and detergents.

(d) Mycoplasma and viruses are not removed by conventional 0.2 μm sterilizing filters.

To understand fully the efficiency of any sterilization method that depends on inactivation, it is important to know the kinetics of the process for relevant micro-organisms, particularly those that are resistant. A graph of the logarithm of the number of surviving organisms against time often gives a straight line. From this the time required to kill any desired number of organisms can be determined. The efficiency of different sterilization methods can be compared by determining whether large numbers, e.g. 1×10^8, can be inactivated or removed and in what time.

3 Wet heat at up to 100 °C

Heat represents one of the most commonly used methods for sterilizing equipment and media. Heat can be used in the dry state but is much more effective in the wet state and is at its most effective at temperatures of 121 °C or greater. In this section the use of wet heat at temperatures of up to 100 °C is considered.

3.1 Theory

Many micro-organisms including bacteria and viruses are relatively easily inactivated at temperatures of 60–80 °C in the wet state. In fact temperatures of 55–60 °C are often adequate. However, there are micro-organisms that are more resistant to heat, such as the spore-forming bacteria and prions. In addition, even for a given susceptible type of micro-organism, mutants or variants may exist that are less heat-sensitive. The exact mode of inactivation will depend on the micro-organism and conditions involved, but denaturation and coagulation of protein is thought to be important. However other targets, such as nucleic acid, may be involved in some situations. With some agents, including the viruses, inactivation kinetics often give rise to a curve with two (or more) components, e.g. an initial phase of rapid inactivation followed by a period when the inactivation is slower. This may suggest that two or more targets are involved in inactivation, or that aggregates or genetic variation exists within the population. For the highly resistant bacterial endospores, the target of wet heat is believed to be the DNA, cell membrane, or the wall precursors and the mechanism of resistance to heat involves dehydration of the cells. The stability of micro-organisms (including viruses) to heat can be influenced by pH, and by levels of salts, protein, and sugar.

In practice the use of wet heat at temperatures of up to 100 °C has only a few limited applications in the cell culture laboratory and it is preferable, where possible, to use other more effective techniques such as autoclaving (Section 4) or dry heating at high temperature (Section 5) for more effective sterilization.

Several points should be noted when considering the use of wet heat below 100 °C for sterilization:-

(a) The degree of sterilization will depend on temperature and time.

(b) Spore-forming bacteria and prions are resistant to all but the most severe conditions, i.e. autoclaving.

(c) It can be useful for heat-sensitive materials that cannot withstand more severe heat-treatment.

(d) It can be used to inactivate viruses in heat-sensitive products (see Section 9.1).

3.2 Pasteurization

Heat-treatment at temperatures of about 60–80 °C has long been used in the food industry. For example milk is treated at 63°C for 30 min or at 72°C for 15 sec. These procedures will kill vegetative micro-organisms and any viruses that are likely to be present, but not bacterial spores. Pasteurization at 60 °C for 10 hours is used in the pharmaceutical industry for certain products derived from human plasma as a method of inactivating viral contaminants (see Section 9.1). Another process that uses wet heat, combined with detergents, is that employed in the washing machines that are used for decontaminating reusable equipment particularly in hospitals.

3.3 100 °C

Boiling or steaming at 100 °C for 5–10 min represents a simple and effective means of sterilizing equipment and fluids. It is very effective and will even destroy some bacterial endospores. However, some types of endospore and also prions are resistant to boiling and require more severe conditions for inactivation. It is possible to inactivate endospores by the use of several 100 °C cycles. This method, known as Tyndallization, involves using three cycles of heat-treatment with intervening periods during which the spores germinate. This type of procedure can be used in the cell culture laboratory for decontaminating incubators that have a heat sterilization option.

4 Wet heat above 100 °C and autoclaving

4.1 Theory

Boiling water can be a very effective sterilizing agent, but in practice more severe sterilization methods that are capable of inactivating bacterial endospores are required for many applications in the laboratory. Using water at a temperature of about 121 °C, i.e. as steam under pressure, is such a method. To produce steam under these conditions special equipment is required, i.e. an autoclave. By having a knowledge of the inactivation kinetics for heat-resistant bacterial endospores, suitable autoclave cycles have been developed that give a high assurance of sterility. Some of the standard autoclave temperature–time cycles that are commonly used are given in *Table 1*. It can be seen that the various cycles are not in fact equivalent and, when survival is compared to the widely used cycle of 121 °C for 15 min, that other recommended cycles are either less severe (115 °C) or more severe (126 °C, 134 °C). Prions are particularly resistant to heat-treatment and require especially extreme conditions for inactivation (see Section 9.2).

Table 1 Standard autoclave cycles

Temperature (°C)	Time (min)	Pressure (bar)[a]	Survival[b]	Equivalent time[c] (min)
115	30	0.7	1 in 10^4	60
121	15	1.0	1 in 10^8	15
126	10	1.4	1 in 10^{17}	4.7
134	3	2.0	1 in 10^{32}	0.8

[a] 1 bar = 10^5 Pa.

[b] For a relatively heat-resistant bacterial endospore such as *Bacillus stearothermophilus*.

[c] Time required to give endospore survival of 1 in 10^8, i.e. time equivalent to 121 °C for 15 min.

4.2 Steam

For steam to be fully effective it must be at its maximum water-holding capacity, but with no droplets to wet wrapped items, and it must not be superheated, i.e. in an over-dry state. This aspect of steam quality can be checked by an autoclave engineer determining the steam-dryness value. The presence of too much air, i.e. non-condensable gas, in the steam supply will also reduce the effective steam pressure in the chamber. This too can be checked by an autoclave engineer in a non-condensable gas test, and may require modifications to the boiler or pipework for improvement. The presence of air in the chamber, either from non-condensable gas in the steam supply or because of poor air removal, will lead to a lower temperature in the chamber. For instance, if only half the chamber contained steam then, under conditions of pressure for which a temperature of 121 °C was expected, only 112 °C would actually be reached. If localized pockets of air were to remain, e.g. in packaged goods or in infectious laboratory discards, this effect could be made even worse. Adequate quantities of steam must be available so that no more than a 10% pressure drop to the supply pressure occurs when the machine is operating. For pharmaceutical applications, clean-steam is used which involves providing the boiler with purified water of a conductivity less than 15 μS.

4.3 Types of autoclave

The types of autoclave available vary widely in size, complexity, and versatility. They range from small, simple pressure cookers, which generate their own steam, to large machines which are fully automated and may possess multiple cycle options. The nature of the load is important for determining which cycle is most appropriate.

4.3.1 Portable bench-top autoclaves

i. Upright models

These are essentially pressure cookers as used in the kitchen at home. Steam is generated in the chamber base, either by an external or internal heat source, and an air–steam mixture discharged from the chamber through a hole in the top. A pressure gauge, possibly a temperature gauge, a pressure control valve, and a safety valve are present. In more sophisticated models there may be an automatic timer, safety features that seal the lid until the cycle has been completed and the contents have cooled to 80 °C, and cycle-stage indicators. A guide to the operation of a basic model is given in *Protocol 1*.

Protocol 1

Operation of a basic upright bench-top autoclave

Equipment

- Upright bench-top autoclave
- Items to be autoclaved

Protocol 1 continued

Method

1. Consult manufacturer's instructions before using instrument. The following protocol is for outline guidance only.
2. Remove lid and fill to required depth with water.
3. Place the items to be sterilized into the machine on shelves or in a basket.
4. Replace lid and seal, heat water to boiling.
5. Allow steam to escape for several minutes before closing the pressure valve (if appropriate).
6. Start timer when required pressure and temperature is reached.
7. At cycle end, turn off heat and allow pressure and temperature to fall.
8. Open the pressure valves and allow to cool to 80°C and atmospheric pressure before removing the contents.

ii. Horizontal models

These are essentially similar to the upright models described above but generally have more features such as both 121°C and 134°C cycles, a drying option using heat (but not vacuum), safety features such as a temperature lock, cycle-stage indicators, and automatic timers. However, steam is generated in an identical way to the upright models. An example of a small horizontal autoclave is shown in *Figure 1*.

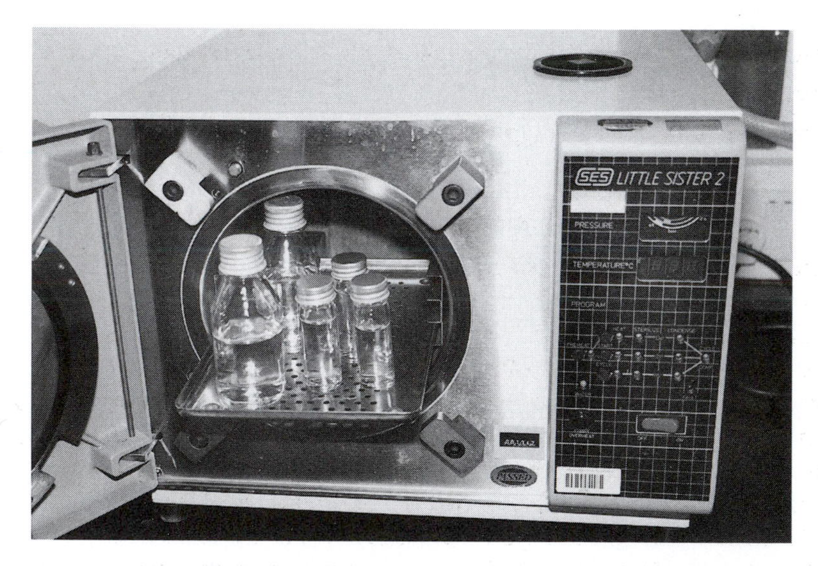

Figure 1 An example of a small horizontal laboratory autoclave. Bottled fluids have been loaded into the chamber.

iii. Uses and limitations

Some of the points that need to be considered when using these small types of autoclave are:-

(a) They have poor air removal and are therefore not suitable for critical porous loads such as wrapped items or laboratory discards.

(b) The use of wet steam and no, or limited, drying capability can lead to wet items.

(c) The temperature in the load is not monitored.

(d) There is no cycle record, i.e. print-out or trace.

(e) The load only cools slowly after sterilization.

(f) Safety features may be limited or absent.

(g) Their use is best limited to small items that have not been wrapped and bottled fluids with loosened caps.

4.3.2 Gravity displacement autoclaves

While there exists a range of larger capacity (e.g.100 litres) floor standing autoclaves, which are either top or front loading, these are often essentially larger versions of the bench-top models already described, although more sophisticated types exist. The next major advance in design involves the use of an external steam supply. In these, air is displaced from the bottom of the chamber by the less-dense steam entering at the top. These machines also possess a jacket for water and steam, or spray-cooling systems in some bottled fluids autoclaves, and thus cool the chamber contents more rapidly. A combination of heat in the jacket, combined with a vacuum, can be used to aid in the drying of the load. This type of autoclave is:-

(a) Useful for items that are easily exposed to steam, e.g. glass or plasticware.

(b) Suitable for bottled fluids.

(c) Not suitable for porous loads and wrapped items because the air removal is not adequate.

4.3.3 Multicycle porous-load autoclaves

Porous-load autoclaves (*Figure 2*) were developed because of the problem of air removal from wrapped items and other loads. These use a vacuum which, when used in pulses combined with steam, efficiently removes air from the load and chamber. It thus helps the penetration of steam into the load. The use of a vacuum also aids in the final drying stage. Such machines tend to be large, with a minimum overall size of about 4 m^3 and a chamber of at least 0.5 m^3, and expensive. Some machines used in hospitals or the pharmaceutical industry are far larger and materials can be directly wheeled in, or loaded via dedicated trolley systems. In the laboratory a multicycle machine is commonly used, with a number of different cycles (*Figure 3*) available for use depending on the nature of the load. This increases versatility in a multi-user situation. Parameters such

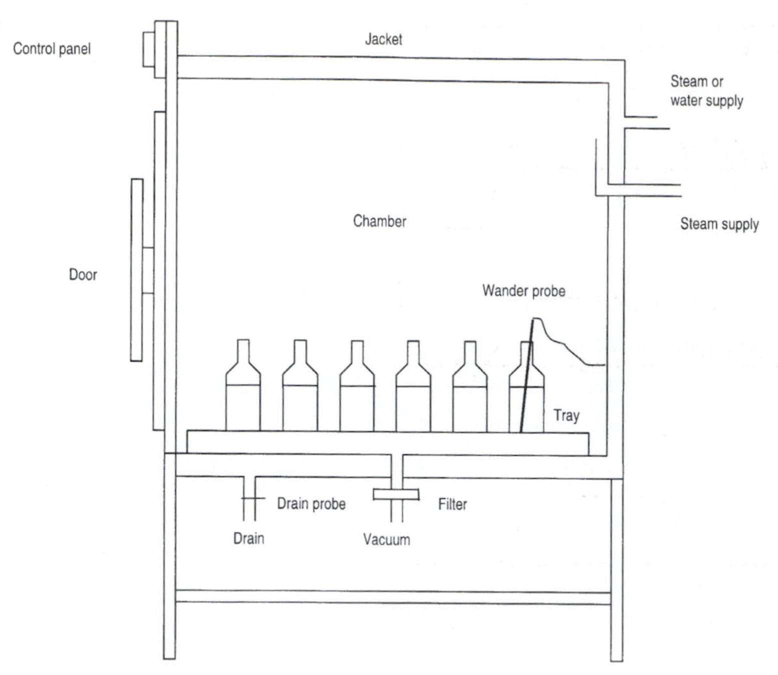

Figure 2 Diagrammatic representation of a large multicycle laboratory autoclave. Temperature probes are located both in the bottled fluids load (wander probe) and in the drain. The control panel contains temperature and pressure gauges, cycle-stage indicators, and a chart recorder/printer. A filter is usually placed on the vacuum line to prevent any micro-organisms from leaving the chamber when autoclaving contaminated discard loads.

as temperature and time can be pre-programmed into the machine and are often not user-adjustable. In addition, a cycle may be available in which any combination of these parameters can be set by the user immediately before use. Pressure and temperature during the run are indicated by dials or digital meters. In addition a printer and/or chart recorder is fitted to allow the temperature throughout the cycle to be recorded. A chart recorder is more easily checked by the user. On more recent models an air detector is usually fitted to confirm that all the air has been extracted from the chamber. Some points to consider with a porous-load autoclave are:-

(a) There is good steam penetration of porous loads, allowing wrapped goods and difficult items to be sterilized.

(b) Different cycles are available allowing flexibility of use.

(c) The chamber has a large capacity, e.g. > 0.5 m^3, and will accommodate e.g. 40 $\times$ 500 ml bottles, several large 5 or 10 litre vessels, or bags of discards.

(d) Sterilized items are free of moisture due to the presence of a heat and vacuum drying stage at the end of the cycle.

Because of the wide range of options and cycles available, this type of autoclave is usually made to order. Thus the user must carefully consider the

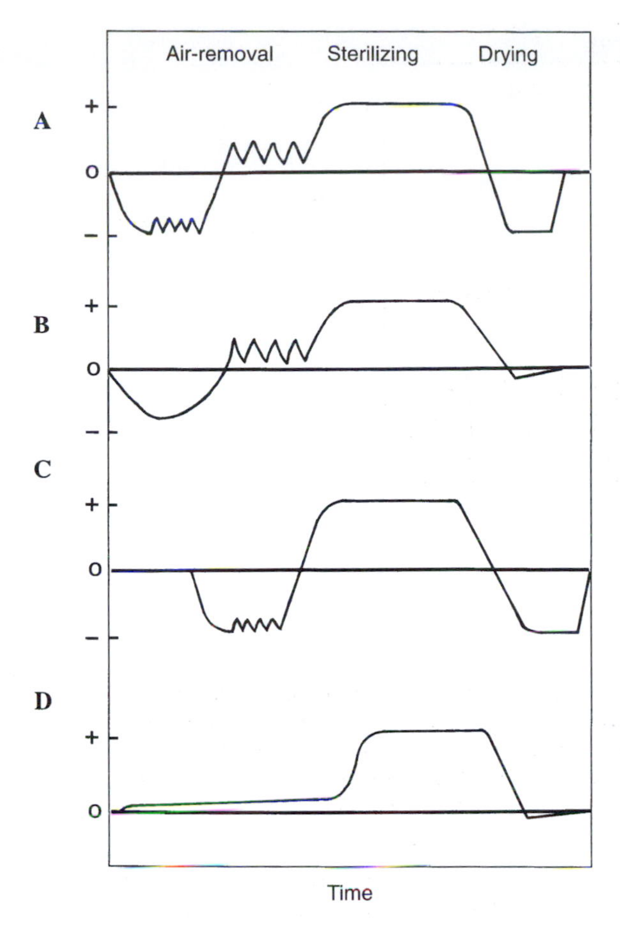

Figure 3 Typical cycles found on a representative multicycle porous-load autoclave. Pressures within the machine, whether positive (+) or negative (–) are indicated. Cycle A, which involves both negative and positive pulsing, is used for fabrics, assembled filter units, and discard loads from which it is particularly difficult to remove air. Cycle B uses a single negative pulse followed by positive pulsing and is used for laboratory plastic and glassware. Cycle C, which uses a series of negative pulses, can be used for discard loads. However, in validation studies, Cycle A has proved more effective for difficult discard loads. Cycle D is a bottled fluid cycle which uses air displacement for air removal.

applications and loads for which the machine will be used, in consultation with a sterilization engineer and the manufacturer.

4.4 Preparation of the load and operation of the autoclave

A basic outline procedure for preparing material to be autoclaved is given in *Protocol 2*. Examples of the types of load used with a large multicycle porous-load autoclave are shown in *Figure 4*.

Protocol 2

Preparation of autoclave loads

Equipment

- Aluminium foil
- Sterilization bags
- Autoclave tape
- Items to be autoclaved

A. Preparation of equipment

1. Wrap items in aluminium foil or place in sterilization bags using autoclave tape. Do not seal completely.
2. Loosen the caps of bottles, and other containers, or wrap up the open tops.
3. Some items, e.g. pipettes or micropipette tips, may conveniently be placed in containers such as tins or beakers which are then covered with aluminium foil or a sterilization bag. Loosen the tops of tins.
4. Micropipette tips are best sterilized in the boxed racks supplied by the manufacturer.
5. The necks of autoclave bags containing mixed discard loads must be opened to ensure good steam penetration.
6. Add a piece of autoclave-sensitive tape to the items in the load.

B. Preparation of fluids

1. For a fluid cycle, bottles may either be sealed or left unsealed.
2. Check caps for their ability to withstand autoclaving. With some types of cap the rubber liner may be sucked into the bottle.
3. Add autoclave-sensitive tape to the items in the load.

Before routine use, the sterilization of typical loads should be validated as described in Section 4.5. The heat-sensitivity of the items should be considered and tested before use. Many plastics can withstand 121 °C, but polystyrene cannot. Fluids such as water, salt solutions, and buffers can also be autoclaved. The heat-sensitivity of more complex formulations containing components such as vitamins, growth factors, antibiotics, or heat-sensitive amino acids, e.g. glutamine, should be considered or tested. Cell culture media and other culture reagents are only autoclavable in a few specific cases, e.g. autoclavable minimal essential medium without glutamine. Sodium bicarbonate solution can also be autoclaved but it is essential that sealed bottles are used. Their integrity can be confirmed after the run if phenol red is included and the solution adjusted to neutrality with CO_2 before sterilization. Phosphate-buffered saline formulations containing calcium or magnesium are best autoclaved at 115 °C to prevent the formation of any precipitate; however formulations lacking these components can be autoclaved at 121 °C. Temperature indicators should be included within the load. Their use on every item allows a clear distinction between items that have been

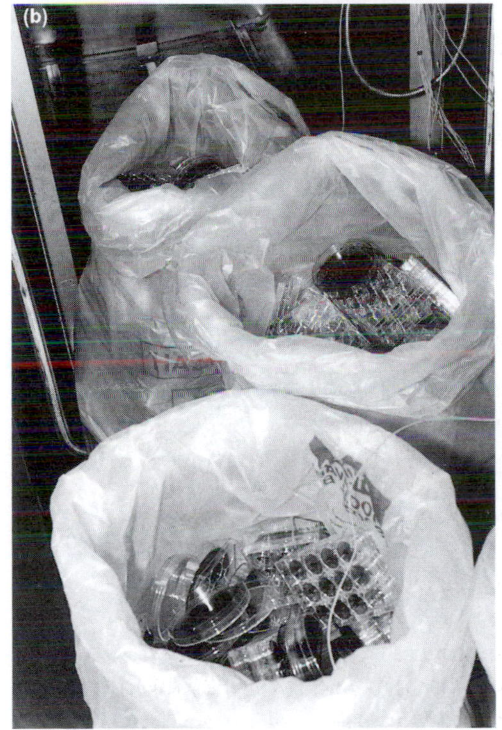

Figure 4 Typical loads for a large multicycle laboratory autoclave. (a) A bottled fluids load. The load temperature probe (wander probe) has been placed in one bottle in the centre. (b) A laboratory discard load. This comprises plastic Petri dishes and plates. The load probe will be placed in one of the bags. For both loads, additional temperature probes have been specially included in order to validate the autoclave cycles.

sterilized and those that have not. Such indicators are essential with the more basic types of autoclave considered earlier which do not record details of the cycle. The simplest system is the use of autoclave-sensitive tape, but autoclave tubes [Browne's tubes (Solmedia)] or indicator strips can be used. Some types of autoclave bag possess integral indicators. Only an approximate indication that sterilizing conditions have been met is given by such indicators. Nevertheless, they do provide a measure of assurance. If there is any problem with the wetness of porous loads at the end of the cycle, consideration should be given to modifying the length of the dry-cooling stage or to simply repeating this stage at the end of the run. *Protocol 3* outlines how the autoclave should be operated.

Protocol 3

Operation of a multicycle porous-load autoclave

Equipment

- Porous-load autoclave

Method

1. Follow the manufacturer's instructions and standard operating procedures.
2. Carry out routine or weekly tasks such as cleaning the door-seal, removing debris/broken glass from the chamber, and cleaning the drain filter.
3. Perform warm-up, leak-rate, or performance tests as necessary (see *Protocol 4*).
4. Load the chamber but do not overfill.
5. Fill in details of run in the log book (see *Protocol 4*).
6. Place the wander probe into the load. In the case of a bottled fluids cycle, place the probe into an identical control bottle with liquid which can be discarded later.
7. Select appropriate cycle and start the run.
8. On completion the cycle-end indicator will come on.
9. Check the chart paper or print-out for a satisfactory run.
10. Open the door and remove items using heat-resistant gloves to prevent injury. In the case of a liquid load that is still hot, it is advisable to stand clear while opening the door as a precautionary measure in case a bottle explodes.
11. Check autoclave tape or other in-load indicators. Label each item to indicate sterility if autoclave tape has not been applied to every item.
12. Complete log book record.

4.5 Autoclave testing

For an autoclave to perform satisfactorily and to ensure the sterility of the load, it is necessary for tests to be carried out by the user and by the sterilization engineer, and for records to be kept. This is also essential in order to meet Good

Laboratory Practice (GLP) and/or Good Manufacturing Practice (GMP) guidelines (see Section 11, and Chapter 10). Such testing goes hand in hand with a regular preventive maintenance programme. An indication of the sort of tests that should be performed (5) is given in *Protocol 4*.

Protocol 4

User tests and records for the autoclave

Equipment

- Autoclave
- Log book

Method

1. Record details of all runs, including pre-use warm-up and leak-rate tests, in a log book with details of date, load, cycle, result (pass or fail), operator, and other relevant information.
2. Carry out a leak-rate test, preceded by a warm-up run. This may be carried out daily on heavily used machines, but weekly is probably adequate in most cases.
3. At the end of every run, check the chart and/or digital recorder for a satisfactory cycle. Relevant parameters include temperature, time, and pulsing profiles.
4. At regular intervals, e.g. daily or weekly, carry out a Bowie–Dick test (see *Protocol 5*), to assess steam penetration into a standardized difficult load (6). This is less critical for modern machines fitted with an air detector.
5. Arrange for a sterilization engineer to carry out servicing at regular intervals, e.g. three to six months, and keep records of test results and work performance.
6. Before the machine is first used and at regular intervals thereafter, e.g. six monthly, arrange for the sterilization of critical worse-case loads to be validated by thermocouple studies carried out by sterilization/quality assurance personnel.

The Bowie–Dick test is designed to confirm that effective air removal and steam penetration into a standard test pack, during a porous load cycle, is both rapid and even. It is performed within one hour of a warm-up run. The standard method that is currently used, together with that originally described (6), are given in *Protocol 5*. An example of a test pack and test sheet is shown in *Figure 5*.

Protocol 5

Bowie–Dick test for porous-load autoclaves

Equipment

- Cotton sheets (approx. 90 × 120 cm), e.g. product code 9053 (Browne Health Care Ltd.)
- Sensor sheets (Browne, code 2385 or similar)

7 Chemical sterilization

7.1 Fumigation

Formaldehyde gas or ethylene oxide can be used for fumigation. Both are effective against all types of micro-organisms including viruses, although bacterial endospores are somewhat more resistant in the case of ethylene oxide. Their activity is greatest at higher temperatures and humidity levels of 75–100%. However, conditions that reduce the accessibility of the micro-organisms to the gas, e.g. being dried in organic or inorganic material, will decrease the effectiveness of fumigation. These fumigants acts as alkylating agents and act on both the nucleic acid and protein components of the micro-organisms.

7.1.1 Ethylene oxide

Ethylene oxide is commonly used for sterilizing items of clean equipment in low temperature autoclaves, or in combination with steam, particularly in hospitals. In the laboratory some items of plasticware may be purchased pre-sterilized with ethylene oxide, e.g. syringes and filters. However, this method of sterilization can leave behind toxic residues, and other methods, e.g. gamma-irradiation, may be preferable (see Section 8.1.1). Some additional points to consider include:-

(a) The gas is toxic and operator exposure needs to be controlled.

(b) Special equipment is needed.

(c) It is not suitable for decontaminating bulk discard loads.

7.1.2 Formaldehyde gas

This gas is commonly used to decontaminate or sterilize laminar airflow cabinets and rooms used for handling cell cultures, and also for small items of equipment. An outline of the procedure used when treating rooms is given in *Protocol* 7. It should be noted that formaldehyde gas is toxic and thus any necessary precautions, including the use of breathing apparatus, should be taken where necessary.

Protocol 7

Fumigation of rooms using formaldehyde

Equipment and reagents

- Formaldehyde generating kettle, or hotplate and a large saucepan
- Masking tape
- Formalin solution

Method

1 Remove all unnecessary items from the room, including those that must not be exposed to the gas, e.g. cell cultures, sensitive electrical equipment, etc.

Protocol 7 continued

2. Clean the room to minimize the level of microbial contamination and to allow good gas penetration.
3. Turn off air-handling system and seal up room with masking tape as far as possible.
4. Place a hot-plate with a large saucepan, connected to a timer, in the room (the time to boil off the liquid should be determined in advance). Alternatively use a dedicated formaldehyde generating kettle.[a]
5. Fill the container with formalin solution (20 ml/m^3 of room volume).
6. Turn on the hot-plate or kettle and leave the room.
7. Lock the door, tape it up fully (including the keyhole), and attach safety warning notice.
8. Leave until the next day.
9. If a total exhaust cabinet or air-extract is present, turn this on by a remote switch. If this is not possible enter wearing breathing apparatus and turn on cabinet and/or air-handling system.
10. Leave until the level of formaldehyde reaches an acceptable level. Testing equipment should be used.
11. The room may require cleaning to remove residues of paraformaldehyde and it may take several days to fully remove the gas.

[a] Other methods for generating formaldehyde gas exist, e.g. heating paraformaldehyde (10 g/m^3) and, for those situations where fumigation is carried out regularly such as in pharmaceutical manufacturing areas, equipment that generates a mist of formaldehyde (e.g. Phagojet, Laboratoires Phagogene).

Microbiological safety cabinets (only models that can be sealed and ducted to the outside) that are used for handling cell cultures may also be decontaminated with formaldehyde gas (*Protocol 8*) on a regular basis, e.g. once a week or once a month, or after it has been found that contaminated cultures have been handled.

Protocol 8

Fumigation of laminar airflow cabinets

Equipment and reagents

- Formaldehyde generator
- Masking tape
- Formalin solution

Method

1. Clean the cabinet.
2. Fill the integral formaldehyde generator attached to the exterior, or a small portable generator placed inside, with *c.* 25 ml of formalin solution.

Protocol 8 continued

3 Also place any items of equipment used in cell culture procedures in the cabinet, e.g. pipette-aids, etc.
4 Replace the door, and seal with masking tape.
5 Switch on formaldehyde generator and place warning notice on the cabinet.
6 Leave overnight before turning on air-exhaust while opening cabinet door.
7 Leave to vent before cleaning cabinet to remove residues of paraformaldehyde.

7.2 Liquid disinfectants

Many types of liquid disinfectant exist and they have a useful role in the cell culture laboratory. The main properties of some of the principal types is summarized in *Table 2*. Disinfectants can either be prepared directly using laboratory reagents or be purchased in the form of proprietary formulations. Some of the factors to be considered when using or selecting a suitable disinfectant are listed below:-

(a) The range and level of anti-microbial activity is less than with other methods of sterilization, e.g. heat.
(b) Spore-forming bacteria and some types of viruses may be resistant.
(c) Some disinfectants are neutralized by organic matter.
(d) The stability of 'working' dilutions varies with the type of disinfectant.
(e) The exposure time required depends on the type of disinfectant and how it is used.
(f) Toxicity to the user should be considered.

Diluted disinfectants can be used in the cell culture laboratory for:-

(a) Routine hygiene and disinfection of items of equipment and surfaces in rooms and cabinets.

Table 2 Effect of liquid disinfectants on micro-organisms

	Effectiveness[a]			
Type	**Fungi**	**Bacteria**	**Endospores**	**Viruses**
Aldehydes	+	+	+	+
Hypochlorites	+	+	+	+
Phenolics	+	+	–	+/v
Alcohol	–	+	–	+/v

[a] Effective (+) and non-effective (–). In the case of non-enveloped viruses, the effect may be variable or partial (v) depending on the particular virus. Examples of different types of disinfectant are given in Section 7.2.

(b) The disinfection of cell culture items after use and prior to washing and re-sterilization, e.g. glassware.

(c) Treatment of used or contaminated cell culture media before disposal.

The most suitable type of disinfectant for a particular application varies (see below). There are a number of methods available for applying the disinfectant. These include using a cloth or hand-held sprayer, or the use of paper towels applied to liquid spills and then soaked in disinfectants. For a full 'wet' disinfection of a room, a 'knapsack' sprayer, or large volume garden sprayer, with a long lance should be used. In this case the necessary safety precautions should be taken to avoid exposure of the face, eyes, and lungs. Items of glassware or plasticware must be fully immersed in the disinfectant to be fully effective.

7.2.1 Aldehydes

Formaldehyde and glutaraldehyde can both be used as liquid disinfectants, although glutaraldehyde is more commonly used and is probably more effective. They both have the advantage that they are not influenced by the presence of high levels of organic matter and they inactivate all types of conventional micro-organisms including bacterial endospores. Formaldehyde is used at about 4%, conveniently prepared by diluting 40% formaldehyde solution (formalin). Glutaraldehyde is used at 2%. Examples of commercial formulations include Cidex (Genesis Service Ltd.) and Gigasept (Sterling Medicare). Treatment times are typically of the order of at least 30 minutes, with longer times required to obtain full sporicidal activity. Proprietary glutaraldehyde-based disinfectants need the addition of an activator before use and then have a recommended lifetime of about one week. Aldehyde vapours, particularly glutaraldehyde, are considered relatively toxic due to their ability to sensitize, and they also have mutagenic and carcinogenic properties. Steps to limit exposure must therefore be taken.

7.2.2 Hypochlorite

One of the main advantages of using hypochlorite is its relatively low cost and ready availability, e.g. household bleach or Chloros (now called Haychlor Industrial; Hays Chemical Distribution Ltd.). However it has the disadvantage that it is not so effective in the presence of high levels of organic matter. It also corrodes metals and thus must not be used with centrifuge rotors or cell culture cabinets. It is not very stable after dilution. The concentration recommended for use when high levels of organic matter are present is 10 000 p.p.m. of available chlorine, although lower concentrations have been recommended for routine hygiene, e.g. 2500 p.p.m. The concentrate is stable but dilute disinfectant should be used within 24 hours. An exposure time of at least 30 minutes, and preferably overnight, is recommended. For treating spills, sodium dichloroisocyanurate powder, e.g. Haz-Tab granules (Guest Medical Ltd.), can be used to release high levels of chlorine rapidly.

7.2.3 Phenolics

Clear phenolic disinfectants, e.g. Hycolin (William Pearson Ltd.), although not inactivated by organic matter, have little activity against bacterial endospores. They are commonly used at a concentration of 2–5%, or as recommended by the manufacturer. They may also leave sticky residues when used for cleaning surfaces.

7.2.4 Alcohol

Ethanol is commonly used for disinfecting surfaces and hands (preferably gloved, but some people treat their hands directly). Its effect is optimal at a concentration of 70–80%, and surfaces must be fully saturated to ensure that the exposure time is sufficient. The ethanol/water mixture is then left to evaporate off naturally. Although a convenient and easy disinfectant to use, and of low toxicity, it is not very effective against fungi, bacterial endospores, or non-enveloped viruses. It is best used as a cleaning agent or disinfectant for less critical applications, e.g. for treating microbiological safety cabinets before and after routine cell culture work (see *Protocol 10*), or in combination with other disinfectants. Appropriate precautions should be taken because it is flammable.

7.2.5 Others

Many other disinfectants can be prepared in the laboratory or purchased as commercial formulations and may have a place in the cell culture laboratory. Examples include hydrogen peroxide at 5–10%, acids, alkalis, ethanol (70%) mixed with 4% formaldehyde or 2000 p.p.m. hypochlorite, or Microzid (Sterling Medicare), an ethanol/propanol/aldehyde mixture. One proprietary disinfectant (Virkon, Antec International), that has been shown to inactivate a wide range of viruses and other micro-organisms, has three components: an oxidizing agent (peroxide), acid (pH 2.6), and detergent. With this agent the recommended conditions are a 1% solution and an exposure time of at least 10 minutes. A face mask should be worn when handling the powder.

8 Filtration

8.1 Filters for bacteria and fungi

The earliest examples of this type of filter were developed in the late 1800s and used such materials as unglazed porcelain, diatomaceous earth, asbestos, or sintered glass. These materials acted as depth filters within which bacteria were trapped. For sterile filtration, such filters have largely been replaced by 0.2 μm membrane filters, first developed in the 1950s. They act more like sieves which trap the bacteria on their surface, although the distinction between the two types of filter is not absolute. Membrane filtration is the method commonly used for solutions that cannot be sterilized by methods such as autoclaving, because they contain heat-labile components. It suffers, however, from the disadvantage that, because sterilization does not take place in the final sealed container as is the case for example with autoclaving, steps must be taken to ensure sterility between filtration and bottling. Uses in the cell culture laboratory may include sterilizing culture media, sera, and other cell culture supplements, and any biological products made by the cells.

8.1.1 Types of filter

A range of filter materials are available which can be used for removing bacteria and fungi (7). For the filtration of liquids, these include materials made by casting processes, such as cellulose acetate, cellulose nitrate, a mixture of both, or nylon or polysulfone. Other materials with low protein-binding properties, e.g. polyvinylidene difluoride, are also available although for most applications in the cell culture laboratory (such as filtering cell culture media) this property is probably not really necessary. In addition, filters can be made from polycarbonate by the irradiation-etch method. These have discrete pores with a much narrower range of sizes. Full details of the various types of filters that are available, together with application and selection guides, and details on chemical compatibility and other aspects, can be found in the extensive literature produced by the various manufacturers such as Gelman Sciences, Nucleopore (Corning Costar), Sartorius, Pall, Millipore, etc.

Membrane filters come in a range of pore sizes, but 0.2 μm is considered the standard for removing bacteria and fungi. Larger pore sizes, and depth pre-filters, e.g. of glass fibre, are useful in serial filtration systems to increase the filtration capacity where significant levels of insoluble material may exist, such as in serum, complete medium containing serum, or cell culture supernatants. For effectively removing mycoplasma, filters with a pore size of 0.1 μm are required. Filters with pores of this size are now used by most commercial processors of serum products. In addition, filtration to this level may also be useful for removing some of the larger viruses that may be present in such products, e.g. infectious bovine rhinotracheitis virus and parainfluenza-3.

Low levels of various extractable materials may be found in membrane filters, e.g. surfactant used during manufacture, or residues of the ethylene oxide gas that is used by some manufacturers as a sterilizing agent (8, 9). For critical cell culture applications, where the complete absence of such residues is necessary, filters with very low extractable levels which have been sterilized with gamma-radiation can be used. Alternatively, where practical, simply discard a small volume of the initial filtrate. Apart from filtering liquid, membrane filters are also used for filtering gases, e.g. CO_2 or O_2 supplied to cell cultures growing in mass culture systems. Some types of cell culture flask (e.g. from Bibby Sterilin, Nunc, or Corning Costar) have filters inserted in the cap. Various types of filter, ranging from membrane filters to simple cotton wool plugs, are also used to protect liquid handling equipment such as pipettes, automatic pipette handling devices, and micropipette tips, from air- or liquid-borne contamination.

8.1.2 Types of filtration unit

In addition to the wide range of filter types that exist, filter housings also come in a range of shapes, sizes, and materials (see *Figure 8*). The filter manufacturers

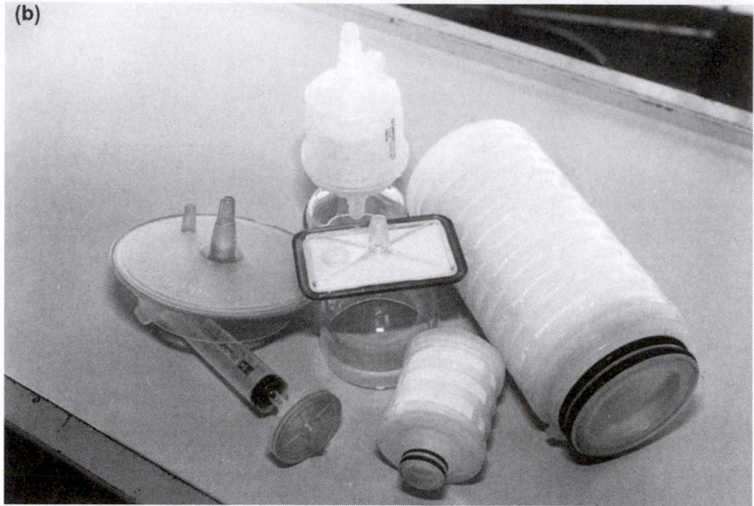

Figure 8 Examples of the range of different filtration units that are available. (a) Disc filters of various sizes, along with the plastic or stainless steel housings into which they are assembled. Note that a rubber washer is placed above the filter disc. After assembly, the unit can be autoclaved. (b) A range of filters in the form of sealed plastic units. Examples of disposable units that come pre-sterilized are shown on the left. They include examples with integral filling bells (designed to protect the filtered product from contamination, see *Figure 10*), a small syringe filter, and a filter designed for vacuum filtration directly into bottles. Examples of filter cartridges, containing pleated filters, are shown on the right. These are assembled in stainless steel holders and autoclaved before use.

should be consulted for details of the physical and chemical properties of both filters and housings. Sizes and formats range from the simple syringe filter of 13 or 25 mm diameter which will filter small volumes, to cartridges of various sizes and shapes, often pleated to maximize filter area, for filtering large volumes. Disposable units designed for filtration directly into a bottle (i.e. bottle-top filters), some with an integral receiving vessel, are also available.

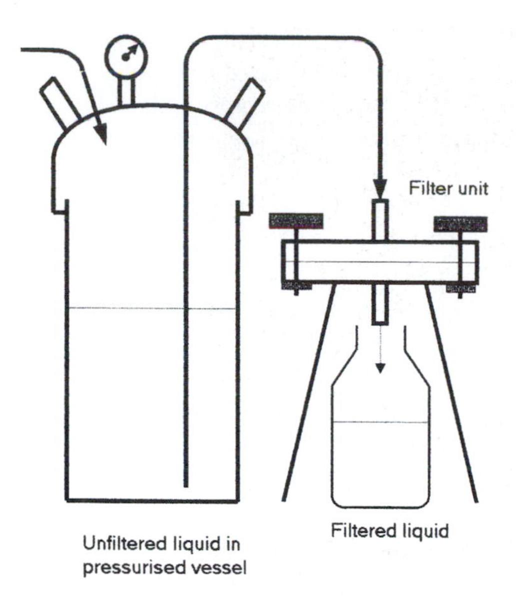

Figure 9 A diagrammatic representation of a system for filter sterilizing liquids. The liquid is held within a pressurized vessel prior to sterilization through a stainless steel housing containing a membrane filter. The direction of the flow is from left to right. Other filter units can be used (see *Figure 8*). An alternative filter system is shown in *Figure 10*.

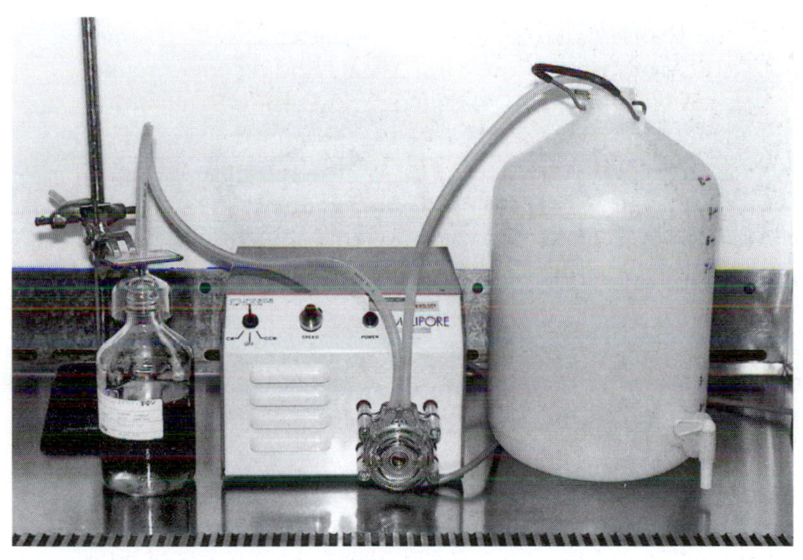

Figure 10 A system for filter sterilizing liquids using a peristaltic pump. A disposable pre-sterilized filter unit with an integral filling bell has been used. The direction of flow is from right to left. Sterile filtration is being carried out in a Class II MSC.

Filter units comprising the filter housing and the filter itself can be purchased either as pre-assembled, disposable units (which may be autoclaved or supplied pre-sterilized) or as a reusable housing which, after fitting the required filter(s), can be sterilized by autoclaving. Some points to consider with regard to the different systems include:-

(a) Disposable units:
- are convenient
- have high unit cost
- have quality control of filter batches, including integrity, carried out by the manufacturer

(b) Reusable systems:
- require assembly and sterilization before use, followed by disassembly and cleaning after use
- carry a greater risk of failure due to improper assembly
- after initial purchase of the housing, have low unit costs

8.1.3 Filtration set-up

Membrane filtration is carried out under positive pressure, e.g. a syringe for small volumes up to *c.* 50 ml, air pressure from a pump or a pressure-line, or a peristaltic pump. Alternatively, negative pressure can be used with bottle-top filters and filter units with integral receiving vessels, but caution must be taken to ensure appropriate vessels are used due to the risks of implosion. Negative pressure filtration has the disadvantage that, with tissue culture medium containing bicarbonate, the pH rise caused by filtration is greater than by positive pressure methods. In addition, when protein is present, frothing and protein denaturation can occur. After filtration, the liquid can either be collected into a single final sterile container or, where smaller aliquots are required, directly into vessels of a convenient size, e.g. 500 ml bottles for basal cell culture medium. Filter manufacturers or suppliers of cell culture equipment can provide various complete filtration systems designed for use with particular liquid volumes.

In *Protocol 9* the typical steps involved in liquid filtration are given. Examples of filtration systems are shown in *Figures 9* and *10*.

Protocol 9

Sterilization of liquids by membrane filtration

Equipment

- Filters
- Filter housing assembly
- Laminar flow cabinet/microbiological safety cabinet

A. Assembly and sterilization of the filter unit

1 Place a 0.2 μm or 0.1 μm filter in the disassembled unit taking care to orient the top of the filter in the correct way.

Protocol 9 continued

2. Where appropriate add additional filters of larger pore size, e.g. 0.45, 0.8, or 1.2 μm, with separators, and/or a depth pre-filter, as required to increase filtration capacity.
3. Assemble the unit taking care to install all supports and O-rings in their correct positions.
4. Attach tubing, if necessary, to inlet/outlet ports. Open any bleed points or air vents, if present, to allow good steam penetration.
5. Wrap the items and sterilize by autoclaving at 121 °C for 15 min using a porous load cycle. Higher temperatures, or dry heat sterilization, must not be used unless recommended by the filter manufacturer.

B. Filtration

1. Assemble any accessories that are required in a laminar flow or microbiological safety cabinet.
2. Remove the sterilized filter from its packaging using aseptic technique, and assemble the complete filtration system.
3. Tighten all connections and attach to the pressure system.
4. Start filtering into the sterile container(s), having first bled off any air in the system via the air vent (if present).
5. Test the integrity of the filter unit and/or carry out sterility tests on the filtered liquid as necessary.

8.1.4 Testing of bacterial filters

To ensure the correct functioning of the filtration system, the assembled unit, i.e. filter and housing, needs to be tested for integrity. This is essential to confirm that there are no defects in the filter that would allow micro-organisms to pass through. The manufacturers carry out various physical tests that are related to the pore size, e.g. bubble point or airflow/diffusion tests. These are then correlated with the actual performance of the filter or filter unit in tests designed to challenge the filters with high levels of a small test bacterium, e.g. *Pseudomonas diminuta* for a 0.2 μm filter. A number of quantitative procedures can be carried out by the user to confirm the integrity of the filter unit. The tests should be carried out after and, where considered necessary, before using the filter. The simplest method is to confirm that significant resistance (i.e. bubble point pressure—see below) is felt when trying to force air forwards or backwards through a wet filter. If air passes through unhindered then the filter is damaged. The quantitative methods are:-

(a) The bubble point method, which is based on determining the pressure required to force air through the wet filter.

(b) The diffusion airflow or forward flow method, which is based on determining the airflow across the wet filter at a specified pressure (*c.* 80% of the bubble point pressure).

(c) The pressure hold test, which is based on measuring pressure decay after pressurizing the upstream of the filter.

In all cases, the filter manufacturers provide information on the pass limits for specific types of filter when used with defined liquids. Dedicated testing equipment, that can be programmed to indicate pass or fail, is also available.

In addition to these physical tests, sterility testing of the final product, by inoculating a range of microbiological growth media, can be carried out. Both integrity testing and sterility testing are routinely carried out in the pharmaceutical industry.

8.2 Filters for viruses

Although 0.1 μm membrane filters can be useful for removing large viruses, particularly when several are used in series, membrane filters of a much smaller pore size are required for the removal of most other viruses. Recently some filter manufacturers have developed filters (10) which fulfil this requirement, e.g. Millipore (11, 12), Asahi Chemicals (13, 14), and Pall Corporation (15, 16). These filters are available in a range of formats, see *Table 3*. Such filters have been shown to remove viruses in a size-dependent fashion. Some filter types come in a range of pore sizes and the most appropriate type that will maximize virus removal without removing significant levels of any important media component or product should be selected. It is essential to test the suitability of such filters for any specific application to ensure that the protein of interest is not removed.

Some aspects to be considered with regard to virus-removing filters include:-

(a) This is a new technology and further developments are likely.

(b) The effect of filtration on the levels of important proteins, e.g. in the media or product, must be tested.

Table 3 Virus filtration systems

Trade name/ Manufacturer	Grade	Material	Mode of use		Mechanism
			Dead-end	Tangential	
Viresolve[a] (Millipore)	70 180	PVDF[b]	–	+	Membrane
Planova[c] (Asahi)	15N 35N	Cellulose[d]	+	+	Depth
Ultipor VF[e] (Pall)	DV-20 DV-50	PVDF[f]	+	–	Membrane

[a] Will allow passage of protein of about 70 kDa (70) or 180 kDa (180).

[b] PVDF, polyvinyldeoxyfluoride.

[c] Pore size of about 15 nm (15N) or 35 nm (35N).

[d] Cuprammonium-regenerated cellulose fibres.

[e] Will allow passage of protein of about 70–160 kDa (DV-20) or about 160–200 kDa (DV-50).

[f] Filter is composed of three discrete layers of PVDF.

(c) Removal of large (> 80 nm) viruses, e.g. infectious bovine rhinotracheitis (IBRV), Epstein–Barr virus, retroviruses, is most effective. Other medium-sized viruses (*c*. 50–60 nm), e.g. parainfluenza-3, and small viruses, e.g. picornaviruses, may also be removed to a lesser extent.

(d) Efficiency can be increased by using multiple units.

(e) Costs are high relative to standard filters.

(f) Units are designed for single use.

(g) Methods for testing the integrity of filters, that can be carried out by the user, have been developed.

Virus-removing filters have found application, in the cell culture field, in eliminating viruses such as bovine viral diarrhoea (BVDV) and other larger viruses such as IBRV and parainfluenza-3 from bovine serum. One manufacturer of bovine serum has used six filter units of 0.04 μm to remove BVDV and other possible viral contaminants (17, 18). Virus filters are currently being evaluated in the pharmaceutical industry for use with a range of cell culture-derived or plasma-derived biological products, and a wider range of filters based on this new technology are likely.

8.2.1 Testing of filters

As with sterilizing filters that are designed to remove bacteria, it is essential that integrity testing is carried out on viral filters. Different test methods are available depending on the type of filter. In addition, the filter manufacturer may use additional test methods during the quality control of the product. The methods that are used include:

(a) Gold particle removal test. This method involves evaluating the removal of gold particles slightly larger than the pore size of the filter. The absorbance of pre- and post-filtration samples are measured on a spectrophotometer. This method is used with the Planova filter system.

(b) Liquid/liquid intrusion test. The filter is first treated with a fluid to lower the pressure required to carry out the test. The pressure required to force a second fluid through the filter is then measured. This method is used with the Viresolve filter system; proprietary fluids are used.

(c) Forward-flow test. In this test method, the filter is first treated with isopropanol in order to reduce the pressure required to carry out the test. The pressure required to force air through the filter is then measured. This method is used with the Ultipor-VF filter system.

In all cases the performance of the filter in the integrity test has been correlated with the removal of a representative virus, usually a bacteriophage which has been selected to have a size near to the pore size. A pass value is recommended by the filter manufacturer. However a pass limit for such tests can usually also be selected by the filter user, based on the graph of correlation between virus removal and integrity test value supplied by the filter manufacturer.

It is essential that the integrity test is carried out as soon as possible after filter use in order that, if a filter is found to have failed, the batch of filtered material can be re-processed. In some cases it is particularly critical to ensure that the test is carried out correctly because it may not be possible to repeat the test; for instance the gold particle test used with Planova filters is destructive.

8.3 HEPA filters

HEPA (high efficiency particulate air) filters are used to filter large volumes of air in sterile or clean rooms, and in laminar flow or microbiological safety cabinets (19, 20). Such filters act as depth-type filters and are capable of removing $> 99.97\%$ of particles of 0.3 μm or larger. They are thus effective against not only bacteria but also viruses which exist in the atmosphere attached to dust particles and liquid droplets. In the pharmaceutical industry standards exist for the quality of air, i.e. the maximum level of particles and viable micro-organisms permitted for a particular class of room or cabinet. For aseptic filling, an environment with no more than 100 particles of 0.5 μm per cubic ft ($3530/m^3$) is required to meet class M3.5 of US Federal Standard 209E. Other clean room classification standards, e.g. BS5295 exist (19). The various types of cabinet that use HEPA filters are listed in *Table 4*. Cell culture should be considered as a potential source of viruses (see Chapters 5 and 9), and it is thus best to handle them in a Class II microbiological safety cabinet (MSC) which provides protection to both the operator and the cells. Laminar flow cabinets with a horizontal flow must **never** be used for handling cells and it is recommended that even models with vertical flow are not used for such purposes (see Chapter 5). Class I MSCs offer no protection to the work but are recommended because they give a high and consistent level of operator protection when handling dangerous pathogens (Hazard Group 3) (21). If it is considered essential that cell cultures and Hazard Group 3 agents are handled together in a Class II MSC, in order to provide protection to the cells, then a case must be made to the Health and Safety Executive or other relevant agency. Additional safety testing of the MSC is likely to be required. Alternatively a fully enclosed Class III cabinet would provide full operator protection and

Table 4 Types of cabinet used for handling cell cultures

	Protection[a]		
Type	**Operator**	**Product**	**Airflow[b]**
Laminar flow cabinet (vertical)	–	+	V
Laminar flow cabinet (horizontal)	–	+	H
Microbiological safety cabinet (Class I)	+	–	X
Microbiological safety cabinet (Class II)[c]	+	+	V, X
Microbiological safety cabinet (Class III)	+	+	E, X

[a] Protection (+), no protection (–).

[b] Vertical laminar airflow (V), horizontal laminar airflow (H), air filtered on exhaust (X), or air filtered on entry and exhaust (E, X).

[c] Also commonly referred to as Class II hood or cabinet.

some level of protection to the work because the incoming air is filtered. A Class II MSC is in fact adequate for work with most common microbiological agents, i.e. Hazard Group 2 pathogens (21). Guidance on the routine use and maintenance of cabinets is given in *Protocol 10*.

Protocol 10

Use and maintenance of Class I and Class II microbiological safety cabinets

1. Turn on airflow while removing door panel. Leave the door in a suitable clean location (do not place on the floor).
2. Spray base and sides of cabinet with 70% ethanol and leave to evaporate.
3. Leave to run for $>$ 5 min.
4. Check airflow dial reading is correct and in 'safe' position.
5. Work in cabinet using standard aseptic technique and limiting rapid body movement.
6. After use, remove all items from the cabinet and clean the interior with disinfectant. Finally spray with 70% ethanol.
7. Turn off cabinet and replace the door.
8. At regular intervals or after handling contaminated cultures etc., fumigate the cabinet if possible (i.e. if extract ducted to the outside)—see *Protocol 8*. Alternatively fully clean and disinfect the interior including underneath the base.
9. Have the performance of the cabinet tested at about yearly intervals.

8.3.1 Performance testing

HEPA filters and cabinets should be tested at least once each year, or every six months in the case of a room or cabinet used for the sterile filling of pharmaceutical products or for handling dangerous pathogens, to ensure they are performing correctly. Cabinet manufacturers and clean room environment specialists are able to carry out the necessary tests. For MSCs these include challenging the HEPA filter with dioctyl-phthalate smoke of about 0.3 μm diameter, and measuring the air inflow and downflow velocity (as necessary for the particular type of cabinet), to confirm they meet the required standard, e.g. BS5276 in the UK. An operator protection test, using potassium iodide generated within the cabinet (KI-Discus test) should also be carried out when the cabinet is first installed. This test is not strictly required again unless the cabinet is moved or relatively high risk pathogens (i.e. those in Hazard Group 3) (21) are handled. The level of particles within a sterile room can be directly monitored by the use of a particle counter. Additional biological tests for product protection in rooms and cabinets used for sterile filling can be carried out by exposing bacteriological agar plates within the cabinet.

9 Viruses and prions

9.1 Virus elimination

Viruses pose a particularly difficult problem in cell culture due to their ability to be latent or persistent without causing any obvious effects. The first approach for controlling such agents is the screening of the cell line. This is dealt with in more detail in Chapter 9. However, there are limitations to this approach. The detection methods may be of limited sensitivity and be too specific with regard to the range of viruses they can detect. An alternative method is to inactivate or remove any virus that may be present (22–25). While this approach is obviously difficult to use with the cells themselves, it can be used with the biological additives added to cell cultures or with products produced by the cultures. However the methods that are commonly used are not as effective as standard sterilization methods such as autoclaving. The best virus reduction methods should be effective against the viruses of main concern, and preferably against as wide a range of virus types as possible. The method should be validated, and shown not to be influenced by process variables, for the specific product involved. For critical situations, multiple virus reduction steps should be considered.

The standard purification method used for a particular product may in itself be effective for removing/inactivating viruses. Manufacturing steps such as precipitation, chromatography, freeze-drying, low pH, ethanol treatment, and product storage, may all contribute significantly to virus reduction. In the case of chromatography, affinity-based methods may be particularly effective.

A wide range of methods have been specifically developed for inactivating and/or removing viruses from biological products. Heat-treatment in the liquid state, i.e. pasteurization, is one commonly used method. For instance, in the case of human albumin, heating at 80 °C for 10 hours is routinely used. This procedure has also been applied to other products; however the addition of specific stabilizers, e.g. sodium octanoate in the case of human albumin, or general stabilizers is required in all cases to prevent protein denaturation. Less severe treatments, e.g. 56 °C for 30 minutes (26) or 56 °C for 3 hours (27) have been shown to be effective for inactivating some types of virus in calf serum. An alternative approach is to heat the protein after freeze-drying. Treatment conditions of 80 °C for 72 hours or even 100 °C for 24 hours have been used with coagulation factors. Chemical methods can also be used to inactivate viruses. The most widely used chemical method is inactivation with solvent/detergent (28). This method was originally developed for inactivating enveloped viruses in plasma products but is also being used for products derived from cell culture. The product is treated with the solvent tri-*n*-butyl phosphate, combined with a suitable non-ionic detergent, e.g. sodium cholate, Tween 80, or Triton X-100. Low pH can also be used for virus inactivation; the use of pH 5 for 30 minutes has been evaluated with calf serum (27). Virus retentive filters are increasingly being used (see Section 8.2).

Various other methods are under development. Others have as yet not found widespread application, for example photochemical methods, in which a chemical is added which is activated by exposure to UV or visible light. Some of these methods may be able to inactivate intracellular as well as extracellular viruses. Ultraviolet light in the UV-C region, i.e. 200–290 nm, is also effective and some manufacturers treat fetal calf serum in this way.

9.2 Prions

Prions or unconventional agents are the agents that cause transmissible spongiform encephalopathies (TSEs) and have become of increasing concern over recent years (4). The diseases caused by these agents include Creutzfeldt–Jakob disease (CJD) and Kuru in man, bovine spongiform encephalopathy (BSE) in cattle, scrapie in sheep, and feline spongiform encephalitis in cats. CJD became of increasing concern when it was realized that human-derived tissue, e.g. cornea, or products like human growth hormone, could transmit the disease. In the case of Kuru, the agent was transmitted during ritual cannibalistic practices in which the brains of dead relatives were consumed. The belief that the TSEs associated with animals were species-specific meant that these agents were of little concern to humans. However, laboratory studies have shown that TSEs can be adapted to new species if large amounts of prion are introduced by an appropriate route, e.g. intracerebrally. In addition, there is now evidence that BSE has been adapted to humans, i.e. the so-called variant form of CJD (vCJD), presumably as a result of the ingestion of BSE contaminated meat (29, 30).

Prions differ considerably from conventional viruses. A nucleic acid component has not been detected and infectivity is always associated with a specific protein, i.e. the prion protein (PrP). This has led to the theory that these are a unique class of infectious agents that are composed of protein alone. The prion protein is in fact a normal cellular component (PrP^c) of unknown function, which can undergo a conformational change to the abnormal form (PrP^{Sc}). This abnormal form is resistant to protease, is tightly membrane-bound and aggregated, and precipitates into characteristic amyloid plaques in the brain.

Unfortunately, prions are extremely resistant to inactivation by conventional sterilization methods (31, 32). In addition, suitable screening methods have yet to be developed that are simple, sensitive, and specific enough for rapid detection. Currently the most sensitive detection methods rely on infectivity titration using mice or hamsters and take about a year to complete. Consequently, at present, health screening and sourcing of biological materials is the approach recommended for controlling the transmission of such agents (33). Thus, in the case of cell culture, it is recommended that alternatives to bovine-derived serum are used. Where this is unavoidable, material from countries that are free of BSE should be used (see Chapter 3, Sections 3.4.2 and 3.5).

Details of the sterilization methods (31, 32) that can be used for prions are summarized in *Table 5*. However, even those methods that have been recommended may only be partially effective. This is due to the fact that there are both a range of different types of prion, as well as different strains of each type. These all vary with regard to their susceptibility to inactivation. Even the most effect-

Table 5 Effectiveness of sterilization methods for inactivating or removing prions

Method		**Effectiveness**[a]
UV and gamma-irradiation		–
Dry heat	160 °C/24 h	–
	360 °C/1 h	+
	Incineration	+
Autoclave	132 °C/1 h[b]	+
	134–138 °C/18 min[c]	+
Potassium permanganate		+
Urea		+
Hypochlorite 20 000 p.p.m./1 h		+
Sodium hydroxide 2 M/1 h		+
Formic acid 96%/1 h		+
Peracetic acid		–
Hydrochloric acid		–
Organic solvents		–
Formaldehyde / formal saline		–
Glutaraldehyde		–
β-Propiolactone		–
Phenolics		–
Hydrogen peroxide		–
Membrane filtration 15 nm		+

[a] Procedures that are effective or partially effective, are indicated (+). Types and strains of prion vary widely in their resistance, making it difficult to generalize.

[b] Using a gravity displacement cycle.

[c] Using an effective porous load cycle.

ive methods do not provide a level of inactivation comparable to that obtained with conventional micro-organisms using standard sterilization methods. One particularly severe approach that has recently been proposed involves treatment with sodium hydroxide followed by, or in combination with, autoclaving. It should be noted that some sterilization methods that are only partially effective have resulted in residual infectivity with an enhanced resistance to other inactivation methods (31). One simple and basic approach is to include the use of vigorous and repeated washing procedures prior to the use of a specific disinfection method.

Treatment with sodium hydroxide or sodium hypochlorite is the most commonly used chemical method. Autoclaving at 132 °C for 1 hour, or 134–138 °C for 18 minutes, are currently the recommended heat-treatment cycles. In the latter case, it is essential to use an autoclave with an effective, i.e. validated, porous load cycle.

Not surprisingly, therefore, it has proved difficult to inactivate prions in biological materials without destroying the material itself. Removal methods, such as the use of virus filters with a pore size of about 15 nm may be effective (34, 35) as long as the prion remains in a highly aggregated form. In addition, prions may be removed by many other manufacturing processes including filtration, centrifugation, chemical precipitation, and chromatography.

10 Safety in the laboratory

Many of the procedures used for sterilization are potentially dangerous. For this reason, appropriate precautions must be taken to prevent injury. The responsibility for safety rests with the laboratory manager as well as those working in the laboratory. In the UK, the Control of Substances Hazardous to Health (COSHH) regulations are designed to control risks from both chemical and biological agents. In addition, there are risks from the equipment being used. The risks associated with all these aspects need to be considered and safe working practices developed. The safety aspects of handling micro-organisms are considered in more detail in specific publications (20, 21). Cell culture should also be considered a risk, both because of the ability of cells to harbour viruses as well as the theoretical potential of cells to colonize individuals (see also Chapter 5).

11 Good Manufacturing Practice and Good Laboratory Practice

Manufacturers of sterile pharmaceutical products, which include those derived from cells, are required by regulatory authorities to follow GMP guidelines (37). These cover all aspects of the manufacturing process, including the operation of equipment/procedures used for sterilization.

GLP may also be relevant in the cell culture laboratory (38). Such guidelines must be followed in studies designed to test the safety of a therapeutic product. In the case of cell cultures, this may involve testing cell lines for adventitious agents or evaluating manufacturing processes for their ability to inactivate/ remove viruses. GLP is considered in detail in Chapter 10.

Acknowledgements

I would like to thank colleagues at BPL for useful discussions and Sebastian Roberts and Karin Roberts for help in preparing the illustrations and manuscript.

References

1. Block, S. S. (ed.) (1983). *Disinfection, sterilization and preservation* (3rd edn). Lea and Febiger, Philadelphia.
2. Russell, A. D., Hugo, W. B., and Ayliffe, G. A. J. (ed.) (1992). *Principles and practice of disinfection, preservation and sterilization* (2nd edn). Blackwell Scientific Publications, Oxford.

3. Threlfall, G. and Garland, S. G. (1985). In *Animal cell biotechnology* (ed. R. E. Spier and J. B. Griffiths), Vol. 1, pp. 123–40. Academic Press, London.
4. Prusiner, S. B. (1991). *Dev. Biol. Stand.*, **75**, 55.
5. Department of Health and Social Security. (1980). *Sterilisers.* Health Technical Memorandum, no. 10. HMSO, London.
6. Bowie, J. H., Kelsey, J. C., and Thompson, G. R. (1963). *Lancet*, **i**, 586.
7. Brock, T. D. (1983). *Membrane filtration*. Springer-Verlag, Berlin.
8. Gelman Sciences. (1993). *Laboratory Solutions*, **1**, 1.
9. Knight, D. E. (1990). *Nature*, **343**, 218.
10. Roberts, P. (2000). *Eur. J. Parent. Sci.*, **5**, 3.
11. DiLeo, A. J., Allegrezza, A. E. Jr., and Builder, S. E. (1992). *Biotechnology*, **10**, 182.
12. DiLeo, A. J., Vacante, D. A., and Dean, E. F. (1993). *Biologicals*, **21**, 287.
13. Manabe, S. (1992). In *Animal cell technology: basic and applied aspects* (ed. H. Murakami, S. Shirahata, and H. Tachibana), pp. 15–30. Kluwer Academic Publishers, Dordrecht, Netherlands.
14. Burnouf-Radosevich, M., Apourchaux, P., Huart, J. J., and Burnouf, T. (1994). *Vox Sang.*, **67**, 132.
15. Roberts, P. (1997). *J. Virol. Methods*, **65**, 27.
16. Oshima, K. H., Evans-Strickfaden, T. T., Highsmith, A. K., and Ades, E. W. (1996). *Biologicals*, **24**, 137.
17. Pall Process Filtration Ltd. (1991). Retention of Viral Contaminants by 0.04 μm Nylon 66 Filters. Pall Scientific and Technical Report STR 1358. Pall Process Filtration Ltd., Portsmouth.
18. HyClone Laboratories. (1987). *Art to science in tissue culture*, **5**, 1.
19. Davis, J. M. and Shade, K. L. (2000). In *The encyclopedia of cell technology* (ed. R. E. Spier), p. 108. John Wiley & Sons, New York.
20. Collins, C. H. (ed.) (1993). *Laboratory-acquired infections* (3rd edn). Butterworths, London.
21. Advisory Committee on Dangerous Pathogens. (1995). *Categorisation of biological agents according to hazard and categories of containment* (4th edn). HSE Books, London.
22. Roberts, P. L. (1996). *Rev. Med. Virol.*, **6**, 25.
23. Prudouz, K. N. and Fratantoni, J. C. (1994). In *Scientific basis of transfusion medicine* (ed. K. C. Anderson and P. M. Ness), p. 852. Saunders, Philadelphia.
24. Burnouf, T. (1992). *Biologicals*, **20**, 91.
25. Foster, P. R. and Cuthbertson, B. (1994). In *Blood, blood products and HIV* (ed. R. Madhok, C. D. Forbes, and B. L. Evatt), p. 207. Chapman and Hall, London.
26. Danner, D. J., Smith, T., and Plavsic, M. (1998). *Bio Pharm*, **50**, 50.
27. Hanson, G. (1997). Handbook of the IBC Symposium on viral clearance, Oct 27–28, 1997, Philadelphia. IBC Conferences Inc., Southborough, MA, USA.
28. Horowitz, B., Prince, A. M., Horowitz, M. S., and Watklevicz, C. (1993). *Dev. Biol. Stand.*, **81**, 147.
29. Bruce, M. E., Will, R. G., Ironside, J. W., McConnell, I., Drummond, D., Suttie, A., *et al.* (1997). *Nature*, **389**, 498.
30. Hill, A. F., Desbruslais, M., Joiner, S., Sidle, K. C. L., Gowland, I., Collinge, J., *et al.* (1997). *Nature*, **389**, 448.
31. Taylor, D. M. (1996). *Br. Food J.*, **98**, 36.
32. Taylor, D. M. (1991). *Dev. Biol. Stand.*, **75**, 97.
33. Committee for Proprietary Medicinal Products, European Medicines Evaluation Agency. (1997). *Note for guidance on minimising the risk of transmitting animal spongiform encephalopathy agents via medicinal products.* CPMP/BWP/877/96.
34. Tateishi, J., Kitamoto, T., Ishikawa, G., Manabe, S., and Yamaguchi, K. (1995). In *Animal cell technology: developments towards the 21st century* (ed. E. C. Beuvery), p. 637. Kluwer Academic Publishers, Dordrecht, Netherlands.

35. Tateishi, J., Kitamoto, T., Ishikawa, G., and Manabe, S. I. (1993). *Membrane*, **18**, 101.
36. Advisory Committee on Dangerous Pathogens. (1998). *Transmissable spongiform encephalopathy agents: safe working and the prevention of infection.* The Stationery Office, London.
37. Medicines Control Agency. (1997). *Rules and guidance for pharmaceutical manufacturers and distributors.* The Stationery Office, London.
38. Organization for Economic Cooperation and Development. (1998). *Principles of Good Laboratory Practice.* OECD Publications, Paris.

Chapter 3
Culture media

T. Cartwright
TCS CellWorks Ltd., Botolph Claydon, Buckingham MK18 2LR, UK.

G. P. Shah
Biomedical Structures, GlaxoWellcome, Gunnels Wood Road, Stevenage SG1 2NY, UK.

1 Introduction

Successful growth and maintenance of animal cells *in vitro* requires culture conditions that mimic *in vivo* conditions with respect to temperature, oxygen and carbon dioxide concentration, pH, osmolality, and nutrition. For this reason, early tissue culture media were based on biological fluids such as plasma, lymph and serum, and tissue extracts especially of embryonic origin.

The culture medium is by far the most important single factor in culturing animal cells. Its function is to provide the pH and osmolality environment essential for cell survival and multiplication and also to provide all the complex chemical substances required by the cell which it cannot synthesize itself. Some of these requirements can be provided by a culture medium composed of simple low molecular weight components generally referred to as a *basal medium*. However, most basal media cannot, by themselves, support the growth of cells and it is common practice to supplement cell culture media with more complex, chemically undefined additives such as animal sera. There have been many attempts to devise media which do not require supplementation of this type, to produce a *serum-free medium*. Growing cells in serum-free media has many practical advantages, but a general purpose serum-free medium has not yet been developed (and is almost certainly an unattainable goal). A complete cell culture medium can thus be considered to be composed of two distinct parts:

(a) A basal medium that satisfies cellular requirements for nutrients, salts, and pH control.

(b) A set of supplements that satisfy other cellular requirements and permit growth of cells in the basal medium.

A nutrient is defined as a chemical substance that enters a cell and is used as either a structural component, as a substrate for biosynthesis or energy metabolism, or in a catalytic role in such metabolism (1). Anything else needed for cellular proliferation is normally classified as a supplement, including all undefined additives such as serum and other biological fluids.

2 Basal media

A range of basal tissue culture media have been developed to cover the main types of cells in common use. Some of these were formulated to include only the minimum number of components essential for cell growth (e.g. Eagle's basal medium) while others were developed to mimic the composition of plasma (e.g. TC199) and contain many ingredients.

2.1 Types of basal media

There are four main categories of basal media for mammalian cells and six main categories for insect cells. These are:

(a) Eagle's medium and derivatives, e.g. BME (basal medium Eagle's), EMEM (minimum essential medium with Earl's salts), AMEM (minimum essential medium with alpha modification), DMEM (Dulbecco's modified minimum essential medium), GMEM (minimum essential medium with Glasgow modification), and JMEM (minimum essential medium with Joklik's modification).

(b) Roswell Park Memorial Institute (RPMI) designed media, e.g. RPMI 1629, RPMI 1630, and RPMI 1640.

(c) Basal media designed for use with serum supplementation, e.g. Fischer's, Liebovitz, Trowell, and Williams'.

(d) Basal media designed for serum-free formulations, e.g. CMRL 1060, Ham's F10 and derivatives, TC199 and derivatives, MCDB and derivatives, NCTC, and Waymouth.

For insect cell culture, the basal media (designed empirically) are:

(a) Grace's medium

(b) Schneider's medium

(c) Mitsuhashi and Maramorosch medium

(d) IPL-41 medium

(e) Chiu and Black medium

(f) D-22 medium

The above basal media are also suitable for the culture of many cell lines derived from cold-blooded vertebrates (when supplemented with an appropriate biological fluid). However, allowance is usually made for differences in salt concentration to obtain the optimum osmolality. Growth at different incubation temperatures may also require adjustments to be made to the pH of the medium, due to alterations in the solubility of CO_2, and the ionization and pK_a of buffers with temperature.

2.2 Constituents of basal media

The standard components required in basal media may be summarized as follows.

2.2.1 Balanced salt solution

Balanced salt solutions (BSSs) have been used since the earliest attempts at culturing cells *in vitro*. A BSS is composed of a combination of inorganic salts that maintain osmotic pressure and buffer the medium at physiological pH. In addition to these effects, the inorganic ions used have other important physiological roles including the maintenance of membrane potential and acting as cofactors in enzyme reactions and in cell attachment. The inorganic ions used are chiefly those of Na^+, K^+, Mg^{2+}, Ca^{2+}, Cl^-, SO_4^{2-}, PO_4^{3-}, and HCO_3^-. When necessary, osmolality can be adjusted by modifying the concentration of NaCl.

Most BSSs do not contain the nutrients required by cells for long-term maintenance or growth, although glucose and sodium bicarbonate may sometimes also be included. There are four main categories of BSSs. These are:

(a) Earle's balanced salt solution (EBSS).

(b) Dulbecco's phosphate-buffered saline (DPBS).

(c) Hank's balanced salt solution (HBSS).

(d) Eagle's spinner salt solution (ESSS).

Balanced salt solutions are of two types:

(a) Those that can maintain the desired pH when equilibrated with air (e.g. HBSS and DPBS).

(b) Those which require equilibration with a gas phase containing CO_2 at a concentration between 5% and 10% (e.g. EBSS and ESSS).

2.2.2 Buffering systems

Culture media generally need to be buffered to compensate for evolution of CO_2 and the production of lactic acid from the metabolism of carbohydrates. Media are frequently buffered with bicarbonate which forms a buffering system with dissolved CO_2 produced in the medium by growing cells. Rapidly growing cells can produce enough CO_2 themselves to maintain pH, but when cells are growing at a low cell density or are in a lag phase, they may produce insufficient CO_2 to maintain the required optimal pH. For this reason, such cultures must be grown in an atmosphere of 5–10% CO_2.

Bicarbonate is both cheap and non-toxic to the cells but its pK_a (6.1 at 37 °C) results in suboptimal buffering in the optimal pH range for cell growth (7.0–7.5). Some media are designed to contain low HCO_3^- but high PO_4^{3-} concentrations and therefore do not require incubation in a CO_2-enriched atmosphere. Sodium β-glycerophosphate is also used as a buffer in some formulations. Each of the basal media has a recommended bicarbonate concentration and CO_2 tension to achieve the correct pH and osmolality (*Table 1*).

Table 1 Recommended CO_2 concentration (gas phase) to use with common basal media

Basal medium	$NaHCO_3$ concentration (mM)	% CO_2 in gas phase
Eagle's MEM (Hank's salts)	4	Atmosphere
Grace's (Hank's salts)	4	Atmosphere
IPL-41 (Hank's salts)	4	Atmosphere
TC100 (Hank's salts)	4	Atmosphere
Schneider's (Hank's salts)	4	Atmosphere
IMDM	36	5
TC199	26	5
DMEM/Ham's F12	29	5
RPMI 1640	24	5
Ham's F12	14	5
DMEM	44	10

For more effective buffering, without the need for elevated CO_2 levels, a range of organic buffers can be employed. The most widely used of these is Hepes (*N*-2-hydroxyethyl piperazine-*N*′-2-ethanesulfonic acid) which has a pK_a of 7.3 at 37 °C. Hepes is therefore a very effective buffer in the pH range 7.2–7.6 and is more resistant to rapid pH changes than bicarbonate. The use of such organic buffers eliminates the need for cell growth in a CO_2-controlled atmosphere. Some media are buffered effectively with both bicarbonate and Hepes. However, Hepes is both expensive and toxic to cells at concentrations above 100 mM. Impurities in Hepes have been shown to cause cytotoxicity at lower effective Hepes concentrations. Other organic buffers, related to Hepes, which have been used are Tes {*N*-Tris (hydroxymethyl)-2-aminoethanesulfonic acid} and Bes {*N,N*-bis (2-hydroxyethyl)-2-aminoethanesulfonic acid}. Although these organic buffers function without CO_2, it should be remembered that bicarbonate is essential to cells as a nutrient independently of its buffering role, and sufficient bicarbonate for this requirement must always be present in the medium. Actively growing cells will usually generate sufficient CO_2 for this themselves.

2.2.3 pH indicators

Unless the cells are being grown in a pH-controlled bioreactor, it is usual practice to add a pH indicator to the medium to facilitate inspection and control of cultures. The most widely used indicator is phenol red which is a cherry red colour in the optimal pH range for the growth of most cells (~ pH 7.0–7.5), but which turns yellow when the culture becomes too acid and purplish when the pH is too high. It should however be noted that most commercially available phenol red preparations contain impurities that have a significant oestrogenic activity (2) and that the use of phenol red-free media is advised when steroid hormone effects on cells are being studied. Phenol red can also interfere with certain fluorescence techniques (see Chapter 4).

2.2.4 Energy sources

Carbohydrates are the major energy source used by cultured cells and glucose is the most frequently used sugar. Other sugars, e.g. maltose, sucrose, fructose, galactose, and mannose may also be included. Glucose is predominantly catabolized by glycolysis which leads to the production of lactate which is toxic to the cells. Some studies indicate that use of other sugars, including galactose and fructose, can substantially reduce the generation of lactate (3). The amino acid glutamine also acts as an important energy source, and in some cell types oxidation of glutamine may account for over 50% of the ATP generated (4).

2.2.5 Amino acids

Most animal cells have a requirement for the essential amino acids, i.e. those which are not synthesized in the body. In the human, these are arginine, cystine, histidine, isoleucine, leucine, lysine, methionine, phenylalanine, threonine, tryptophan, and valine. Cysteine and tyrosine are also included in this group to compensate for inadequate synthesis. Most animal cells also have a high requirement for glutamine. Glutamine acts both as an energy source and as an amine donor in the synthesis of nucleic acids and other compounds (5). Other amino acids are often added to compensate either for a particular cell type's incapacity to make them or because they are made but lost into the medium.

2.2.6 Vitamins

Several vitamins of the B group are necessary for cell growth and multiplication. Many of these are precursors for essential metabolic cofactors. The vitamins most commonly added to basal media are *para*-amino benzoic acid, biotin, choline, folic acid, nicotinic acid, pantothenic acid, pyridoxal, riboflavin, thiamine, and inositol.

2.2.7 Hormones and growth factors

Hormones and growth factors exhibit a variety of different effects on the survival and proliferation of cells. They are included in some media (especially serum-free media) at relatively low concentrations. Insulin and hydrocortisone are the most widely used, but growth factors like NGF (nerve growth factor) and EGF (epidermal growth factor), as well as certain interleukins, colony stimulating factors, and acidic and basic fibroblast growth factors have also been used (for examples, see Chapter 7).

2.2.8 Proteins and peptides

Although an absolute requirement for proteins and/or peptides by cells in culture has not been established, relatively few media have been formulated in which cells grow rapidly in the total absence of protein or polypeptides. Common examples of protein supplements are fetuin, α-globulin, fibronectin, albumin, and transferrin. These may serve a variety of functions as will be discussed in Section 3.3.

2.2.9 Fatty acids and lipids

As with the proteins and peptides, there is little direct evidence of a general essential nutritional role for lipids although some cells are known to have specific requirements; for example the NS-1 and the related NS-0 myeloma lines are strict auxotrophs for cholesterol. Fatty acids and other lipids are important components of several serum-free medium formulations, but their role is generally poorly understood and lipid requirement is an important parameter in medium design which has yet to be optimized.

2.2.10 Accessory factors

This is a blanket term used to cover other additives that are necessary for effective cell growth. These include the 'trace' elements, especially iron, zinc, copper, and selenium. Addition of other compounds including nucleosides and citric acid cycle intermediates may be advantageous with some cell types.

2.2.11 Antibiotics

Although antibiotics are used widely for protection in laboratory-scale tissue culture, their use should be avoided since resistant micro-organisms may develop, and the antibiotics may also adversely effect cell growth and function (6). Wherever possible, antibiotics are not used in industrial-scale cell culture where reliance is placed on plant which is correctly designed and of appropriate quality to maintain culture sterility. Robots are becoming increasingly used in industry to avoid microbial contamination from human sources. The use of antibiotics in biopharmaceutical production is generally not acceptable from a regulatory viewpoint.

When antibiotics are to be used, the key factors governing their selection are:

- absence of cytotoxicity
- broad anti-microbial spectrum
- acceptable cost
- minimum tendency to induce formation of resistant micro-organisms

Mixtures of penicillin (100 IU/ml) and streptomycin (50 μg/ml) are the most frequently used anti-bacterial agents. Gentamycin (50 μg/ml) is more expensive but is widely used to treat persistent contamination. Amphotericin B (Fungizone) at 2.5 μg/ml is the most commonly used anti-fungal agent, but is cytotoxic for some insect cells. Nystatin (25 μg/ml) is also an effective anti-fungal agent in tissue culture medium. For further information on antibiotics, see Chapter 9.

2.3 Choice of basal medium

The choice of medium is not always obvious and is still frequently made empirically despite many years of exhaustive research into matching particular media to specific cell types and culture conditions. Guideline information is usually available in the literature or from the originator of the cells. As a general guide, EMEM and BME will usually support the growth of primary mammalian and avian cells and human diploid cells as well as continuous cell lines such as HeLa, L-cells, and BHK-21. RPMI medium is intended primarily for cultures of human haemopoietic cells but may also support the culture of a wide range of mammalian cells including hybridomas. Fischer's medium is intended primarily for mouse leukaemic cells. Iscove's modified Dulbecco's medium (IMDM) is widely considered best for most cells of haemopoietic origin and supports the growth and differentiation of primary human and murine bone marrow cultures. Most insect cells, e.g. *Sf9*, *Sf21*, *Bombyx mori*, *Trichoplusia ni*, and *Drosophila* will grow effectively either in Grace's medium supplemented with fetal calf serum or in purpose-designed serum-free media such as EX420 (JRH) and ESF921 (Expression Systems).

2.4 Preparation of basal medium

Contamination of medium with micro-organisms (e.g. bacteria, yeast, fungi), and noxious chemical substances (e.g. traces of heavy metals), is the greatest hazard in medium preparation. For this reason, particular care must be taken in the selection and preparation of materials. High purity apyrogenic water should be used (see Chapter 1). Biochemicals used should always be of analytical grade. Items of glassware used in dispensing and storage of reagents and media must be cleaned very carefully to prevent traces of toxic materials from contaminating the inner surfaces of vessels and thence becoming incorporated into the medium. Ideally they should also be treated to remove or inactivate endotoxin. Basal media are frequently prepared by diluting a series of concentrated stock solutions, e.g. amino acid and vitamin concentrates. These stock solutions are stored in conditions appropriate for the individual components. Incompatible substances are kept separate until they are mixed together to make the complete medium formulation.

The complete medium is usually sterilized by filtration (0.1–0.2 μm). Alternatively, some media can be sterilized by autoclaving (e.g. Eagle's MEM) but care must be taken to stabilize the B vitamins, and labile additives such as glutamine should be substituted by glutamate or added after autoclaving. Powdered media are prepared by dissolving the powder in the recommended amount of water and ensuring that all the constituents are completely dissolved. Unstable constituents (e.g. sodium bicarbonate, ascorbate) are usually added as a sterile concentrate, just before use. Similarly glutamine, which is a key metabolite for growth of animal cells, is also relatively unstable and is usually added just before use. The dipeptide glycylglutamine is more stable than glutamine and has comparable biological activity (7).

2.4.1 Preparation of media from powder

For large-scale tissue culture applications, it is often economical and practical to make up single strength media from powder as outlined in *Protocol 1*.

Protocol 1

Preparation of media from powder

Equipment and reagents

- Graduated cylinder of appropriate capacity[a]
- pH meter
- Magnetic stirrer unit (and hot plate, if required)
- Filtration equipment
- Sterile bottles for dispensing medium
- Labels for bottles
- Class II microbiological safety cabinet
- Osmometer (if available)
- Analytical balance
- Appropriate quantity of tissue culture grade water
- Appropriate quantity of medium powder
- Appropriate quantities of buffer and additives
- 1 M sodium hydroxide and 1 M hydrochloric acid as required

Method

1. Using a graduated cylinder,[a] dispense 90% of the required volume of tissue culture grade water. The temperature of the water should be 15–20 °C.
2. Add the appropriate amount of powdered medium whilst gently stirring the water. Stir until all the powder has dissolved.
3. Rinse the powdered medium container with a small amount of tissue culture grade water and add to the bulk volume.
4. Add the required amount of buffer and other additives.
5. Adjust the pH to 0.2 to 0.3 pH units below the desired pH using 1 M NaOH or 1 M HCl. Stir gently whilst adjusting the pH. The pH will normally rise 0.1 to 0.3 pH units during filtration (in bicarbonate-buffered media).
6. If an osmometer is available, check osmolarity of a sample of the medium – this provides a rapid check that no gross errors of formulation have occurred.
7. Assemble filtration system to deliver filtered medium into the microbiological safety cabinet, place the sterile medium bottles in the cabinet.
8. Sterilize by filtration using a membrane of porosity 0.22 μm or less. A positive pressure filtration system is recommended in order to minimize the loss of CO_2 and consequent rise in pH.
9. Fill the filtered medium into the medium bottles in the cabinet using aseptic technique. Label the medium bottles with description, date, and batch number.
10. Store the bottled medium at 2–8 °C in the dark.

[a] When preparing larger volumes of medium, it may be more practical for some purposes to measure weights rather than volumes. A large capacity balance will be required in such cases.

2.4.2 Preparation of single strength medium from 10× concentrate

Single strength medium can be prepared from a 10× concentrate by following the steps in *Protocol 2*.

Protocol 2

Preparation of medium from 10× concentrate

Equipment and reagents

- See *Protocol 1*
- Medium concentrate

Method

1. Using a graduated cylinder, add the appropriate volume of the concentrate to 80% of the required volume of tissue culture grade water and mix gently.
2. Add the required amount of buffer.
3. Add the appropriate volume of 200 mM L-glutamine and any other required additives.
4. Follow *Protocol 1*, steps 5–11.

2.4.3 Precautions to be taken during media preparation

(a) Because of the hygroscopic nature of many medium components, avoid using partial quantities of pre-packaged powdered media.

(b) Pre-filter water (0.1 μm) prior to use.

(c) Avoid excess acid/base additions.

(d) Mixing vessels should be properly cleaned (depyrogenation may also be appropriate for some applications).

(e) Filtration should be performed as soon as preparation of the medium is complete.

2.4.4 Quality control of basal medium

Complete basal medium should satisfy the following criteria:

(a) The solution should be clear.

(b) It should have the correct pH at room temperature: this is specific to individual media.

(c) Osmolality should be correct: specific to individual media but usually in the range 280–300 mOsmol/kg.

(d) Amino acid composition (analysed by HPLC) should conform with the medium specification.

(e) The concentrations of key elements should be confirmed by chemical analysis to conform with the medium specification.

(f) Sterility of the medium should be checked by a standard protocol.

(g) Endotoxin level should be less than 1 ng/ml.

(h) Basal medium must also be able to support the growth of the required cells through at least two subculture generations (serum supplementation is usually necessary to achieve this).

2.4.5 Storage of medium

As a general rule, medium is best kept in the dark at 4 °C and storage time at this temperature should not exceed three months unless otherwise specified by the manufacturer. Glutamine is a relatively unstable amino acid that decomposes spontaneously in the medium to produce ammonia which is cytotoxic. Once glutamine is added, the shelf-life is generally reduced to two to three weeks although studies on individual media may show the storage time at 4 °C, or even room temperature, to be much longer than this. The dipeptide glycylglutamine has been shown to be an adequate substitute for glutamine and to have comparable biological activity (7). Glycylglutamine is much more stable than glutamine during both autoclaving and storage of medium. Media which contain labile constituents should either be used within two to three weeks of preparation or stored at −20 °C.

3 Serum

3.1 Why use serum?

When Eagle and others in the 1950s (8) produced basal media containing all the amino acids, carbohydrates, vitamins, and minerals thought to be essential for successful cell culture, it became apparent that such media were insufficient and that other, unidentified factors were required. It was found that supplementing the basal medium with animal body fluids could repair this deficiency and it became common practice to supplement basal medium with up to 20% of animal serum. Because of its rich content of growth factors and its low content of gamma globulin, fetal bovine serum (FCS), most frequently used at 10% (v/v) concentration, has become adopted as the 'standard' medium supplement. Lower concentrations of FCS may be desirable for some applications. Extensive study of serum has shown that it enhances the growth of cells by a number of distinct mechanisms:

(a) Serum contains a cocktail of most of the growth factors required for cell growth and maintenance and is therefore an almost universal growth supplement effective with most types of cells. Using serum-supplemented medium therefore reduces the need to spend time developing a specific optimized medium for every cell type under investigation.

(b) Serum buffers the cell culture system against a variety of perturbations and toxic effects such as pH change, presence of heavy metals, proteolytic activity, and endotoxin.

These points are discussed in more detail in Section 3.3. The use of serum also imposes a number of difficulties (discussed in Section 3.4) which impact on the safety, reproducibility, and cost of biopharmaceuticals. These difficulties can be minimized by careful selection and validation of serum sources. Although almost all new manufacturing processes using animal cells are designed for use with serum-free medium in order to avoid these difficulties, many existing processes still use FCS-supplemented medium. This situation is unlikely to change fundamentally in the near future since regulatory constraints generally make it impractical and uneconomic to alter existing processes. Current world demand is in excess of 500 000 litres of FCS per year.

3.2 Types of serum

Despite its high cost, FCS remains the most frequently used serum for medium supplementation. Several other types of serum have been proposed as cheaper alternatives to FCS. Calf serum is quite widely used industrially and is available either as 'new-born' calf serum (which has high levels of biotin) or as mature calf serum. New-born calf serum contains high levels of γ-globulin as a result of the ingestion of colostrum immediately after birth. Adult bovine serum is also used occasionally but is not so effective as calf serum. Horse serum is also sometimes used, especially for some human cell lines. The use of human serum has been proposed for some fastidious human cell lines but it is not clearly established that human serum performs better than FCS.

Whereas FCS is collected at the abattoir, generally by aseptic cardiac puncture, calf serum and horse serum can be collected in better-controlled environments from 'donor' animals. In the donor system, herds of virus-screened animals are kept separate from other stock and used exclusively for serum production. The health of every animal is constantly monitored with special reference to viral infection. Animals are bled at regular intervals using aseptic venepuncture as for human blood donors. Advantages of this system over slaughterhouse collection include:

(a) Better control of animal husbandry and hygiene.

(b) Comprehensive knowledge of the animals' health status.

(c) Full control of blood collection and processing in an integrated Good Manufacturing Practice (GMP) process.

(d) Improved consistency of serum quality, since animals remain in the donor herd for several years.

(e) Full traceability from bottled serum back to the individual animal if required.

The availability of serum guaranteed negative for specific animal viruses is of critical importance in the production of animal vaccines. Serum from donor animals is available from a number of suppliers including CanSera Inc., South Pacific Sera, Salzman Corporation, and TCS CellWorks Ltd.

3.3 Constituents of serum

Serum is an effective growth-promoting supplement for practically all types of cell (for some exceptions, see Chapter 7) because of its complexity and the multiplicity of growth-promoting, cell protection, and nutritional factors that it contains. These can be divided into specific polypeptides that stimulate cell growth (growth factors), carrier proteins, cell protective agents, cell attachment factors, and nutrients such as trace elements (some of which may be attached to carrier proteins). Some serum macromolecules can fill more than one of these roles.

3.3.1 Growth factors

Polypeptide growth factors are of particular importance in serum. These 5–30 kDa proteins act via cell surface receptors to generate signals which stimulate cell proliferation or differentiation. In some cases, the presence of certain growth factors may not be stimulatory as such but may still be essential as 'survival factors'. Deprivation of these factors initiates a pre-programmed auto-destructive sequence of events (apoptosis) which results in cell death even though the cells may be fully provided with nutrients and maintained under otherwise optimal culture conditions (9).

Different cell types have different growth factor requirements and a given growth factor may stimulate or inhibit depending on the cell type and the concentration of the factor (10). Different types of serum (and different batches of the same serum type) may contain different absolute and relative levels of various growth factors. This is one of the main reasons why performance testing of serum batches is necessary to ensure satisfactory results with the specific cell line of interest.

3.3.2 Albumin

Albumin is the major protein component of serum and has several functions which contribute to the growth and maintenance of cells in culture. Albumin acts as a carrier protein for a range of small molecules, particularly lipids. Transport of fatty acids is an important function since these are essential for cells but are toxic in the unbound form and are also very poorly soluble in water. Steroids and fat soluble vitamins may also be transported bound to albumin. (Other lipids such as cholesterol, cholesterol esters, triglycerides, and phospholipids are transported in serum as micelles formed with specific lipoproteins.)

Albumin also has specific binding sites for thyroxin and for metal ions such as Ni^{2+}, Zn^{2+}, and Cu^{2+}. There is evidence that albumin may also transport other metals and also carry other, unidentified factors which support cells in culture. The absorptive capacity of albumin also enables it to play a detoxifying role by binding toxic metal ions and other inhibitory factors.

Albumin also functions as a pH buffer and protects cells against damage by the shear forces that occur in pumped or stirred culture systems. This latter effect appears to be entirely mechanical and to be related to the hydrodynamic properties of the medium since cells become protected immediately after albumin addition, before the albumin has time to exert any metabolic effects (11).

3.3.3 Transferrin

Iron is an essential trace element for cultured animal cells but can be toxic if presented in an inappropriate form in the medium. Iron salts are also sparingly soluble in medium. Transferrin (siderophilin) is the major iron-transporting protein in vertebrates, where it represents 3–6% of total serum protein. Practically all animal cells express specific cell surface transferrin receptors. The transferrin/ Fe^{3+} complex is taken up via these receptors and, after release of the iron, apo-transferrin is liberated from the cells and recycled. It is unclear whether iron transport is the only role of transferrin; some reports suggest that it may also transport vanadium and manganese, and others that it may have a wider role in heavy metal detoxification.

Copper may also be transported by a serum carrier protein (ceruloplasmin), and also as a chelate with small peptides such as Gly-His-Lys (GHL, liver growth factor).

3.3.4 Anti-proteases

Serum contains two classes of wide spectrum protease inhibitors, α_1-antitrypsin and α_2-macroglobulin, each representing about 2% of total serum protein. Proteases are secreted by many cells in culture (to an extent that depends on culture conditions) and are used in the subculture of anchorage-dependent cells. The powerful anti-protease action of serum prevents proteolytic damage to cells and their products.

3.3.5 Attachment factors

Serum also provides a source of attachment factors that facilitate the binding of anchorage-dependent cells to the substratum. Cells will generally synthesize these factors themselves, but culture is enhanced if they are provided in the medium. The major serum protein involved in attachment is fibronectin. Laminin, fetuin, and collagens also play a role. The importance of attachment factors goes far beyond supplying a simple 'glue'; amongst other effects, their binding with cell surface receptors links directly into the cell's signal-transduction cascade to influence survival, differentiation, and proliferation of the cells.

A summary of the main components of serum that are known to be important in culture medium is given in *Table 2*.

3.4 Potential problems with the use of serum

There are a number of serious disadvantages incurred when serum is used to supplement culture medium. These have different impacts depending on the intended use of the cultured cells. In the production of biopharmaceuticals, compliance with rigorous regulatory controls concerning potential contamination by viruses and other adventitious agents is a primary concern as is the need to purify the final product away from contaminating proteins. These concerns may be of limited relevance in research studies where an accurate knowledge of growth factor content may be more critical. The main difficulties encountered when using serum are discussed below.

3.4.1 Lack of reproducibility

Serum batches vary considerably depending on the characteristics of the source animals used, on the feed stuffs employed, on the time of year, etc. Different

Table 2 Nutritional and protective factors which may be supplied by serum

Factor	Concentration
Specific growth factors	
EGF, PDGF, IGF, FGF, IL-1, IL-6, insulin	1–100 ng/ml
Trace elements	
Iron	1–10 μM
Zinc	0.1–1 μM
Selenium	0.01 μM
(also Co, Cu, I, Mn, Cr, Ni, V, As, Si, F, Sn)	
Lipids	
Cholesterol	*c.* 10 μM
Linoleic acid	0.01–0.1 μM
Steroids	
Polyamines	
Putrescine	0.01–1 μM
Ornithine	0.01–1 μM
Spermidine	0.01–1 μM
Attachment factors	
Fibronectin	1–10 μg/ml
Laminin	–
Fetuin	–
Mechanical protection	
Albumin	17–35 mg/ml
Buffering capacity	
Albumin	17–35 mg/ml
Neutralization of toxic factors	
Albumin	17–35 mg/ml
Transport of metals	
Transferrin/Fe^{3+}	2–4 mg/ml
Ceruloplasmin	–
Protease inhibitors	
α_1-antitrypsin	1.5–2.5 mg/ml
α_2-macroglobulin	0.7–2.0 mg/ml

batches contain different absolute and relative levels of growth factors. Certain factors may be deficient in some batches while others may be present at excessive levels that would be inhibitory for some cells. Variations of this sort are not tolerable in manufacturing processes and are countered by the batch reservation system where batches are held on reserve by the serum producer while the would-be user completes testing of samples for efficacy in his system. The situation is also improved if the size of reserved batches is large enough to permit production over a considerable period.

In experimental cell biology, it is also important to be aware of the inherent variability of serum which renders it very difficult to study the specific effects of molecules such as growth factors, hormones, cytokines, adhesion molecules, or matrix components, all of which are present in serum at undefined and variable levels.

The presence of specific antibodies in serum may also profoundly affect the results obtained. This is particularly true in the production of viruses. Antibodies in the serum may result from a natural infection with the relevant virus or a related species (which may be transmitted transplacentally in some cases), or from prior vaccination of the animals used. It should be remembered that some antibodies may also cross the placenta and that even FCS may contain significant inhibitory activity to infectious agents to which the mother has been exposed. Specific virus-free herds of donor animals may be the preferred serum source in this situation.

Serum also varies depending on the quality and reproducibility of the procedures used for its collection. For instance, the length of time between collection of blood and removal of cells is critical if the lysis of cells and the release of cell contents (possibly including viruses) is to be kept low. Sterility of the operation, exclusion of pyrogens, and several other process parameters are also critical. Reputable serum producers will have invested heavily in the quality of their operation, and production of serum to GMP standards is a more complex process than many users realize. Reliable tissue culture quality serum is accordingly an expensive product.

3.4.2 Risk of contamination

Serum represents a major potential route for the introduction of adventitious agents including bacteria, fungi, mycoplasma, and viruses into cell culture. This could be disruptive in research projects and dangerous in pharmaceutical manufacture. Possible virus contamination is the major current regulatory concern. In order to minimize risks of contamination, suppliers must apply rigorous health checks to the animals used, use GMP facilities for the collection and processing of serum, apply thorough quality control testing, and ensure rigorous batch documentation to permit verification of all the procedures. The process employs aseptic collection of blood by cardiac puncture or by venepuncture, aseptic clotting and clot removal, clarification by centrifugation, and sterilization by filtration terminating in double 0.1 μm filters followed by sterile filling. Viruses will not be effectively removed by 0.1 μm filtration and several companies are now producing filters of 40 nm pore size and below to provide enhanced clearance of virus (see Chapter 2). However, it should be recognized that some of the larger serum proteins (e.g. IgM) may also be removed by such filters. After filtration, serum is tested for microbial sterility, for contamination by a panel of viruses that commonly occur in the donor species, and for the capacity to support the growth of a range of test cells. An example of a typical Certificate of Analysis reporting the results of the quality control tests routinely performed on serum is shown in *Figure 1*.

As the outbreak of bovine spongiform encephalopathy (BSE) in the UK illustrates all too clearly, this approach cannot eliminate all risks of contamination, particularly that of the presence of unsuspected infectious agents. As a further line of defence, regulatory authorities now require that only serum from specified countries of origin, where particular viruses are thought not to occur, can be used in the production of pharmaceutical, veterinary, and sometimes diagnostic

FOETAL CALF SERUM - CERTIFICATE OF ANALYSIS

COUNTRY OF ORIGIN ....

PRODUCT DESCRIPTION:

BATCH NUMBER:

Sterility Testing

Bacteria
Yeasts and fungi
Bacteriophages
Mycoplasma
Viruses:
Bovine Viral Diarrhoea (BVD)
Parainfluenza 3 (PI3)
Infectious bovine rhinotracheitis (IBR)

Physical and Biochemical Analysis

pH at 37°C	
Osmolality	mOsmol/kg
Albumin	mg/ml
Beta globulin	mg/ml
Gamma globulin	mg/ml
Haemoglobin	µg/ml
Total protein	mg/ml
Electrophoretic profile	
Endotoxin	EU/ml
Visual check	

Functional Testing

Diploid fibroblast growth capacity	% of control
Human epithelial cell growth capacity	% of control
Myeloma/Hybridoma growth capacity	% of control
Relative cloning efficiency	% of control
Relative plating efficiency	% of control
Cytotoxicity check	

Documentation Approval .... Signed .... Date

Product Release .... Signed .... Date

Figure 1 An example of the type of quality control testing regime which should be applied to batches of serum for tissue culture use. Certificates of analysis like this are normally issued by the serum supplier. The original documentation used to generate this summary sheet should be available for examination on request. The precise specification values on the certificate depend on the specific type of serum in question. All sera tested should, of course, show no detectable contamination in the sterility tests. Tests for other viruses may be necessary for particular applications. (NB: for serum of non-bovine origin, virus testing will be based on other virus types.)

products. These restrictions are unlikely to affect the research user, but are mandatory for manufacturers of biopharmaceuticals. This has resulted in the creation of a 'league table' of acceptable countries with New Zealand, Australia, and the USA occupying the highest places. Not surprisingly, this adds considerably to the cost of pharmaceutical-grade serum and because of this, an extensive 'black market' has grown up supplying relatively low price serum of doubtful origin which, in addition, may not have been processed and tested to the necessary rigorous standards (12).

To further reduce the risk of viral contamination, serum can be subjected to virus inactivation procedures such as gamma-irradiation, UV-irradiation, or treatments with chemicals such as beta-propiolactone. Heat inactivation (usually at 56 °C for 1 h) also inactivates some viruses. In general, all of these processes also usually result in decreased growth-promoting activity and increased cost.

3.4.3 Availability and cost

Serum is a by-product of the meat industry. As such, its supply (particularly that of FCS) depends on the agricultural policies in the different supplying countries. The supply of genuine New Zealand FCS is very limited and this product therefore commands a very high price (four to six times that of 'research grade' material). Whatever the origin, very significant investment and operating costs are incurred by any manufacturer who produces FCS according to GMP principles. Correctly collected, processed, and validated FCS will always therefore contribute greatly to the cost of a tissue culture process.

3.4.4 Influence on downstream processing

The presence of serum in tissue culture medium presents particular difficulties when purifying products secreted by cells. At 10% concentration, serum contributes about 4–8 mg of protein per ml while recombinant proteins are frequently expressed at levels of tens of micrograms per ml. In this situation, efficient purification of the required protein may be difficult, and in some cases it may even be impossible to devise an economically acceptable purification process. Monoclonal antibodies may be secreted at higher levels (400–600 μg/ml in the best cases) but are particularly difficult to purify from serum-containing medium because of the high level of endogenous immunoglobulins (significant even with FCS).

Downstream processing may account for over 80% of the cost of commercial production of pure proteins and may determine the commercial viability of the whole process. In such circumstances, dependence of the upstream process on serum supplementation can be a critical disadvantage. One approach which is sometimes used to reduce this problem is to change to a low serum concentration (e.g. 2%) when cell growth is completed so as to reduce the concentration of serum during the product generation phase of the process.

3.5 Sourcing and selection of serum

3.5.1 Sources of serum

For reasons already discussed the geographical source of serum for some applications may be imposed by regulatory agencies. Because serum (especially FCS) is such an expensive commodity there is a temptation for unscrupulous suppliers to misrepresent the origin and quality of the material (12). It is important therefore for users who require high quality serum of defined origin to use a reputable supplier who has direct sources in the required production location. Such suppliers will be pleased to supply comprehensive and reliable original documentation for every serum batch and to allow audits of all stages of the production process. This approach is important both for validation of the quality of individual batches and for assuring continuity of supply. A recent additional safeguard pioneered by the New Zealand government has been the Internet publication of export certificates for FCS so that potential buyers can check these documents at source. Another new development has been the requirement for serum processing and filling to be completed in the country of origin so as to limit the movement of bulk raw serum. This further limits the number of companies that can supply fully compliant pharmaceutical-grade serum. It should also be noted that, independent of the needs of the biotechnology industry, movement of serum between countries is also subject to restrictions by governments because of concerns over animal health (although movement between different EU countries is in principle unrestricted).

Continuity of supply and consistency of quality are, of course, also extremely important to the research user who may not be operating under strict regulatory constraints. Again this is best achieved by dealing with reputable suppliers who have the physical and logistical resources to provide this service.

3.5.2 Selection of serum

Selection of serum type and serum batch are essentially based on empirical evaluation by the user. When a given cell type is first grown, it may be worthwhile to test its capacity for growth on cheaper serum types, possibly including mixtures of different types of serum.

Although serum offered commercially will have been subjected to thorough tests (*Figure 1*), it is still routine for many users to 'batch test' serum in their own system before buying. Serum suppliers will generally provide samples from several different batches free of charge for the user to test in his own laboratory, the quantity of serum potentially required by the user being held on reserve until testing is complete. User testing is primarily to check performance (*Protocol* 3) but may also involve verification of the absence of adventitious agents that are of particular concern for the user.

It is important to bear in mind that tests of serum batches for growth or productivity should always be performed in an identical system (in terms of basal medium, cell conditions, and culture configuration) to that in which the serum will finally be used, and should always include a previously evaluated reference serum as control. It is recommended that the cells are subcultured at least three times as indicated below. At each passage, growth results should be normalized relative to the reference serum. Yields within ± 20% of the reference serum would normally be considered satisfactory. In addition to cell and product yield, cells should also be examined at each passage for any indication of cytotoxicity or abnormal morphology. Tests of cell growth from very low seeding levels may be important for some specific applications. Tests of cloning efficiency or of plating efficiency may also be useful (see Chapter 8). The procedure described in *Protocol 3* is used to test the capacity of serum to support growth of the required cells over several passages.

Protocol 3

Functional testing of serum batches

Equipment and reagents

- 25 cm^2 sterile tissue culture flasks
- Haemocytometer and coverslips (or other cell counting system)
- Aliquots of serum batches to be tested
- Aliquot of reference serum batch
- Basal medium and supplements
- Test cell line
- Trypsin-Versene or EDTA solution

Method

1. Prepare flasks containing the appropriate basal medium supplemented with either 5% of the serum to be tested or 5% of a previously tested reference serum (use of 5% serum gives a more sensitive indication of the growth-promoting capacity of the serum than does the use of higher percentages).
2. Seed the flasks with a number of cells appropriate to the cell line used (generally 1000–1500 cells/ml for adherent cells).
3. Incubate the cells for five to seven days, harvest, and count.
4. Subculture the cells using a 1:5 split ratio (or other ratio appropriate to the cells used), seed cells from the test serum into fresh medium supplemented with 5% test serum, and the reference serum cells into 5% reference serum.
5. Incubate flasks for five to seven days as before, harvest, and count.
6. Repeat steps 4 and 5.
7. Calculate test serum growth-promoting capacity relative to the reference serum (generally accept reference serum value ± 20%).

3.6 Serum storage and use

Serum is frozen rapidly by the supplier immediately after bottling and is held frozen at −20 °C. Little data has been published on the shelf-life of serum held in this way but two years has become accepted as a rule of thumb. Measurement of serum stability in real time is difficult (due mainly to the difficulty of standardizing cell culture over the long test period required) and time-consuming, while accelerated degradation tests cannot be meaningfully applied to the frozen material. However, the one rigorous real time study of which we are aware suggests that two years at −20 °C is a very conservative estimate and that frozen serum remains effective after at least five years.

When required for use, serum should be thawed rapidly and with gentle mixing to minimize protein damage due to salt concentration effects. An agitated water-bath at 37 °C is best although serum should be removed as soon as fully thawed and not allowed to warm up. Thawed serum should be clear without any significant precipitation. Once thawed, sterile serum can be held at 4°C for a maximum of two to three weeks. Thawed serum can be aliquoted and refrozen but should not be thawed and refrozen more than once.

4 Replacement of serum in media

Much of the present understanding of the nutritional requirements of cells in culture stems from Eagle's work (8) on the fundamental requirements for growing mammalian cells. Based on this information, several attempts have been made to replace serum in part or in full by serum-derived factors or by completely synthetic media. One approach followed is to reduce the serum requirement by supplementing culture medium with processed serum products. Controlled process serum replacements (CPSR) are prepared by processes that yield defined products with much higher batch to batch consistency than serum. CPSR products are derived from bovine plasma and have lower protein and endotoxin levels than serum. Natural serum can also be replaced by supplemented/fortified serum. Serum may be fortified with mitogens, growth factors, hormones, proteins, and protein stabilizers and trace elements. Such fortified serum can often be used at a much lower concentration than natural serum.

4.1 Reduced-serum media for early-passage normal human cells

There has been a major increase recently in the use of primary or early-passage normal human cells as models for drug discovery, metabolism, and toxicology studies. This approach is based on the premise that these will yield more directly useful information on the behaviour of potential drugs in man than would the use of either transformed human cells or animal cells, since they represent the real targets for drug action (e.g. human coronary artery endothelial cells for the study of coronary artery disease, or lung smooth muscle cells for the study of potential anti-asthma drugs, etc.).

It is highly desirable to eliminate serum and other undefined protein extracts from cultures of this sort, but the cells are extremely fastidious and this approach has met with very limited success. Thus the commercially available media for such applications contain a small amount of serum, undefined tissue extracts, and a number of recombinant growth factors (*Table 3*).

A special consideration in the case of such early-passage cells is the need to achieve effective growth of the required cell type while suppressing the growth of other cell types that will inevitably be present at low levels at this stage of culture. This can be approached by modifications to the basal medium (a well known case is the use of D-valine to favour the growth of epithelial cells since only this cell type express D-amino acid oxidase). However selective growth is most frequently accomplished by supplementing the low serum medium with cell type-specific polypeptide growth factors.

Table 3 A selection of commercially available low serum media intended for use with primary and low-passage normal human cells[a]

Medium supplier	Cell type							
	Keratinocytes	**Large vessel endothelial cells**	**Microvascular endothelial cells**	**Fibroblasts**	**Melanocytes**	**Smooth muscle cells**	**Skeletal muscle cells**	**Corneal epithelial cells**
Cascade Biologics Inc.	Medium 154	Medium 200	Medium 131	Medium 106	Medium 154	Medium 231	–	Medium 165
Clonetics Corporation	Keratinocyte BM	Endothelial cell BM	Endothelial cell BM MV	Fibroblast BM	Melanocyte BM	Smooth muscle cell BM	–	–
Gibco BRL	Keratinocyte SFM	Endothelial SFM	–	–	–	–	–	–
Promocell GmbH	Keratinocyte BM	Endothelial cell BM	Microvascular endothelial cell BM	Fibroblast BM	Melanocyte BM	Smooth muscle cell BM	Skeletal muscle BM	–
Sigma	Keratinocyte BM	Endothelial cell BM	Microvascular endothelial cell BM	Fibroblast BM	Melanocyte BM	Smooth muscle cell BM	–	Corneal epithelial BM
TCS CellWorks Ltd.	Keratinocyte BM	Large vessel endothelial cell BM	Microvascular endothelial cell BM	Fibroblast BM	Melanocyte BM	Smooth muscle cell BM	Skeletal muscle BM	Corneal epithelial BM

[a] The media included in the table are basal media (BM) that are specifically optimized for the cell type in question. These are supplied together with a pack of frozen supplements which may include cell type-selective recombinant growth factors, tissue extracts, and a low level (typically below 2%) of FCS. The precise composition of the BMs and their corresponding supplements is usually proprietary information.

4.2 Serum-free media (SFM)

For the reasons discussed in Section 3.4, whenever possible, it is desirable to culture cells in serum-free medium. A properly designed SFM:

- is reproducible
- is not reliant on the economics of the world cattle market
- simplifies downstream purification
- contains no unknown factors, e.g. viruses or growth inhibitors

Table 4 Serum-free media for established cell lines

Cell line	Source	Basal medium	Supplements	Substratum modification	Refs
GH3	Rat pituitary carcinoma	F12	ins, trf, T_3, TRH, PTH, som C, FGF	None	13
HeLa	Human cervical carcinoma	F12	ins, trf, FGF, EGF, hc, trace elements	None	14
PCC.4aza-1	Mouse embryonal carcinoma	F12	ins, trf , 2-ME, fetuin	None	15
M2R	Mouse melanoma	DMEM/F12	ins, trf, test, FSH, NGF, LRH	None	16
TM4	Mouse testes	DMEM/F12	ins, trf, FSH, som C, GH, RA	None	17
RF-1	Rat ovarian follicle	DMEM/F12	ins, trf, hc	fn	18
M1	Mouse myeloid leukaemia	F12	ins, trf, trace elements	None	19
B104	Rat neuroblastoma	DMEM/F12	ins, trf, prog, putr, sel	fn, polylysine	20
C62 BD	Rat glioma	DMEM/F12	ins, trf, T_3, hc, PGE_1, sel	None	21
MCF-7	Human mammary carcinoma	DMEM/F12	ins, trf, EGF, $PGF_{2\alpha}$	fn	22
BHK-21	Hamster kidney	DMEM/F12	ins, trf, EGF, FGF, BSA	fn	23
3T6	Mouse embryo fibroblasts	DMEM/ Waymouth	ins, $FeSO_4$, EGF	None	24
116NS-19	Mouse hybridoma	MEM or RPMI 1640	inf, trf	None	25
HL60	Human promyelocytic leukaemia	DMEM/F12	ins, trf, sel	None	26
MPC-11	Mouse plasmacytoma	DMEM/F12	trf, LH, sel, LRH, PGE_1, EGF, T_3, glucagon, NGF, $PGF_{2\alpha}$	None	27
Flow 2000	Human embryo fibroblasts	MCDB 108	ins, EGF, dex	Polylysine	28
WI38	Human embryo fibroblasts	MCDB 104	ins, trf, EGF, dex, PDGF	None	29
K562	Human erythroleukaemia	RPMI 1640	trf, sel, BSA	None	30
U-251 MG	Human glioma	DMEM	trf, FGF, hc, sel, biotin	fn	31
NCI-H69	Human small cell lung carcinoma	RPMI 1640	ins, trf, sel, hc, oestradiol	None	32
LA-N-1	Human neuroblastoma	DMEM/F12	ins, trf, prog, putr, sel	Polylysine	33
MDCK	Dog renal epithelium	DMEM/F12	ins, trf, T_3, hc, PGE_1, sel	None	34

Abbreviations: BSA, bovine serum albumin; dex, dexamethasone; EGF, epidermal growth factor; FGF, fibroblast growth factor; fn, fibronectin; FSH, follicle-stimulating hormone; GH, growth hormone; hc, hydrocortisone; ins, insulin; LH, luteinizing hormone; LRH, luteinizing hormone releasing hormone; 2-ME, 2-mercaptoethanol; NGF, nerve growth factor; PDGF, platelet-derived growth factor; PGE, prostaglandin E; PGF, prostaglandin F; prog, progesterone; PTH, parathyroid hormone; putr, putrescine; RA, retinoic acid; sel, selenium; som C, somatomedin C; T_3, triiodothyronine; test, testosterone; trf, transferrin; TRH, thyrotropin releasing hormone.

A number of cells have been grown successfully in SFM, usually in a medium specifically developed for one cell line. The requirements of cell lines differ greatly and success with one cell line does not guarantee success with other, even closely similar cell lines. A great deal of effort has gone into developing SFM, but until recently, success has been limited. However, with the identification of essential growth factors and nutrients required by different cells (see Chapter 7), several very effective serum-free media have been formulated. *Table 4* lists the established cell lines which are able to proliferate in defined medium without adaptation; those requiring a serum 'weaning off' period are not included. A variety of tissue sources and species are represented; clearly there are unique combinations of growth factors and hormones that promote optimal proliferation of specific cell types (see also Chapter 7). The most consistent requirements appear to be for the polypeptide hormone, insulin, and the iron-transport protein transferrin. Other supplements include polypeptide and steroid growth hormones, polypeptide growth factors, trace elements, reducing agents, diamines, vitamins, and albumin complexed with unsaturated fatty acids. An important consideration for some applications is that any animal-derived supplements or proteins can pose similar contamination risks to those of serum (see Section 3.4.2).

Several commercially produced, ready-to-use SFM are now available which have been designed for the main 'workhorse' cell types used in biopharmaceutical manufacture such as CHO, BHK, and hybridomas. These are summarized in *Table 5*.

Table 5 A selection of the serum-free media currently available commercially

Medium supplier	**Cell types**				
	Hybridomas	**CHO**	**Insect cells**	**Lymphoid cells**	**General purpose**
Biowhittaker	UltraDoma UltraDOMA-PF*	UltraCHO	Insect XPRESS	*Ex vivo* range	UltraCulture
Life Technologies	Hybridoma-SFM Hybridoma PFHM-II*	CHO-S-SFM II CD CHO*	SF900II EXPRESS FIVE Drosophila-SFM	AIM V	EPISERF
Invitrogen			DES High Five SFM		
Hyclone	Hyq CCM1 Hyq SFX-Mab Hyq PF-Mab*	Hyq CCM5 Hyq SFX-CHO Hyq PF-CHO*	Hyq CCM3		
CSL-JRH	EX-CELL TM 610-HSF EX-CELL TM 620-HSF	EX-CELL™ 301 EX-CELL™302 EX-CELL™ 325-PF*	EX-CELL™ 400 EX-CELL™ 401 EX-CELL™ 420 EX-CELL™ 405		
Expression Systems		ESF CHO*	ESF 921*		
Sigma Biosciences	QBSF S2 QBSF S5 Hybri-Max	CHO-SFM CHO-PFM* CHO medium CD*			
PAA Laboratories	Cytoferr™*	Cytoferr™ CHO*			
Irvine Scientific	HB101 HB GRO/PRO* IS MAB-V™*	HB CHO IS CHO IS CHO-V™*	IS BAC™*		

* Designates protein-free medium.

It should however be remembered that different strains of the same cell type may have different medium requirements, and that 'fine-tuning' of these commercial media may be necessary to obtain optimum results with an individual specific strain or construct.

4.3 Design of SFM

A defined SFM is one in which a group of components of known purity, present at a known concentration, are formulated together to optimize performance for a given cell type. Several important factors must be considered to achieve this goal. Amongst these are the origin of the cell line (i.e. species and tissue), compatibility of media components and their ratio, any interactions between components, and the specific application for which the cell line is being cultured, e.g. production of biomass or generation of secreted product. The two main approaches generally followed in designing a serum-free medium are:

(a) **Reduced serum**. In this approach, the concentration of serum in the basal medium is progressively reduced whilst other components, e.g. growth factors and hormones, are added to identify the factor(s) capable of restoring growth to the level obtained in the presence of serum. This process can be very lengthy because at each change, growth assays using the serum-supplemented control and repeat verification assays need to be done (35).

(b) **Basal medium**. A different approach is to add components (singly or in combinations) to a basal medium in a stepwise manner until a medium is progressively 'built up' to give a similar or equivalent cell growth to the serum-supplemented medium.

For either of these approaches, the following critical factors need to be considered in designing an efficient, defined SFM.

4.3.1 Basal medium

The selection of basal medium can be extremely important in terms of energy sources, buffers, and inorganic ions. Generally, the starting basal medium formulation is chosen on the basis of known preferences of the required cell line or cell type. Accordingly it is recommended to start with the same basal medium in which the cells have been growing with serum. Alternatively, start with commercially available serum-free media which have been developed for a particular cell type (for examples, see *Table 5*).

4.3.2 Lipids

These include ethanolamine, phosphoethanolamine, sterols, fatty acids, and phospholipids. In serum-supplemented media, these are usually carried on macromolecules, principally proteins. In SFM, fatty acids are usually provided in a bound form (either to albumin or other serum proteins) or in the form of phospholipid-enclosed vesicles (liposomes). If serum albumin is used directly as a lipid source, it should be noted that the endogenous lipid content of albumin may be dependent on the methods used for its purification; the solvent precipitation frequently used may result in substantial stripping of lipid from the protein. Also, pasteurized human albumin will have been stabilized with octanoic acid or another hydrophobic stabilizing agent prior to heating, and it may be important to replace this with more physiologically relevant lipid before use in cell culture. Recent developments have permitted the use of totally synthetic hydrophilic carriers such as cyclodextrins for the transport of lipids (36).

4.3.3 Buffering

Buffers maintain a proper environment for the metabolism, growth, and functioning of cells. Major ions (Na^+, K^+, HCO_3^-, and HPO_4^{2-}) are usually regarded as the principal components in pH control, along with H^+ and OH^-, which enter into the ion balance. Other components, including amino acids, if present in high concentrations, can contribute to the buffering power of a medium. Besides bicarbonate, zwitterionic organic buffers like Hepes, Bes, and Tes may be used in systems in which strict control of the gas phase is not required. However, careful consideration must be given to the concentration of these buffers since they can be toxic to the cells (37). In this connection it may be useful to screen batches of, for example, Hepes for the presence of toxic by-products before incorporating them into the medium. It should also be noted that some of these zwitterionic organic buffers can chelate biologically important cations (38). A useful buffer for use in the presence of low or no bicarbonate is sodium β-glycerophosphate (39).

4.3.4 Trace elements

The major ions - Na^+, K^+, Ca^{2+}, Mg^{2+}, Cl^-, HPO_4^{2-}, and HCO_3^- are principally involved in maintaining electrolyte balance and contributing to the osmotic equilibrium of the system. Trace elements are also included in many serum-free media because of their beneficial effects. Interrelationships exist between Fe^{2+}, Zn^{2+}, and Cu^{2+} ions which are needed for many cells. Most SFM also include Co^{2+} and SeO_3^{2-}. Cells derived from heart and kidney tissue have a high requirement for K^+ whilst Ca^{2+} is required for control of mitosis (40) and the Ca^{2+}/Mg^{2+} ratio is important in controlling cell proliferation and transformation (41). Selenium is proving to be important for many cell types (1). Other trace elements include Sn, V, Al, and As (42). Iron is frequently added as a transferrin complex but can also be added in other forms such as ferric citrate, ferrous sulfate, or ferrous nitrate.

4.3.5 Mechanical stabilizers and adhesion factors

For optimal growth, cells grown in suspension culture require protection from shear due to agitation, bursting air bubbles, etc. Shear damage can be reduced by increasing the viscosity of the medium. Carboxymethyl cellulose and polyvinyl pyrrolidone have been used for this purpose. The most widely used shear protectant is Pluronic F-68. This is a non-ionic block copolymer with an average molecular weight of 8400 daltons, consisting of a central block of polyoxypropylene (20% by weight) and blocks of polyoxyethylene at both ends. Pluronic F-68 has been demonstrated to have a significant effect in protecting animal cells grown in suspension in sparged or stirred bioreactors. The protective effect is thought to be exerted through the formation of an interfacial structure of adsorbed molecules on the cell surface. It is thought that the hydrophobic portion of the molecule interacts with the cell membrane, while the polyoxyethylene oxygen may form hydrogen bonds with water molecules to generate a hydration sheath, which provides the protection from laminar shear stress and cell–bubble interactions (43, 44).

Cell attachment and growth of anchorage-dependent cells can be improved by pre-treatment of the substrate in a variety of ways. Treatments used include coating with adhesive glycoproteins such as fibronectin, laminin, chondroitin, epibolin, or serum spreading factor (see also Chapter 5, Section 5.4). Recently an engineered fibronectin substitute called Pronectin® (available from Protein Polymer Technologies Inc.) has been developed for this purpose. This has the advantage of promoting better attachment than the natural matrix proteins since it contains multiple copies of the sequence Arg-Gly-Asp (RGD) recognized by the integrins. It also has the advantage of not being derived from animal sources and therefore being free of potential viral contamination (45).

4.3.6 Selection of components

The following checklist is a useful starting point for consideration of components for inclusion in an SFM formulation:-

(a) Transport proteins, e.g. transferrin, bovine serum albumin, lactoferrin.

(b) Stabilizing protein(s), e.g. aprotinin, bovine serum albumin, fetuin, soya bean trypsin inhibitor.

(c) Growth regulators, e.g. insulin, hydrocortisone, triiodothyronine.

(d) Growth factors, e.g. EGF, FGFs, NGF, PDGF, IGFs (somatomedins).

(e) Attachment proteins, e.g. fibronectin, collagen, laminin, fetuin, serum spreading factor.

(f) Crude extracts, e.g. bovine pituitary extract, brain extract, liver extract, tissue digests.

(g) Essential nutrients, e.g. cholesterol, linoleic acid, ethanolamine, trace elements.

It may be useful to refer to the many specific examples given for different cell types in Chapter 7.

4.3.7 Practical hints on solubilizing specific components

(a) Riboflavin, folic acid, tyrosine, and cystine need dissolving in NaOH.

(b) Insulin needs to be dissolved in HCl.

(c) Fatty acids, lipids, and fat soluble vitamins can be dissolved in alcohol solutions or attached to either protein carriers, e.g. BSA, or cyclodextrins, or surfactants, e.g. Tween 80 or Pluronic F-68.

(d) Pluronic F-68 is more soluble in cold water and, when making Pluronic F-68 solutions, it should be added to water and not vice versa.

(e) Hypoxanthine dissolves easily on heating.

4.3.8 Adaptation of cells to low serum/serum-free medium

The following protocol describes a generalized procedure for adapting cells to low serum-supplemented medium or SFM. The procedure should ensure that cell viability and protein synthesis are not compromised at any adaptation stage.

Protocol 4

Adaptation of cells to low serum/serum-free medium

Equipment and reagents

- 24-well plates
- Erlenmeyer flasks
- 25 cm^2 tissue culture flasks
- Haemocytometer and coverslips
- Shaking platform
- Cryovials
- Liquid nitrogen freezer
- Appropriate cell line
- Appropriate basal media and supplements
- Appropriate batch of serum
- Tissue culture grade dimethyl sulfoxide (DMSO)

Method

1. Determine the optimal seeding density in the serum-supplemented medium which allows for 5- to 15-fold growth during the experiment/culture period.
2. Using the above optimal starting density, replace the basal medium with the serum-free medium. Set up a series of cultures in this medium with varying concentrations of serum (e.g. 1–10%). Ensure enough replicate cultures are set up to give meaningful interpretation of results. Include selective agents to maintain gene copy number if recombinant cell lines are used.
3. Select the condition that gives 60–80% (or more) of the cell growth of the control cultures, and using this condition expand the cells to a larger scale (e.g. 24-well plate to 10 ml suspension to 50 ml suspension to 100 ml suspension, or 24-well plate to 25 cm^2 flask to 75 cm^2 flask to 175 cm^2 flask). Check for expression of desired product/ characteristics. If none of the conditions achieve the 60–80% growth of controls, consider alternative serum-free media or the addition of supplements.
4. Make a frozen bank of cells at this stage (e.g. 5×10^6 to 10^7 cells/ml) from an exponentially growing culture.
5. Set up a second series of cultures as in step 2, reducing the serum supplementation further (e.g. 5% going down to 0.1%). Grow cells for three to six days.
6. Repeat steps 4 and 5 until serum supplementation is eliminated.
7. Prepare a frozen bank of the serum-free adapted cells and check for the absence of mycoplasma.

4.3.9 Difficulties that may be encountered with SFM

When cells are grown in serum-free conditions, they no longer benefit from the multiple protective and nutritional effects that serum provides (*Table 2*). The robustness of the process in serum-free medium depends on attention to the following points:

(a) Cells appear more fastidious in the absence of serum - design of a dedicated medium for each cell type is usually necessary for optimal results.

(b) Culture conditions become more critical in SFM - better control of key process parameters (pH, oxygenation, etc.) is therefore necessary.

(c) SFM has a reduced capacity to inactivate or absorb toxic materials (e.g. heavy metals, endotoxin, etc.).

(d) Greater attention to the purity of components and to depyrogenation is required.

(e) Specific shear-protection agents may need to be added.

(f) A significant adaptation period may be required before cells are fully weaned into SFM. This makes the design and testing of serum-free medium a long and labour-intensive process. It should also be noted that, even after an appropriate period of adaptation, the cell growth rate, cell density, and viability of cultures are almost always lower in SFM than in the initial serum-containing medium.

4.3.10 Animal protein-free medium

The development of completely protein-free culture medium is an important objective for biopharmaceutical manufacturers. The main reason for this is to minimize the possibility of the introduction of adventitious viruses or prion agents into the product by elimination of all substances of animal or human origin. The same general design considerations apply as for SFM, but the difficulties faced in producing a satisfactory medium are considerably greater since the protein fractions routinely used in SFM (e.g. serum albumin and transferrin) must obviously be avoided.

As discussed in Section 3.3.2, albumin is the most important protein component in most SFMs. In principle, recombinant serum albumin can be produced in yeast and other micro-organisms, but recombinant albumin does not generally support cell growth effectively, presumably because some of the essential factors carried by albumin in the serum are not present in yeast culture supernatants (46).

Protein hydrolysates (peptones) have been shown to replace albumin effectively in some cases and it is likely that, like serum albumin, these bring to the medium a range of trace elements, oligopeptides, supplementary amino acids, and lipids. However such hydrolysates are most frequently of animal origin and may still represent an unacceptable risk of contamination with viruses or prion agents. Recently, it has been shown that medium formulated with plant protein-derived peptone can be effective for some cell types (47).

It should be noted that the medium used to produce recombinant proteins such as growth factors in micro-organisms itself frequently contains large quantities of animal-derived peptones. Thus the use of recombinant growth factors in medium does not necessarily eliminate them as a potential source of contamination with infectious agents. In the same way it is worth noting that most of the lipid supplements in commercial use are derived from bovine tallow, and that several amino acids are still frequently derived from the hydrolysis of animal proteins. Similarly, attachment factors are usually produced from serum or from cell culture. In this case though, the problem can be avoided by the use of Pronectin® (see Section 4.3.5).

Dedicated animal protein-free media are in wide use for some the types of transformed producer cells commonly used by biopharmaceutical manufacturers. Design of suitable animal protein-free media for other cell types remains a major technical challenge.

4.3.11 Component interaction and factorial experimental design

In the past, the design of SFM was predominantly empirical. The most common approach was to reduce the serum supplementation progressively ('weaning off') and to determine the critical factors involved in cell growth and protein expression.

Based on this approach, a database is now becoming established for the nutritional requirements of the most commonly used cell types (see Chapter 7) and this information can be applied generically. For example, NS-1 myelomas but not their hybridomas, are strict auxotrophs for cholesterol. In general, transformed cell lines have a lower requirement for growth factors than untransformed cells and in some instances, protein requirements can be completely eliminated from the medium (*Table 5*).

The advent of rapid and sensitive analytical methods such as HPLC has facilitated the measurement of the utilization of low molecular weight nutrients during cell culture. Based on this data, medium optimization can proceed by supplementation of rapidly utilized components and reduction of under-used components. However, interpretation of such studies is complicated by the dynamic interaction between utilization of different components. This has led several groups to base optimization studies on factorial experimental design aimed at accommodating complex systems involving multiple interacting factors (43). Such statistical optimization approaches based on Plackett-Burman (48) designs have been used to develop serum-free media for the production of erythropoietin and IFNγ by suspension culture of CHO cells (49, 50). From the statistical analyses, positive medium determinants for protein production were determined and then added to the basal medium. Additional complications can arise since cellular metabolism may alter during the course of a fermentation and different nutrients may become critical at different phases of the culture (51). Therefore, reductionist approaches to medium design should be applied with caution.

For further discussion of the design of serum-free media, see ref. 52.

5 Influence of cell culture systems on the choice of medium

The efficiency of an animal cell culture system depends on the interaction of many different factors. The support that the medium supplies for the cells is a crucial element, but this in turn is influenced by the type of cell culture system in which the cells are propagated.

Cultures may be operated in a simple batch mode, or as fed batches, or as a perfused system. The cells may be grown in suspension culture or attached to surfaces, or they may be grown at very high density (in excess of 10^8 cells/ml in various types of plug flow reactor) or at densities around 10^5 to 10^6 cells/ml in simple cultures.

Cells may be enclosed in compartments which favour the development of local micro-environments (as in hollow-fibre bioreactors or macroporous microcarrier cultures) or they may be fully exposed to the bulk medium. They may experience significant shear forces (either by pumped medium flow or by stirring) or they may be completely isolated from shear forces (e.g. in the interior of macroporous microcarriers or when encapsulated). In these different situations, the environment and stresses experienced by the cells modulate the support that they require from the culture medium.

5.1 Batch or perfusion cultures

When cells are grown as batch cultures, all the nutrients required for the duration of the culture must be present in the initial medium. The two major energy sources, glucose and glutamine, need to be present at unphysiologically high concentrations which may lead to the accumulation of high levels of the toxic metabolites, lactate and ammonia. In the fed batch approach, glucose and glutamine can be added at intervals as they become depleted, thus improving efficiency and limiting toxicity.

In perfusion systems, i.e. culture systems where cells are retained in the bioreactor while medium flows through, fresh medium is continuously supplied and spent medium removed. Ideally, the perfusion rate and the concentration of each component in the medium should be adjusted to match the consumption rate of each nutrient. Careful design of the medium is therefore needed to match medium composition to the cells' metabolic needs. Another point that requires consideration is that the perfusion rate should not be so rapid as to flush out autologous growth factors.

5.2 Anchorage-dependent cells

Many of the cell types which are currently the focus of intense research activity, such as endothelial cells and epithelial cells, will only grow and function when attached to surfaces. Some of the cells commonly used in vaccine manufacture, such as human diploid fibroblasts, Vero cells, and MDCK cells, also exhibit anchorage dependence.

The need for cells to be attached to surfaces imposes at least two special requirements on the medium. First, although some of these cells can synthesize their own attachment factors, attachment is generally accelerated and viability improved if factors such as fibronectin and laminin, which form part of the ECM, are incorporated into the medium formulation. These factors bind to the surfaces to which the cells will attach and act as ligands for specific cell surface receptors called integrins. Serum-containing medium carries sufficient of these factors naturally but they may need to be specifically added to serum-free formulations.

An alternative approach is to pre-coat the culture surfaces with ECM components such as collagen or fibronectin (see Chapter 5, Section 5.4), or a synthetic matrix substitute such as Pronectin®.

Polypeptide growth factors, particularly those of the FGF family, bind to elements of the ECM which may then act as a slow release pool for these factors. The dynamic interplay between growth factors, the matrix proteins, and the attached cells may be an important aspect of cellular physiology (53).

The second requirement that anchorage dependence imposes on the medium is the need to inhibit the proteolytic enzymes that are used for liberating the cells from the substrate during subculture, in order to minimize cell damage. In the presence of serum, this is achieved by the endogenous protease inhibitors. In serum-free medium, protection can sometimes be achieved by rinsing the cells in medium containing soya bean trypsin inhibitor after trypsinization. It should be noted, however, that the 'crude' trypsin preparations commonly used in cell culture contain proteolytic activities other than trypsin and that a single, purified trypsin inhibitor may not give adequate protection. [It will be found that highly purified trypsin is much less efficient at releasing cells from surfaces than the impure preparations usually employed (54).] A useful alternative to trypsin called Accutase is available from Innovative Cell Technologies Inc. Accutase is reported to cause less cell damage than trypsin and also has the advantage of containing no mammalian- or bacterial-derived products.

5.2.1 Aggregate cultures

When cells that are normally anchorage-dependent are grown in conditions where attachment factors are limiting, they tend to form aggregates which can sometimes be propagated in suspension. Such cultures can be useful for vaccine production or other large-scale operations, but a limitation may be the heterogeneity of the cultures and a tendency for necrotic regions to develop in the centre of the larger aggregates with subsequent cell lysis and liberation of cell contents and debris into the medium.

5.3 Stirred suspension cultures

A primary concern in suspension cultures is the protection required against damage to cells by shear forces. In serum-containing medium this protection is provided by serum proteins, mainly albumin. In serum-free conditions, several synthetic polymers (usually at around 1 g/litre) have been used to fill this role (see Section 4.3.5).

Recent studies have shown that shear damage occurs when turbulent eddies of similar size to the cells are produced in the medium (55). This phenomenon appears to be more associated with the formation and collapse of bubbles induced by cavitation and air entrainment by the impeller or by bubble disengagement from the liquid surface when sparging rather than by shear forces in the body of the liquid. The surfactant Pluronic F-68 is particularly effective in protecting cells from damage of this sort and is now widely used in serum-free medium as a protective agent against shear (see Section 4.3.5).

Limitation of oxygenation is the main factor that currently restricts the scale and density of many animal cell cultures, and cell damage by sparging or by high speed agitation precludes the use of these methods to increase oxygenation of the culture. Improved shear protective agents may help to minimize this problem. An alternative approach which is widely used is to perform oxygenation in a compartment of the bioreactor that is physically separated from the cells in order to avoid contact between the cells and bubbles. This implies an efficient separation system and an adequate circulation system to ensure that oxygenated medium reaches all cells in the bioreactor equally.

In some bioreactor configurations (some spin-filter devices and hollow-fibre systems) cells are completely separated from the vessel in which oxygenation occurs. In others such as encapsulated cells or cells in macroporous carriers, the oxygenation may occur in the same vessel as the cells, but they are protected from direct contact with bubbles or liquid turbulence.

An important consideration when Pluronic F-68 or other protective agents are added to products intended for therapeutic use is the need to remove these agents effectively during downstream processing. Anti-foaming agents (particularly silicones) are sometimes added to culture medium, but their use is best avoided since such materials are notoriously difficult to eliminate completely from the final product. Since the BSE crisis, another (largely theoretical) concern has arisen since many surfactants are ultimately derived from bovine tallow and it may be a regulatory requirement to show that this came from a 'safe' source.

5.4 High density culture systems

As already indicated, several of the cell culture systems used to manufacture biologicals operate at very high cell densities, often in excess of 10^8 cells/ml. This has the advantage that higher concentrations of product can be generated but imposes the need for better control of culture conditions to ensure that the cells are kept within the specified environmental limits. High cell density systems can be divided into two types: homogeneous systems where the cells are continually mixed with the bulk medium, and flow systems where the cells are held immobile while medium flows past them. Many configurations of these types of bioreactor exist – we will only consider two which represent the best-performing examples of their class.

5.4.1 Macroporous carrier systems

These systems employ porous particles with pores large enough to allow cells to enter the particle and colonize the internal space. The particles may be of neutral buoyancy and used as suspended microcarriers or they may be of higher density (typically 1.3–1.6 g/cm^3) and employed in a fluidized bed configuration. This configuration permits very efficient mass transfer, an homogeneous cell culture, and high cell density (56).

A particular feature of this and other high density systems is that the cells are able to form their own local micro-environment while still being able to receive nutrients from the bulk medium and to release toxic waste products. An important consequence of this is the reduced need for growth factors and matrix components since these are secreted by the cells and maintained in their own micro-environment. In cases where cells do not produce the required factors they may relatively easily be engineered to do so.

In addition, since the cells inside the particles are not directly exposed to the bulk medium, shear forces are much less of a problem. However, the protease levels secreted by cells in high density cultures may be significant, and particular attention should be paid to the need to include protease inhibitory capacity in the medium to protect product from degradation.

In high density cultures, particular components of the medium may become depleted very rapidly, and analysis of the input and output medium streams should be performed to determine whether any of these become limiting and to permit adjustment of their concentration in the medium and/or adjustment of the medium flow rate. It should be noted that deficiency of certain amino acids is one factor that can induce the secretion of proteases by animal cells (57). Another important consideration with high density cultures is the provision of adequate instrumentation to ensure that pH, pO_2, and temperature remain within the required limits. Note that medium buffering alone cannot maintain pH in these high density cultures and that some form of automatic pH control is essential.

5.4.2 Hollow-fibre systems

In hollow-fibre systems, the cells are held in the bioreactor separated from the medium flow by the walls of capillary tubes (hollow fibres) through which the medium flows. Typically, the capillary walls are impervious to macromolecules but allow the passage of low molecular weight nutrients including oxygen (58).

In common with other plug flow systems, hollow-fibre bioreactors may suffer from the development of concentration gradients of nutrients, particularly oxygen, along the length of the medium flow path. Adequate medium flow is therefore essential to limit the heterogeneity of the culture that this may generate. Unlike the situation with macroporous carriers, protein products produced in hollow-fibre systems are not released into the bulk medium but are retained by the capillary membrane in the cell compartment from where they can periodically or continuously be harvested in relatively concentrated form (see also Chapter 5, Section 5.8).

5.5 Very low density cultures

Cloning of cells and recovery of some cell types from primary isolates both require that cellular proliferation is achieved at very low initial cell densities. Not surprisingly, here the opposite situation pertains to that discussed above and the cells require maximum support from the medium. Serum supplementation is often not sufficient and 'conditioned medium' harvested from actively growing cultures of the same or another cell type is frequently added (at concentrations up to 25%) as a source of additional factors. Although the basic nutrients will be partially depleted, conditioned medium contains a further cocktail of undefined macromolecules including growth factors, hormones, and detoxifying factors and also low molecular weight compounds such as tricarboxylic acid cycle intermediates which may help with establishment of the culture. In extreme cases, it may be necessary to use 'feeder layers' of metabolically active but non-proliferating cells (a state usually attained by irradiation) which are co-cultured with the required cell to generate supporting factors *in situ* (see Chapter 6, Section 2.4.4 and Chapters 7 and 8).

More recently, efficient media for clonal growth have been developed which are based on the use of the most appropriate standard basal medium supplemented with the recombinant growth factors known to be needed by the required cell type. Effective pH control is also critical in this situation and can be achieved using buffers such as Hepes or Mops. CO_2 is also an essential requirement and, since cells at low density may not produce enough themselves, incubation in an atmosphere containing CO_2 at an appropriate concentration for the medium in use is essential (see *Table 1*). It is particularly important for cultures at low cell density that the medium used should be thoroughly equilibrated with the correct concentration of CO_2 before contacting the cells.

6 In-house medium development and production versus commercial supply

6.1 Economic considerations

Medium production from the basic raw materials is a major undertaking involving as it does the sourcing and quality control of a large number of individual components. In addition, the technology involved in milling and mixing powders and ensuring homogeneity and compatibility of the different components in the mixture is complex and not likely to exist within most biotechnology companies or cell culture laboratories.

For most users, the real choice to be made is between the different types of medium formulation and presentation that are available commercially. These include ready-to-use single strength (15) medium, medium concentrates (10× or in some cases 50×), or pre-mixed powder. Liquid medium is supplied in a variety of packaging ranging from small bottles (usually 500 ml) to flexible plastic containers with volumes from one to several hundred litres.

Laboratory-scale operations are likely to favour bottled single strength medium. For small process operations, single strength medium in 10 litre or 20 litre bags is a particularly convenient presentation. On larger scales, the choice depends on technical and economic factors that are specific to each manufacturer. The use of large containers of single strength medium permits a minimum investment in preparation, quarantine, and quality control facilities and equipment, and staff requirements may also be reduced. However, it is expensive both in terms of the litre cost of medium and also possibly in the cost of delivery. Storage of large quantities of medium in acceptable conditions (usually at 4 °C) may also incur significant costs (note that this is also true for the storage during quarantine of medium prepared in-house).

The use of medium concentrates may represent a suitable compromise. However, in this case it is necessary to invest in a plant for the production of tissue culture quality water and the vessels and pipework required for dilution. Continuous flow dilutors are available for on-line dilution. The cheapest approach in terms of material costs is to buy fully formulated powdered medium and reconstitute this for use on site. This requires investment in an adequate medium kitchen with ancillary facilities and appropriate personnel to operate it. When evaluating these options, it is important from an economic viewpoint that the true costs of the installation and operation of the required medium kitchen facilities are fully taken into account. These include the installation itself, clean room facilities and disposables, trained operators, and the necessary quality control back-up.

6.2 Development of a dedicated medium

Commercial decisions regarding the in-house development or subcontracting to a commercial organization may also be important when optimization of medium for a specific cell type or process is required. Here the key question is how frequently it is necessary to develop a new dedicated medium and whether the maintenance of in-house medium development expertise and facilities is economically justified by this frequency. Several of the major medium production houses offer confidential services in which a team that is engaged full time on medium development will produce an optimized medium formulation specifically for the customer's application.

6.3 Quality assurance and control

Critical elements in quality assurance for medium are the choice of supplier, evaluation of the documentation provided by the supplier, and audit of the supplier's facilities when appropriate. For basal and serum-free media, many suppliers have registered a Drug Master File with the relevant regulatory authorities which provides the necessary guarantees concerning the manufacturing process. For serum, confidence must be based almost entirely on the batch documentation provided.

To ensure the consistent performance of a medium, standard procedures must be set up for all operations concerning the production of the working medium from bought-in components. This particularly includes standardization of the storage conditions and the duration of storage of the complete medium before use, since the complete medium frequently has only a limited shelf-life.

Local quality control testing by the user before committing a medium batch to the production process is still of critical importance. Testing of serum batches has already been discussed in Section 3.5.2. Key tests to be applied to batches of powdered or liquid medium are summarized in Section 2.4.4. Simple tests such as verification of osmolality and pH should routinely be applied to every medium preparation (including those from the same batch of powder or concentrate) to ensure that no error of solution or dilution has been made.

Verifying sterility of the final medium before use is a less simple question because in some cases, the time required for completion of the sterility tests (typically seven to ten days) may not be compatible with the demands of the production process. Some operators assume sterility if all preparation procedures have been performed without incident and then verify sterility in parallel with production. The speed and sensitivity of sterility testing can be improved by the use of filter concentration methods to detect very low levels of contaminating micro-organisms.

References

1. Bettger, W. and Ham, R. (1982). *Adv. Nutr. Res.*, **4**, 249.
2. Shin, S. H. and Milligan, J. V. (1998). In *Endocrine cell culture* (ed. S. Bidey), p. 38. Cambridge University Press, Cambridge.
3. Leibowitz, A. (1983). *Am. J. Hyg.*, **78**, 173.
4. Reitzer, L. J., Wice, B. M., and Kennell, D. (1980). *J. Biol. Chem.*, **254**, 2669.
5. McKeehan, W. L. (1982). *Cell Biol. Int. Rep.*, **6**, 635.
6. Cooper, M. L., Boyce, S. T., Hansborough, J. F., Foreman, T. J., and Frank, D. H. (1990). *J. Surg. Res.*, **48**, 190.
7. Roth, E., Ollenschlager, G., Hamilton, A., Langer, K., Fekl, W., and Jaksez, R. (1988). *In Vitro Cell. Dev. Biol.*, **24**, 96.
8. Eagle, H. (1955). *Science*, **122**, 501.
9. Williams, G. T., Smith, C. A., Spooncer, E., Dexter, T. M., and Taylor, D. R. (1990). *Nature*, **343**, 76.
10. Moses, H. L., Coffrey, R. J., Leof, E. B., Lyons, R. M., and Kesi-Oja, J. (1987). *J. Physiol.*, Suppl. **5**, 1.
11. Van der Pol, L. and Tramper, J. (1992). In *Animal cell biotechnology: developments, processes and products* (ed. R. E. Spier, J. B. Griffiths, and C. MacDonald), p. 192. Butterworth-Heinemann, Oxford.
12. Hodgson, J. (1993). *Bio/Technology*, **11**, 49.
13. Hayashi, I. and Sato, G. (1976). *Nature*, **259**, 132.
14. Hutchings, S. and Sato, G. (1978). *Proc. Natl. Acad. Sci. USA*, **75**, 901.
15. Rizzino, A. and Sato, G. (1978). *Proc. Natl. Acad. Sci. USA*, **75**, 1844.
16. Mather, J. and Sato, G. (1979). *Exp. Cell Res.*, **120**, 191.
17. Bottenstein, J., Hayashi, I., Hutchings, S., Mather, J., McClure, D., Ohasa, S., *et al.* (1979). In *Methods in enzymology* (ed. R. Wu, L. Grossman, and K. Moldave), Vol. 58, p. 94. Academic Press, London.

18. Orly, J. and Sato, G. (1979). *Cell*, **17**, 295.
19. Honma, Y., Kasukabe, T., Okabe, J., and Hozumi, M. (1979). *Exp. Cell Res.*, **124**, 4.
20. Bottenstein, J. and Sato, G. (1979). *Proc. Natl. Acad. Sci. USA*, **76**, 514.
21. Bottenstein, J. (1981). *Cancer Treat. Rep.*, **65**, 671.
22. Barnes, D. and Sato, G. (1979). *Nature*, **281**, 388.
23. Maciag, T., Kelley, B., Certundolo, J., Ilsley, S., Kelley, P., Gaudreau, J., *et al.* (1980). *Cell Biol. Int. Rep.*, **4**, 43.
24. Dicker, P., Heppel, L., and Rozengurt, E. (1980). *Proc. Natl. Acad. Sci. USA*, **77**, 21.
25. Chang, T., Steplewski, Z., and Koproswski, H. (1980). *J. Immunol. Methods*, **39**, 369.
26. Brietman, T., Collins, S., and Keene, B. (1980). *Exp. Cell Res.*, **126**, 494.
27. Murakami, H., Masui, H., Sato, G., and Raschke, W. (1981). *Anal. Biochem.*, **114**, 4.
28. Walthall, B. and Ham, R. (1981). *Exp. Cell Res.*, **134**, 301.
29. Phillips, P. and Cristofalo, V. (1981). *Exp. Cell Res.*, **134**, 297.
30. Pessano, S., McNab, A., and Rovera, G. (1981). *Cancer Res.*, **41**, 3592.
31. Michler-Stuke, A. and Bottenstein, J. (1982). *J. Neurosci. Res.*, **7**, 215.
32. Simms, E., Gazdar, A., Abrams, P., and Minna, J. (1980). *Cancer Res.*, **40**, 4356.
33. Bottenstein, J. (1980). In *Advances in neuroblastoma research* (ed. A. Evans), p. 161. Raven Press, New York.
34. Taub, M., Chuman, L., Saier, M., and Sato, G. (1979). *Proc. Natl. Acad. Sci. USA*, **76**, 3338.
35. Ham, R. (1982). In *Growth of cells in hormonally defined medium* (ed. G. H. Sato, A. B. Pardee, and D. A. Sirbaska). Cold Spring Harbor Conference on Cell Proliferation, Vol. 9, p. 39. Cold Spring Harbor Laboratory Press, NY.
36. Yamane, I., Kan, M., Minamoto, Y., and Amatsuji, Y. (1982). In *Growth of cells in hormonally defined medium* (ed. G. H. Sato, A. B. Pardee, and D. A. Sirbaska). Cold Spring Harbor Conference on Cell Proliferation, Vol. 9, p. 87. Cold Spring Harbor Laboratory Press, NY.
37. Poole, C., Reilly, H., and Flint, M. (1982). *In Vitro*, **18**, 755.
38. Waymouth, C. (1981). In *The growth of vertebrate cells in vitro* (ed. C. Waymouth, R. Ham, and P. Chapple), p. 105. Cambridge University Press, New York.
39. Ling, C., Gey, G., and Richters, V. (1988). *Exp. Cell Res.*, **52**, 469.
40. Perris, A. and Whitefield, J. (1967). *Nature*, **216**, 1350.
41. Whitefield, J., Perris, A. and Rixon, R. (1969). *J. Cell. Physiol.*, **74**, 1.
42. Bettger, W. and Ham, R. (1982). In *Advances in nutritional research* (ed. H. Draper), p. 249. Plenum Press, New York.
43. Murhammer, D. and Goochee, C. (1990). *Biotechnol. Prog.*, **6**, 391.
44. Murhammer, D. and Goochee, C. (1990). *Biotechnol. Prog.*, **6**, 142.
45. Capello, J. and Crissman, J. W. (1990). *ACS Polymer Reprints*, **31**, 193.
46. Keenan, J., Dooley, M., Pearson, D., and Clynes, M. (1997). *Cytotechnology*, **24**, 243.
47. Noe, W., Schorn, P., Bux, R., and Berthold, W. (1994). *In Animal cell biotechnology: products of today, prospects for tomorrow* (ed. R. E. Spier, J. B. Griffiths, and W. Berthold), p. 413. Butterworth-Heinemann, Oxford.
48. Plackett, R. I. and Burman, J. P. (1946). *Biometrica*, **33**, 305.
49. Yoon, S. K., Ahn, Y., Kwon, I., Han, K., and Song, J. Y. (1998). *Biotechnol. Lett.*, **20**, 101.
50. Castro, P. M. L., Hayter, P. M., Ison, A. P., and Bull, A. T. (1992). *Appl. Microbiol. Biotechnol.*, **38**, 84.
51. Bell, L. K., Bebbington, C. R., Bushell, M. E., Sanders, P. G., Scott, M. F., Spier, R. E., *et al.* (1991). In *Production of biologicals from animal cells in culture* (ed. R. E. Spier, J. B. Griffiths, and B. Meignier), p. 304. Butterworth-Heinemann, Oxford.
52. Karmiol, S. (2000). In *Animal cell culture: a practical approach* (3rd edn) (ed. J. R. W. Masters), p. 105. Oxford University Press.

53. Gospodarowicz, D., Ferrara, N., Schweigerer, L., and Neufeld, G. (1987). *Endocrinol. Rev.*, **8**, 95.
54. Dickerson, C. H., Birch, J. R., and Cartwright, T. (1980). *Dev. Biol. Stand.*, **46**, 67.
55. McQueen, A., Meilhoc, E., and Bailey, J. E. (1987). *Biotechnol. Lett.*, **9**, 832.
56. Looby, D. and Griffiths, J. B. (1990). *Trends Biotechnol.*, **8**, 204.
57. Froud, S. J. K., Clements, G. J., Doyle, M. E., Harris, E. L. V., Lloyd, C., Murray, P., *et al.* (1991). In *Production of biologicals from animal cells in culture* (ed. R. E. Spier, J. B. Griffiths, and B. Meigneir), p. 110. Butterworth-Heinemann, Oxford.
58. Knazek, R. A., Guillino, P. M., Kohler, P., and Dedrick, R. L. (1972). *Science*, **178**, 65.

Chapter 4
Microscopy of living cells

Ian Dobbie and Daniel Zicha
Imperial Cancer Research Fund, 44 Lincoln's Inn Fields, London WC2A 3PX, UK.

1 Introduction

The ability to observe living cells has always been dependent on the availability of appropriate technology. Despite the fact that the principle of how to make powerful compound microscopes was known as far back as the 16th century, the quality of actual devices from this period was very poor due to optical aberrations. Indeed, the first observation of individual bacteria and protozoa - by Anton von Leeuwenhoek - was not reported until 1674 and even this was achieved using a single lens microscope.

The discovery by Joseph Jackson Lister of a method for combining lenses to improve resolution by reducing aberrations opened up the field of microscopy, and enabled Matthias Schleider and Theodor Schwann to make the observations which led them, in 1838, to propound the cell theory, i.e. the idea that the nucleated cell is the unit of structure and function in plants and animals. Later progress in basic principles, such as Ernst Abbe's theory of light diffraction in 1873, allowed Carl Zeiss to achieve resolution at the theoretical limits of visible light in 1886.

Following the development of tissue culture by Ross Harrison in 1907, the techniques of microscopy were soon applied to the observation of living cells in culture, and by the 1920s Ronald Canti had pioneered the use of time-lapse cinemicroscopy for recording the behaviour of cultured cells. However, the straightforward application of the compound microscope provided very limited contrast with living cells. Consequently, much effort was expended to develop suitable contrast-enhancement techniques, leading to the invention of interference microscopy by A. A. Lebedeff in 1930, phase-contrast by F. Zernike in 1935, and differential interference contrast by P. H. Smith in 1955 (with further improvements by G. Nomarski in 1969).

The availability of these contrast-enhancement techniques finally allowed live cells to be studied in detail using both qualitative and quantitative techniques. In the 1950s, Michael Abercrombie pioneered the application of rigorous quantitative methods to the analysis of cell behaviour, at a time when microscopic images were still captured on celluloid film. The introduction of video microscopy greatly improved the convenience of dynamic data acquisition (see the comprehensive review by Inoué) (1) and the more recent advent of computers and digital imaging technology has added further impetus to progress in this field.

Another important factor in the recent dramatic increase in the popularity of light microscopy has been progress in fluorescence microscopy, which has allowed dynamic observation of specific intracellular structures in live cells. This has been allied to powerful techniques of molecular biology, to provide important information about the relationship between specific intracellular structures and the molecular mechanisms of cellular functions. Such studies will become even more important in the 'post-genomics' period and new techniques are continually emerging to support the process. Two important new developments in light microscopy for live cells illustrate this trend:

(a) The enhancement of 3D observation by confocal microscopy and deconvolution techniques.

(b) Techniques for the visualization of specific molecular interactions, such as fluorescence resonance energy transfer (FRET) and fluorescence correlation spectroscopy.

2 The modern light microscope

2.1 Basic scheme

The modern light microscope is based on the principle of the compound microscope (2) and consists of four main parts (*Figure 1*):

- a light source with a collector lens (A)
- a condenser (B) to focus the light on a sample (C)
- an objective (D) to collect the light from the sample and project it through a tube lens (E)
- an eyepiece (F) to produce a focused image in the observer's eye or a camera

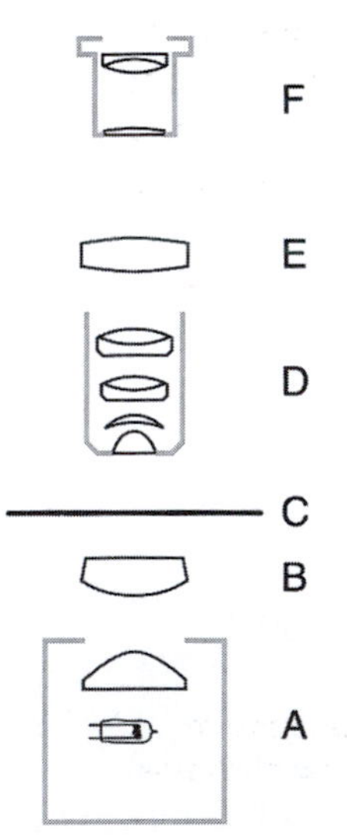

Figure 1 A schematic of a compound microscope showing a light source with collector lens (A), condenser (B), sample (C), objective (D), tube lens (E), and eyepiece (F).

Additional magnification, usually ×1.5 or ×2.5, can be achieved by insertion of an additional lens (an optovar) located between the objective and the eyepiece. An optovar reduces the field of view and does not increase resolution.

All segments in this chain are essential for the production of quality images utilizing the full resolution of the microscope. The main producers of light microscopes are Leica, Nikon, Olympus, and Zeiss.

2.2 Properties of light

In order to understand the different microscopic techniques it is necessary to have a basic understanding of some of the properties of light (3). Visible light is an electromagnetic wave with a wavelength between about 400 nm (violet light) and 700 nm (red light). Light comes in 'packets' (quanta) called photons, with energy inversely proportional to the wavelength; violet light has higher energy than red light.

The electric and magnetic fields which make up light change sinusoidally as the light propagates. The position of these fields within this cycle is known as the phase of the light. With most light sources, such as tungsten or arc lamps, the phase of the light at different times and in different regions of the source is not related in any way. These are referred to as incoherent sources. Lasers emit light that is in phase and are known as coherent sources.

When two coherent light beams combine they add in a manner dependent upon their relative phase, an effect called interference (*Figure 2*). Using light from different regions of an incoherent source will produce no interference, as there is no fixed relationship between the waves. However, coherent sources, such as lasers, will easily produce interference. If two interfering beams are exactly in phase then they will add and the amplitude will be doubled (constructive interference, *Figure 2A*), whereas if the two beams are exactly out of phase then they will cancel each other out and produce zero amplitude (destructive interference, *Figure 2B*). The observed intensity of light is equal to the square of its amplitude.

The speed of light in vacuum is 2.997925×10^8 m/sec, but in different materials light travels at different speeds. The refractive index is the ratio of the speed of light in a given material to that in vacuum. When light travels over a

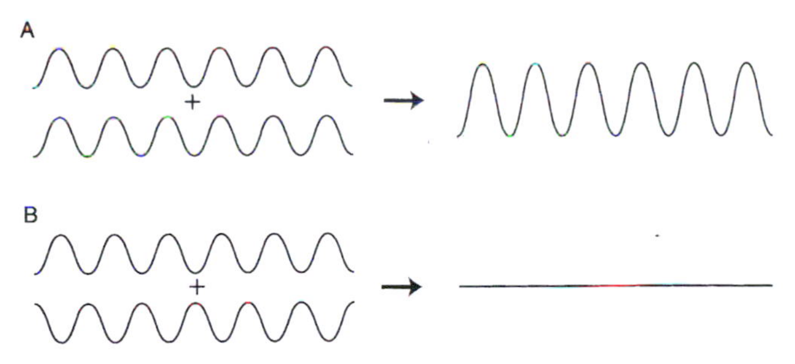

Figure 2 (A) Constructive interference occurs when two coherent beams combine exactly in phase and results in double the amplitude. (B) Destructive interference occurs when the phase of the two beams is shifted by a half wavelength and results in zero amplitude.

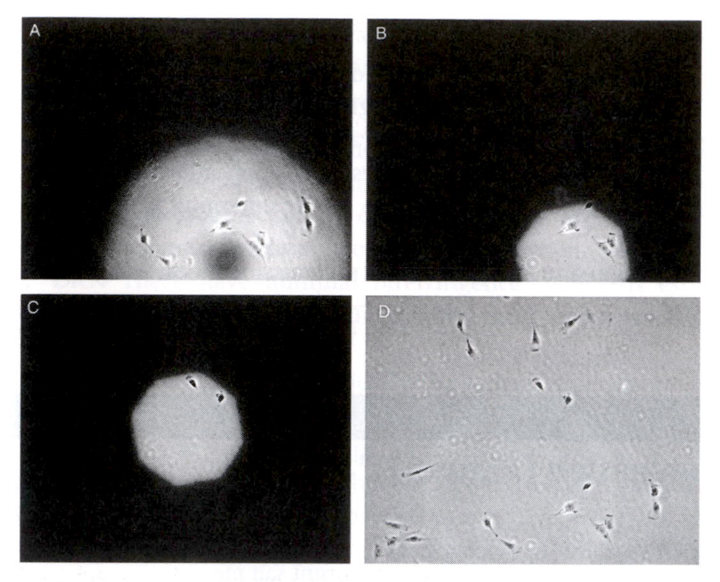

Figure 7 Setting Köhler illumination. (A) Initially the field iris is shown off centre and out-of-focus. (B) The iris is first brought into focus, and then centred (C). (D) Finally the field iris is opened to just overfill the field of view. At all times the sample is in-focus. (These images were taken using phase-contrast on a Nikon Diaphot 200 microscope with a $\times 10$, NA = 0.25, objective and a Princeton Instruments RTE/CCD-1300-Y/HS MicroMax cooled CCD camera.)

Protocol 2

Tungsten lamp alignment[a] for Zeiss Axiovert microscope[b]

Equipment

- Microscope with a lamp housing
- Centring telescope or a Bertrand lens

Method

1. Remove the lamp housing and the diffuser from the microscope and project an image of the filament on a wall about 3 m away.
2. Adjust the focus and position screws so that the filament and its mirror image overlap and are both in-focus. The two filament images should be adjusted so that they each fill in the gaps in the other, to cover a central rectangular region.
3. Mount the lamp housing and perform final adjustment by observing the images of the filament using a centring telescope or Bertrand lens. The filament image should be central, in-focus, and interlaced with its mirror image.
4. Remove the lamp housing, replace the diffuser, and remount the housing on the microscope.

[a] Usually required only when the lamp is changed. Follow manufacturer's instructions to replace the bulb.

[b] Not all makes require tungsten lamp adjustment. Zeiss Axiovert is given as an example.

In order to generate the best image it is important to readjust the illumination for every change of objective. Usually this will just require the repetition of *Protocol 1*, steps 4–7.

Proper alignment of a mercury or xenon arc lamp (*Protocol 3*) is essential in order to achieve good fluorescence imaging where the illumination light is reflected to the specimen using a dichroic mirror (more details are given in Section 3.11). As these arc lamps have relatively short lives it is often necessary to replace the bulbs and then realign completely.

Protocol 3

Arc lamp alignment[a]

Equipment

- Fluorescence microscope
- UV protective glasses
- White paper

Warning: Arc lamps emit a large amount of UV radiation. This can seriously damage your eyes and give you severe skin burns. Wear UV eye protection and take care not to shine unfiltered light on your skin.

Method

1. Follow manufacturer's instructions to replace the arc lamp. It should not be touched with bare hands, but if this happens, the bulb should be cleaned with ethanol. Generally arc lamps should be handled cold since otherwise there is a danger of explosion.
2. Remove one of the objectives and place the sheet of white paper at the sample position.
3. Using a dichroic mirror, reflecting visible light, illuminate the paper with the arc through the empty objective mount.
4. Focus the direct image of the arc on the paper using the lamp adjusters. The lamp housing contains a reflector, which produces a mirror image of the arc. Unless the manufacturer instructs otherwise, move the mirror image of the arc next to the direct image, using the mirror adjusters. Centre the arc images within the circular field.

[a] Usually required only when the lamp is changed.

It is possible to buy fibre-optic scramblers (Technical Video, Q in *Figure 8*) that allow the mounting of the arc lamp away from the microscope. These systems remove the large heat source of the arc lamp from direct contact with the microscope, allow filter wheels and shutters to be mounted remotely, reducing vibration at the microscope, and are able to provide a more even field of illumination.

In order to achieve good microscopic images it may be necessary to illuminate with high intensity light, which is not well tolerated by most cells. In order to protect the cells, the transmission illumination can be restricted to a single wavelength by an interference filter (M in *Figure 8*). Further protection of cells amplitude. The undiffracted background light beam hits the objective phase plate and develops the λ/4 phase shift while the light beam diffracted by cells passes outside the phase plate and is unaffected. When the two beams recombine to form an image by interference, the λ/4 phase shift introduces a dramatic contrast enhancement. An example of a cell taken with phase-contrast is shown in *Figure 9A*. Phase-contrast is the most popular contrast enhancement technique for live cells because it is affordable, robust, and produces high-contrast images. However it is not suitable for accurate determination of cell edges especially by automatic image processing since thick edges produce a bright 'halo', obscuring them, and thin edges vary irregularly in their intensity. Setting up a microscope for phase-contrast microscopy is described in *Protocol 4*.

Protocol 4

Setting up phase-contrast illumination

Equipment

- Microscope with condenser phase annulus and matching objective phase plate
- Centring telescope or Bertrand lens attachment on the microscope

Method

1. Adjust the microscope for Köhler illumination as described in Protocol 1, leaving the microscope focused on your sample.
2. Ensure that you are using a phase objective with a matching phase annulus in the condenser, each designated as Ph1, Ph2, or Ph3.
3. Insert the centring telescope or Bertrand lens, which enables observation of the phase annulus and phase plate. Focus the centring telescope or Bertrand lens so that the phase plate is in-focus. The phase annulus should be clearly seen in the centre of the image.
4. Move the phase annulus using its adjusters so that it is central within the bright phase plate.

Simple and relatively cheap microscopes, such as the Eclipse TS100 (Nikon), CK 30 / CK 40 (Olympus), or Axiovert 25 (Zeiss), are available for phase-contrast microscopy. These simple microscopes have adjustable size of the field iris but they do not require centring of phase rings. Different microscopic techniques, including fluorescence (see Section 3.11), are possible with optional upgrades.

3.5 Differential interference contrast (DIC)

In DIC microscopy a Wollaston prism splits polarized incident light into two components that are shifted relative to one another by a very small amount (about 1/3 of a wavelength). The two beams pass through the sample, are recombined by another prism, and then pass through an analyser. The analyser is a rotating

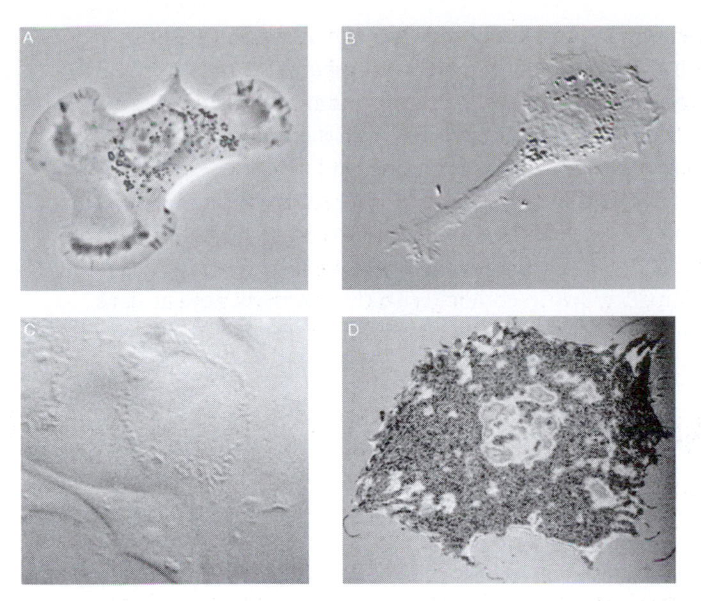

Figure 9 Images of rat sarcoma cells obtained using a selection of microscopic techniques. (A) Phase-contrast. (B) Differential interference contrast. (C) Hoffman modulation contrast. (D) Reflection contrast. (A and B) Taken with a Zeiss inverted LSM 510 microscope with a ×25, NA = 0.8, water immersion objective, and the field width was 120 μm. (C) Taken on a Nikon Diaphot 300 microscope with a ×20, NA = 0.55, objective, a Hamamatsu Orca I cooled CCD camera, and the field width was 110 μm. (D) Taken with a Zeiss upright LSM 510 microscope with a ×63, NA = 1.4, oil immersion objective, and the field width was 70 μm.

polarization filter. Contrast is generated by the interference of the two beams that have passed through closely neighbouring regions of the sample. DIC produces a shadow-cast image that is especially good for picking out small variations in refractive index such as the edges of cells, vesicles, or narrow filaments (*Figure 9B*). The contrast can be adjusted by lateral movement of the Wollaston prism or by rotating the analyser.

3.6 Hoffman modulation contrast

The Hoffman modulation contrast system uses an illumination slit in the condenser and the objective aperture holds a second complementary mask, called a modulator, which consists of three parallel segments – a dark segment, an intermediate segment in the centre, and a bright segment at the other side. Light deviated by the sample to one side is thus blocked whereas light deviated to the other side is unchanged and the background is partially suppressed by the middle segment. A shadow-cast image, similar to DIC, is therefore fairly easily produced (*Figure 9C*).

3.7 Reflection contrast

Reflection contrast microscopy uses reflected light backscattered from the cells to form an image. Illumination as well as observation is achieved by means of the objective. In order to eliminate unwanted reflections, the objective needs to have a highly efficient antireflection coating. Zeiss designates these objectives 'Antiflex' and produces a polarization filter set for separation of the illumination and observation beams. Reflection contrast with live cells is especially useful for visualization of focal adhesion contacts. This technique uses reflections from the coverslip and from the ventral surface of cells to very sensitively measure the separation of the cell from the coverslip. The light beams from these two different reflections interfere and produce a high-contrast image based on the separation of the cell from the coverslip. Regions in contact with the coverslip appear dark and $\lambda/4$ away from the coverslip they reach a maximum intensity (*Figure 9D*).

3.8 Varel contrast

Varel contrast is a new low-cost technique for imaging live cells. It combines phase-contrast and inclined unilateral illumination. The objective has an additional Varel plate, which is bigger than the phase plate and arranged concentrically. The illumination is restricted by a stop with an aperture shaped like a ring sector. The aperture can be positioned coinciding with the image of the Varel plate producing Varel contrast (*Figure 10A*), between the images of the plates producing inclined bright field illumination (*Figure 10B*) and outside the image of the Varel plate producing unilateral dark field (*Figure 10C*). The technique is especially useful in cases where phase-contrast fails, for example because of the curved bottom of culture vessels. Varel contrast is available from Zeiss and a similar system called Relief Phase Contrast is available from Olympus.

3.9 Polarizing microscopy

Plane-polarized light passing through a cell is rotated by variable amounts due to the birefringence of different areas in the cell. If a fixed polarizer is mounted before the condenser and a second rotatable polarizer is mounted after the objective then the intensity of the output image is dependent upon the rotation of plane-polarized light through the sample, and the angle of the rotating polarizer, called the analyser. As the analyser is rotated different features in the

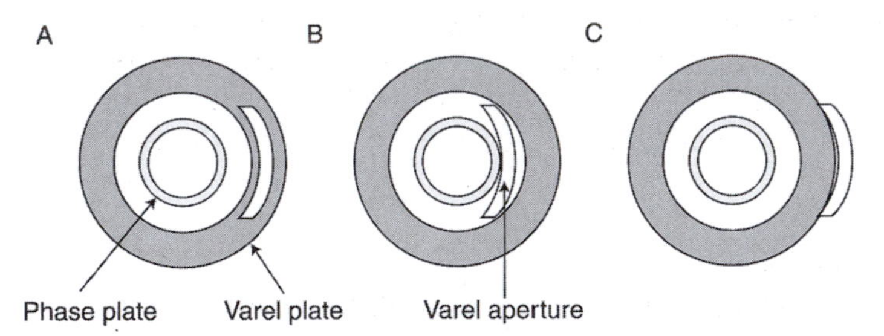

Figure 10 A schematic diagram of the objective and condenser elements of Varel contrast. Illumination is restricted to the Varel aperture in the condenser, and is shown superimposed on the image of the phase plate and the Varel plate in the objective. These can be used in three different configurations by moving the Varel aperture: (A) Varel contrast, (B) inclined bright field illumination, and (C) unilateral dark field illumination.

sample can be visualized. A compensator is added in front of the analyser to enhance contrast.

3.10 Interference microscopy

Although all types of microscopy rely on interference phenomena, a group of microscopy techniques is referred to as interference microscopy (4). Microscopes employing these techniques visualize the light phase shift introduced by the cells on the basis of interference of the object beam passing through a cell and a reference beam, which does not. The beams are separated by a semitransparent mirror before the sample and recombined after. This technique is not currently commercially available but it has a considerable potential for computer assisted imaging of live cells (5).

3.11 Fluorescence microscopy

In fluorescence microscopy, a fluorescent dye in the sample is illuminated with high intensity light of one wavelength (e.g. blue), and produces a weaker emission at a longer wavelength (e.g. green). Fluorescence microscopy is generally used in the so-called epifluorescence mode where illumination and observation both occur through the objective. This has the great advantage that only the reflected excitation light (a relatively small amount) needs to be filtered out in order to observe the emitted fluorescence. In general light from a high intensity arc lamp passes through an excitation filter which only allows through the required excitation wavelengths. It is reflected by a dichroic mirror through the objective onto the sample. Light from the sample is collected by the objective, the longer wavelength emissions pass through the dichroic mirror and then pass through an emission filter in order to further reduce reflected excitation light before passing to the eyepieces or detector. With multi-band (dual, triple, or quad) filters and dichroic mirrors it is possible to view several fluorescent dyes simultaneously or consecutively. The main manufacturers of fluorescence filters, dichroic mirrors, and complete filter sets are Chroma and Omega.

A wide range of fluorescent dyes has been used to specifically stain different structures within cells. Recent development of a range of fluorescent proteins has allowed the application of molecular biology techniques to produce fluorescently labelled structures within live cells. As an example, a sequence of images of a live cell producing green fluorescent protein (GFP) is shown as a montage in *Figure 11A*.

3.11.1 Confocal microscopy

Confocal microscopy (6) provides 3D information about cells by optical sectioning. Illumination is restricted by a pinhole projected onto a limited region in the sample. Observation is restricted by a second pinhole, placed in the position where the image of the illuminated region is formed. This arrangement of the illumination and observation pinholes is said to be confocal and eliminates light originating from features above and below the focal plane in the specimen,

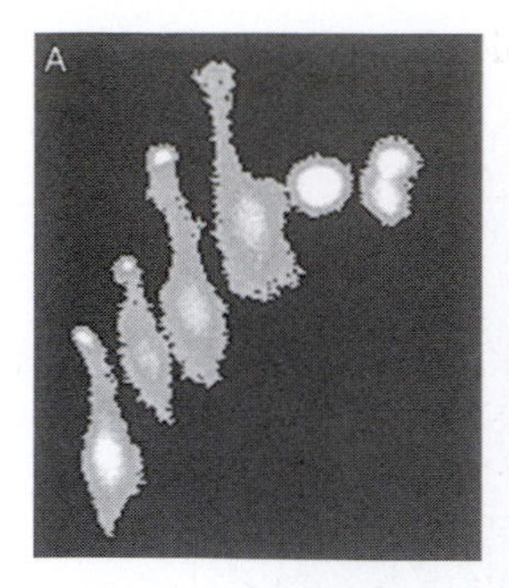

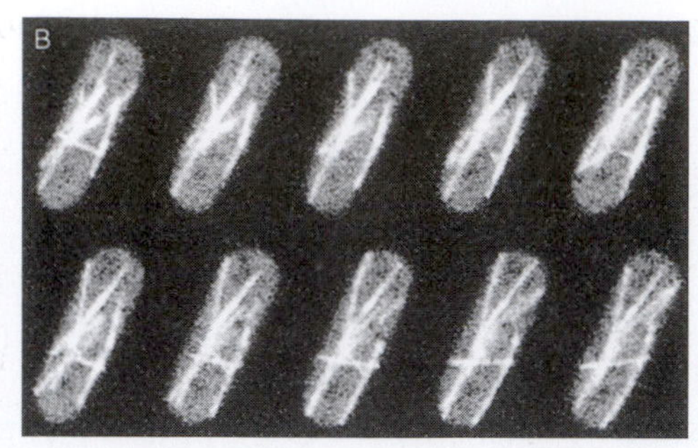

Figure 11 (A) Conventional and (B) confocal fluorescent images. (A) A superimposition of time frames from conventional fluorescence time-lapse images of a rat sarcoma cell producing a GFP fusion protein. The motile cell is progressing upwards and ends in a division. Each time frame is displaced laterally to enhance visibility. The total field of view is 170 μm across. (The images were taken on a Zeiss Axiovert 135TV microscope with a ×10, NA = 0.32, objective and a Princeton Instruments RTE/CCD-1300-Y/HS MicroMax cooled CCD camera.) (B) A gallery of time frames taken at 15 sec lapse-intervals from confocal fluorescence imaging of yeast (*S. pombe*) containing a GFP-tubulin fusion protein. The microtubules can be seen extending and shrinking between time frames. The yeast cell is approximately 10 μm in length. (These images were taken with a Zeiss LSM 510 inverted confocal microscope with a ×63, NA = 1.4, oil immersion lens, and were kindly provided by Damian Brunner, Imperial Cancer Research Fund, London.)

producing optical sectioning. At the same time the maximum achievable resolution increases by almost a factor of two. An image is produced by scanning the sample with the illumination and observation pinholes. The original confocal microscopes used a spinning disk with small holes in it for both illumination and observation. An improved version of the original design has recently been implemented by Perkin Elmer. Current confocal microscopes mainly work by scanning a single laser spot across the sample using galvanometer mirrors. An example of confocal imaging is presented as a sequence of projections in *Figure 11B*. The main producers of laser scanning confocal microscopes are Bio-Rad, Leica, Nikon, Olympus, and Zeiss.

3.11.2 Two-photon excitation

This recent development achieves excitation of a fluorescent molecule by a simultaneous interaction with two coherent photons, each of them having exactly a half of the required energy. The excitation light has therefore twice the wavelength normally required to excite the dye. In order to achieve two-photon excitation, pulsed lasers with femto-second pulses are used. The photon density is only high enough to achieve two-photon excitation within the focal spot. Consequently no out-of-focus material is excited, lowering the background and reducing photobleaching. It is also possible to excite UV dyes using visible wavelengths, eliminating the necessity for UV lasers and UV optics. The infrared light needed to excite visible wavelength dyes has greater penetration depth in thick samples. This technique can be extended to three or more photon excitation.

Two-photon confocal microscopes are commercially available from Bio-Rad, Leica, and Zeiss.

3.11.3 Fluorescence resonance energy transfer (FRET) and fluorescence correlation spectroscopy

Recent advances in microscopy have lead to developments enabling extremely sensitive measurements of molecular interaction using fluorescence techniques in live cells. In FRET a single molecule is labelled with two different fluorescent dyes in different places, or two interacting species are separately labelled with different fluorescent dyes. One dye is excited and the emission of the other is measured showing how much energy is transferred between them. The separation and relative orientation of the two fluorochromes strongly affects the amount of energy transferred. In fluorescence correlation spectroscopy a single fluorescently labelled species is monitored, via a confocal pinhole, in a focus-limited spot. Time correlation between changes in emitted intensity is used to calculate the rate at which single fluorescent molecules move through the volume of interest. These correlations can then be used to estimate the size of the molecule based on its diffusion rate. Two fluorescent dyes can be monitored simultaneously to produce cross-correlations and hence information about the interactions between the two labelled molecules. Zeiss manufactures an auto-correlation system and has announced availability of a cross-correlator extension to their LSM 510 confocal microscope.

4 Specimen preparation

Microscopy of living cells needs not only good microscopic conditions but also the correct environment for the cells. Matching these two aspects requires care. In order to produce quality high-resolution images, cells have to be cultured on glass coverslips. Most objectives used for biological work are designed for a number 1.5 (170 μm thick) coverslip. Some objectives are adjustable, allowing correction over a range of coverslip thicknesses. Zeiss designates these objectives 'Korr'. Dipping objectives are designed to be immersed directly into the culture medium to view cells from above with no coverslip, although evaporation may be a problem (see Section 5).

4.1 Culture flasks and Petri dishes

In general both culture flasks and Petri dishes are now made of plastic. Visualization of live cells in these requires objectives and condensers with long working distances. Standard ×10 objectives will generally be suitable. However at higher power (×20 or ×40) special objectives will usually be required. At even higher magnification, the plastic bottom of culture flasks or Petri dishes may be too thick to allow satisfactory focus on the cells. Due to high birefringence, plastic vessels are generally unsuitable for any technique using polarized light. Furthermore their thickness reduces resolution and the surface is generally not very

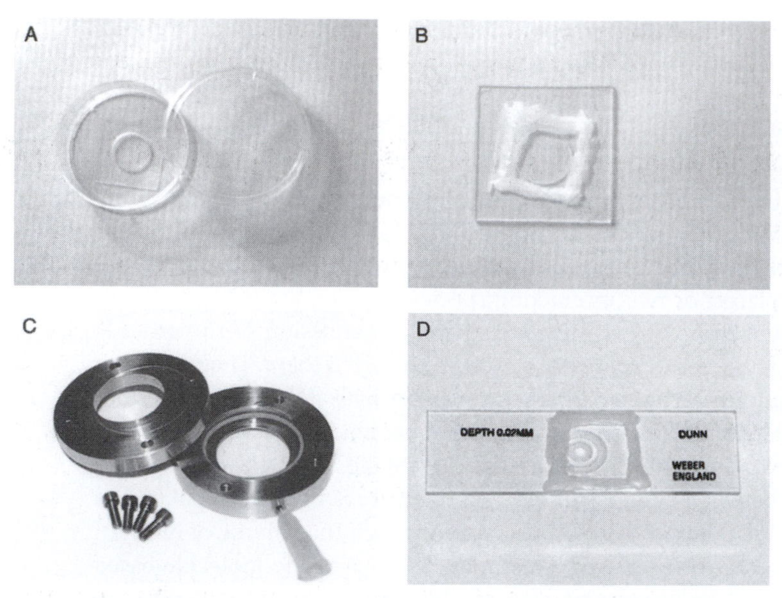

Figure 12 A selection of chambers for microscopy of live cells. (A) A 35 mm Petri dish with a standard thickness coverslip attached to a window in the base (MatTek). (B) A glass slide with a hole drilled through it and a coverslip glued to one side. A second coverslip (22 mm × 22 mm) with cells cultured on it is attached to the other side with a wax mixture. (C) A chamber for two round coverslips clamped on either side of a rubber O-ring. The cells are cultured on one coverslip. The stainless steel body of the chamber is machined to fit a standard 35 mm culture dish holder and has holes on either side for the insertion of narrow hypodermic needles (one is shown) allowing perfusion of medium through the chamber. (D) A Dunn chamber (Weber Scientific) for chemotaxis assays with direct viewing of cells. The cells are cultured on a coverslip which is attached with the wax mixture (see Section 4.3).

flat, and consequently they should be avoided for any critical microscopy. If using an inverted microscope, one method by which these problems can be avoided is to culture the cells in a dish which has a standard thickness coverslip mounted in a central window cut in the bottom (*Figure 12A*, MatTek).

4.2 Coverslips

It is often preferable to culture cells directly onto coverslips (BDH) and then mount these coverslips on a chamber of some kind for imaging. Cells can easily be cultured on a cleaned small coverslip (see *Protocol 5*) in a 35 mm culture dish (Corning) by adding cell suspension and allowing the cells to settle and adhere. This will deposit cells on both the coverslip and the culture dish around the coverslip. If only small numbers of cells are available then it is possible to place a small drop of cell suspension onto the coverslip (100–200 μl is a reasonable amount) in the culture dish and allow the cells to settle in an incubator. After the cells have started to adhere (the time is cell type dependent; usually between 30 minutes and several hours), gently add more culture medium to flood the whole dish to prevent evaporation damaging the cells.

It is important to properly clean the coverslips before culturing cells on them (*Protocol 5*). In addition it may be beneficial to pre-coat the coverslips with, for example, fibronectin or gelatin.

Protocol 5

Acid cleaning of coverslips

Warning: Take care when handling the acids as they are highly corrosive and dangerous. Wear appropriate protective clothing.

Equipment and reagents

- Teflon coverslip racks (Molecular Probes)
- 300 ml of a mixture of 2 vol. of concentrated nitric acid to 1 vol. of concentrated HCl
- Distilled water
- 70% (v/v) ethanol in ultrapure water produced by a Milli-Q Plus (Millipore Corporation) or similar system

Method

1. Place the coverslips on a Teflon rack and immerse in the nitric acid/HCl mixture (this should be done in a fume cabinet). Leave for 2 h, swirling occasionally.
2. Remove the rack from acid mixture and immerse in distilled water. Leave the rack in the water for 15 min and repeat three times.
3. Store in 70% ethanol.
4. Remove ethanol by either flaming, or blowing off with clean compressed air. Since extensive flaming can easily bend the coverslips it is preferable to use clean air.

4.3 Chambers

Once cells are cultured on coverslips it is necessary to construct some kind of chamber in order to observe the cells on the microscope. For basic observation it is possible to create a very simple chamber from a glass slide and two coverslips. A 15 mm hole is drilled through the slide with a diamond burr and a coverslip is permanently glued to one face of the slide with Araldite epoxy resin (RS). A second coverslip with adherent cells can be attached using a wax mixture once the chamber is filled with medium (*Figure 12B*). The wax mixture is made from beeswax (Fisher), soft yellow paraffin (Fisher), and paraffin wax (melting point 46°C; Fisher) in the ratio 1:1:1. The chamber can be cleaned and reused. If medium changing is required then a more complicated type of chamber can be constructed. Two stainless steel plates can be machined to hold coverslips compressing a rubber O-ring between them. Two holes drilled through the plates allow the insertion of fine hypodermic needles through the rubber seal so that the medium can be replaced (*Figure 12C*).

A bacteriological counting chamber, 'Z-special unruled' (Weber Scientific), can be used for mounting a coverslip with pre-cultured cells which is then sealed in place using the wax mixture (see above). This arrangement produces particularly good images since microscopes are usually optimized for samples of this size. A modification of this chamber (the Dunn chamber), specifically designed for assessment of chemotaxis (7), is also available from Weber Scientific (*Figure 12D*).

4.4 Interaction with microscopic techniques

For high-resolution microscopy it is necessary to use cells on coverslips and oil, glycerol, or water immersion lenses as discussed in Section 2.4.

Fluorescence microscopy often needs extra precautions since the light emitted by the fluorescent dye is often weak and live cells are sensitive to the excitation light. It can often be very advantageous to culture cells directly onto a coverslip and observe them using an oil immersion lens since its higher numerical aperture gathers significantly more light. It should be noted that phenol red pH indicator, often used in culture medium, is fluorescent and should be avoided in order to reduce background. To avoid cell damage, it is essential to minimize exposure. It is also beneficial to use longer wavelength excitation whenever possible. For example use dyes which absorb at green wavelengths such as Cy3 (Amersham Pharmacia Biotech) rather than dyes which absorb at blue wavelengths such as Alexa488 (Molecular Probes).

5 Environmental control for microscopic specimens

Environmental control requires that the sample is held at the correct temperature, evaporation of the cell culture medium is prevented, and that the medium has the correct pH. The last of these may often require a CO_2-enriched atmosphere. However, it may be easier to work with a sealed chamber and a CO_2-independent medium (such as a Hanks' based culture medium - see Chapter 3, Section 2.2.2).

Temperature controlled stage heaters can be purchased from microscope manufacturers. Heated stages may have a hole under the sample position to allow access for the condenser or objective. This means there may be unwanted cooling of the sample in the position that is observed. The effect can be much more dramatic if there is immersion liquid between the objective and the sample, as this will act as a heat conductor. Such cooling through the immersion liquid can be prevented by using an objective heater (Bioptechs). Alternatively, a Perspex box with hatches allowing access to the stage position and focus controls can be constructed and heated using a fan-heater with a thermostatic controller (for example made from controller 208 2739, T probe 219 4674, heater element 224 565, low noise a.c. fan 583-325 - RS Components).

If a CO_2-rich atmosphere is required, it may be easier to encase the microscope in a box. An air/CO_2 mix at the correct concentration can be pumped into the box and will only slowly need to be replenished in order to preserve the concentration within the box. If such a heated box is used with open culture dishes then the atmosphere should be humidified to minimize evaporation of the culture medium. Alternatively an open-top culture dish used with either a dipping objective or an inverted microscope can have a layer of mineral oil (designed for tissue culture, Sigma) poured on top of the culture medium to prevent evaporation.

If time-lapse imaging is required then very careful temperature control is required since variations of temperature of less than 1 °C can lead to significant shifts in the image due to thermal expansion and contraction of the microscope body or stage. In these circumstances not only the specimen but the microscope as a whole must be temperature controlled, which may be achieved in a warm room.

6 Recording of microscopic images

Image recording by celluloid film has become obsolete because of the awkward development process and difficult handling of images for reproduction, presentation, and publication. Digital recording is becoming more widely used than video-tape recording because it provides sufficient speed, resolution, and sensitivity. Digital image acquisition techniques for microscopy have, in fact, superseded even 35 mm film in quality (8).

Choice of camera is determined by the microscopy technique. Standard CCD video cameras are adequate for contrast enhancement techniques such as phase-contrast and can be purchased from Cohu, Hitachi, Pulnix, and Sony. Weak fluorescence signals require sensitive scientific low-light level CCD cameras such as Hamamatsu Orca and Princeton Instruments MicroMax. A standard CCD video camera can be connected to a video printer, video recorder, or to a computer equipped with a digitization board. Scientific CCD cameras are designed to be directly connected to computers by digital interfaces. Commercially available computer-based recording systems are listed in Section 6.1.

An important consideration when choosing a microscope is the camera port. Maximum light efficiency is achieved by an arrangement which does not require mirrors and minimizes the number of optical components in the light path to the camera. This type of direct camera port is easily achieved with an upright microscope whereas an inverted microscope requires a port on the under surface of the microscope. Since rotation of the camera on the bottom port is difficult, a rotating microscope stage on the inverted microscope is useful.

A computer-controlled shutter between the light source and the sample is recommended to eliminate unnecessary illumination, which can cause photodamage.

6.1 Time-lapse imaging

Imaging of live cells requires a suitable microscopic technique, environmental control for the cell culture on the stage of the microscope, and recording equipment. This equipment is also essential for time-lapse recording (see *Protocol* 6) (9). Recent progress in digital imaging and computing has dramatically simplified time-lapse recording. The environmental control of tissue culture conditions needs to be provided as described in Section 5. For long-term recording it becomes much more critical, as does the elimination of unnecessary illumination. An automatic illumination shutter is recommended for the observation of live cells, whereas for time-lapse recording it is a necessity. Suitable software is then used to acquire images in time-lapse mode and to operate light shutters and wheels with fluorescence filters for multi-channel recording. Complete imaging systems can be purchased from Applied Precision, Kinetic Imaging, Princeton Instruments, Till Photonics, or Universal Imaging.

Protocol 6

Time-lapse imaging

Equipment (letters refer to *Figure 8*)

- Microscope (A)
- Anti-vibration table (X)
- Heater (U)
- Humidified CO_2 supply for open tissue culture dishes (V)
- Automatic light shutter (K)
- Automatic filter wheel for fluorescence multi-channel recording (L)
- Camera (P)
- Computer (T) with software for time-lapse imaging

Method

1. Set up the microscope and choose the observation field.
2. Focus using the monitor.
3. Choose exposure, time-lapse interval, and number of frames for recording.
4. Start the time-lapse programme.

6.1.1 Four-dimensional imaging

The development of computer technology combined with digital image capture has lead to the increased possibility of collecting data in more than two dimensions. Many modern microscopes, especially confocal microscopes, are able to capture stacks of 2D (x, y) images at different depths through the specimen (z positions) leading to 3D information about structures. This can be performed in time-lapse mode to produce 4D information about the structure in x, y, z, and time.

7 Analysis of microscopic images

Studies of the molecular mechanisms of cellular functions require reliable detection of subtle changes in cell behaviour against a background of high intrinsic cell-to-cell heterogeneity and variability in time. Quantitation is an obvious solution, especially combined with automatic image processing providing the necessary large amount of data required for eliminating the effects of heterogeneity. Pre-processing often enhances images; however it can be detrimental to quantitative analysis.

7.1 Image enhancement by deconvolution

Acquired 3D fluorescence information can be processed by deconvolution in order to eliminate blurring resulting from structures which are out-of-focus. The procedure can be used to improve confocal images but it gives more dramatic improvement with images acquired from conventional microscopy. Globular and filamentous structures are especially suitable for this technique. There is a range of commercially available programs for this purpose, all of which usually require considerable computing power. Huygens (SVI) is a flexible deconvolution program allowing many parameters to be defined for optimum performance and is designed for some Unix machines (SGI). Autodeblur (AutoQuant Imaging) is a simpler program for personal computers.

7.2 Image processing – extraction of morphometric data

Vertebrate cells in tissue culture visualized using standard contrast-enhancement techniques pose enormous problems for automatic image processing. Interactive semi-automatic techniques can be used for acquisition of limited amounts of data. Many image-processing programs, such as NIH *Image* (http://rsb.info.nih.gov/nih-image/), allow measurements of, for example, position or spread area of cells (10). Automatic algorithms can be used in special cases such as highly refractive cells, e.g. neutrophil leucocytes or *Dictyostelium discoideum*. Fluorescence images with good contrast can also be used for automatic image processing. Special interferometric techniques are ideally suited for automatic analysis of cell images but these are not currently commercially available (11).

Measurements based on cell outline can be used to measure various parameters such as spread area, total amount of fluorescence intensity, displacement in a time-lapse recording, and morphometric parameters such as elongation or polarity.

7.3 Data analysis – statistical evaluation

Data derived from microscopic observations of live cells usually form a hierarchical structure. The hierarchy of the data will typically contain some of the levels shown in *Figure 13*. Statistical tests on such data require analysis of variance (ANOVA) with a nested model (12, 13). In a general case, the nested model also has to be unbalanced because the number of data points at different levels are not equal since they depend on chance, e.g. number of cells in the observation field.

The ANOVA test calculates variability in the data introduced at individual levels and assigns statistical significance to the variability at individual levels taking into account interactions between the levels. The level of control-treatment is then usually expected to show a significant additional difference on top of the

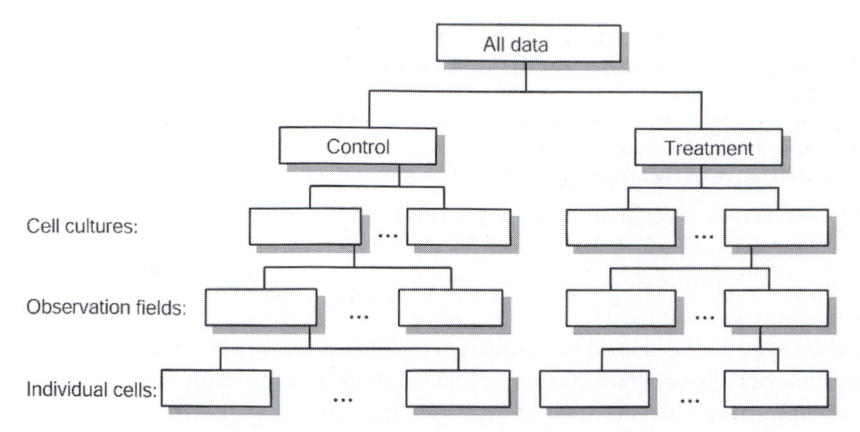

Figure 13 An example of a nested hierarchical data structure for ANOVA. More sublevels of nested data can be added when more than one measurement is acquired from each cell, e.g. time-lapse or 3D data.

variability at the lower levels. When experimental conditions are well controlled and reproduced then the level immediately below, the culture level, does not introduce significant variability. This means that individual cell cultures are reproduced with the same internal variability without additional variations between cultures. The cell level usually gives a high significance since individual cells are highly variable even within individual cultures.

7.4 Data interpretation – simulation modelling

Microscopic observation of live cells reveals a great deal of information about dynamic intracellular structures and their interaction with cell behaviour. Because of the complexity of these interactions it is difficult to determine their consistency, and computer simulation becomes an attractive technique. Successful models of well understood biological systems have been developed using established simulation techniques (14). More complicated systems with partial understanding require the development of a powerful simulation environment (15) with flexible representation schemes to achieve theoretical predictions which can be tested by further experiments.

Acknowledgements

We are grateful to Dr Colin Gray and Dr Alastair Nicol, Imperial Cancer Research Fund, London, UK, for helpful discussions during the preparation of the manuscript.

References

1. Inoué, S. and Spring, K. R. (1986). *Video microscopy: the fundamentals*. Plenum Press, New York.
2. Slayter, E. M. and Slayter, H. S. (1993). *Light and electron microscopy*. Cambridge University Press, Cambridge, UK.

3. Jenkins, F. A. and White, H. E. (1981). *Fundamentals of optics*. McGraw-Hill, Singapore.
4. Pluta, M. (1993). *Advanced light microscopy: measuring techniques*, Vol. 3. Elsevier, Amsterdam, London, New York, Tokyo.
5. Dunn, G. A. and Zicha, D. (1997). In *Cell biology: a laboratory handbook* (ed. J. E. Celis), Vol. 3, p. 44. Academic Press, San Diego.
6. Inoué, S. (1989). *Handbook of biological confocal microscopy* (ed. J. B. Pawley), p. 1. Plenum Press, New York.
7. Zicha, D., Dunn, G. A., and Brown, A. F. (1991). *J. Cell Sci.*, **99**, 769.
8. Entwistle, A. (1998). *J. Microsc.*, **192**, 81.
9. Zicha, D. (2000). In *Freshney's culture of animal cells* (ed. R. I. Freshney), p. 429. Wiley-Liss, New York.
10. Cammer, M., Wyckoff, J., and Segall, J. E. (1997). In *Basic cell culture protocols* (ed. J. W. Pollard and J. M. Walker), Vol. 75, p. 459. Humana Press, Totowa.
11. Zicha, D. and Dunn, G. A. (1995). *J. Microsc.*, **179**, 11.
12. Snedecor, G. W. and Cochran, W. G. (1989). *Statistical methods* (8th edn), p. 217. Iowa State University Press, Ames, Iowa, USA.
13. Milliken, G. A. and Johnson, D. E. (1992). *Analysis of messy data: designed experiments*, Vol. I. Chapman and Hall, New York and London.
14. Bray, D. and Lay, S. (1994). *Biophys. J.*, **66**, 972.
15. Cooper, R. and Fox, J. (1998). *Behaviour Research Methods, Instruments and Computers*, **30**, 553.

Chapter 5
Basic cell culture technique and the maintenance of cell lines

James A. McAteer
Department of Anatomy, Indiana University School of Medicine, 635 Barnhill Drive, Indianapolis, IN 46202-5120, USA.

John M. Davis
Research & Development Department, Bio Products Laboratory, Dagger Lane, Elstree, Hertfordshire WD6 3BX, UK.

1 Introduction

The protocols and descriptions of technique presented in this section are intended to provide an introduction to some of the more general and routine aspects of cell culture methodology. The reader may wish to consult a number of excellent publications which give detailed procedures and instructive critical analysis of other general methods and many specialized cell culture techniques (1–19).

1.1 The terminology of cell and tissue culture

Before proceeding to the practical aspects of cell culture methodology it may be useful to examine the terminology of cell and tissue culture. The following definitions help explain how cultured cell populations are derived, and how cells in culture come to express some of their distinguishing characteristics. For additional information please see a fundamental reference of the tissue culture literature, entitled 'Terminology associated with cell, tissue and organ culture, molecular biology and molecular genetics' (20).

1.1.1 The origin of cell lines

i. Primary culture

Primary cultures are derived from intact or dissociated tissues or organ fragments. A culture is considered a primary culture until it is subcultured (or passaged), after which it is termed a cell line (*Figure 1*).

ii. Subculture (also called passage)

To subculture is to transfer or transplant cells of an ongoing culture to a new culture vessel so as to propagate the cell population or set up replicate cultures

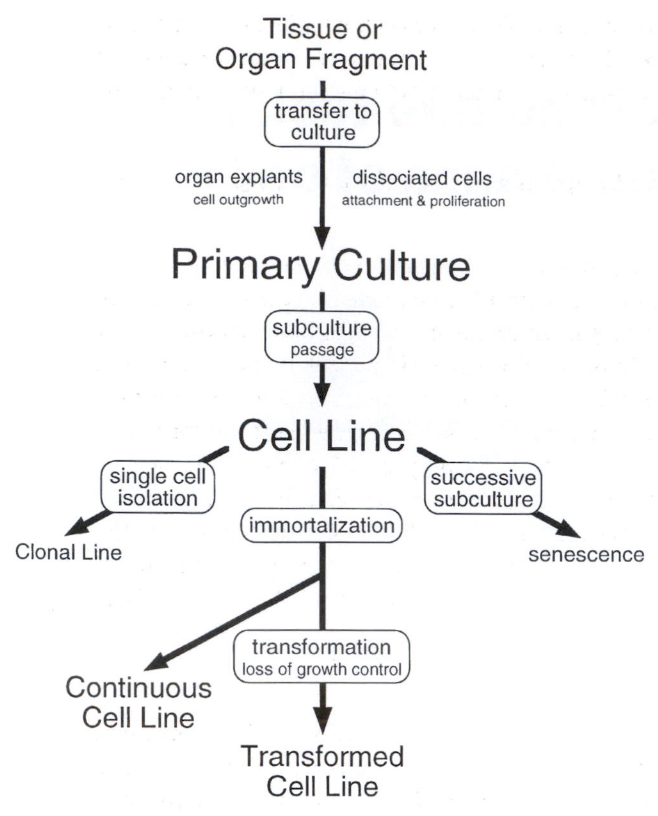

Figure 1 Scheme depicting the origin of cell lines.

for study. When starting with a primary culture, the product of the first subculture is also called a secondary culture. The term tertiary culture is rarely used.

iii. Cell line

A cell line is a cell population derived from a primary culture at the first subculture. Modifiers such as finite (indicating a finite *in vitro* lifespan) or continuous (indicating unlimited generational potential) should be used only when these properties have been determined. The term cell line does not imply homogeneity or the degree to which a culture has been characterized.

iv. Cell strain

This term is used to describe a subcultured population selected on the basis of its expression of specific properties, functional characteristics, or markers.

v. Clonal culture (clone)

Clonal culture or clonal selection is the establishment of a cultured cell population derived from a single cell. Intuitively, one might expect that a culture derived by the successive mitoses of a single cell would consist of identical cells.

This is not necessarily the case, and clonally derived cell populations commonly come to express some degree of heterogeneity. Thus, clonal culture does not imply absolute homogeneity or long-term stability of characteristics within the cell population (see Chapter 8).

1.1.2 The characteristics of normal and transformed cells

i. Normal cell

Although one may describe the properties of 'normal' cells, the term defies precise definition. For primary cultures, 'normal' implies that the cells were derived from a normal, healthy tissue or organ (or fragment), not a tumour or other pathological lesion. Following subculture and the generation of cell lines, normal cells commonly (but not always) exhibit a characteristic set of properties including (but not limited to) anchorage dependence, density-dependent inhibition of growth, and a finite lifespan (number of cell doublings attainable). The converse of normal is 'transformed'. In defining these terms one usually considers properties of cell behaviour commonly expressed by cells of one category and not the other. Such a list of features (*Table 1*) is helpful in characterizing cultures, but it must be recognized that no single property is necessarily diagnostic of the normal versus transformed condition, and exceptions do exist.

ii. Anchorage dependence

Anchorage-dependent (also termed substrate-dependent) cells require attachment to a surface (substrate) in order to survive, proliferate, or express their differentiated function. In common usage, anchorage dependence implies the requirement for attachment in order to support proliferation. Most normal cells must attach to a substrate in order to proliferate. The alternative to anchorage dependence is the capacity to proliferate in fluid suspension (see Section 6) or within a semi-solid, three-dimensional matrix such as semi-solid agar medium (21). Many transformed cell lines can form colonies in soft agar and this ability along with the lack of

Table 1 Characteristics of normal versus transformed cells

Normal cells
Anchorage-dependent
Density-dependent inhibition of proliferation
Finite lifespan
Altered characteristics with increased *in vitro* age
Transformed cells
Continuously culturable, with infinite lifespan
Reduced requirement for serum or growth factors for optimal growth
Shorter population doubling time
Reduced substrate adhesion
Genetic instability (e.g. show heteroploidy and aneuploidy)
'Altered growth control'
loss of contact inhibition
loss of density-dependent inhibition of proliferation
loss of anchorage dependence
increased colony formation in soft agar

density-dependent inhibition of proliferation is considered an expression of 'altered growth control'. Anchorage dependence does not, however, mean that a cell line is normal, since most transformed cells also grow well on a rigid substrate.

iii. Monolayer culture

The term monolayer culture describes the distribution of cells as a single layer on a substrate such as glass or plastic. The ability of cells to grow as a monolayer does not necessarily mean that they are normal or that they are anchorage-dependent. The phrase 'substrate-dependent monolayer culture' is appropriate only when it has been determined that the cells are indeed attachment-dependent and that the culture is composed of a single layer of cells.

iv. Density-dependent inhibition (or limitation) of proliferation

Most normal cell lines exhibit a marked reduction in proliferative activity that correlates with the attainment of confluency, that is, occupancy of all available attachment surface. In most cases this is a population density effect, not a cell–cell contact effect (i.e. contact-inhibition of growth). Density-dependent inhibition of proliferation can occur before confluence is reached, and reflects diminished nutrient supply and the release of cell-derived factors (including waste products) into the medium.

v. Saturation density

The saturation density of a culture is its population density (cells/cm^2) at the point when it reaches density-dependent inhibition of growth. Transformed cell lines commonly have a higher saturation density than normal cells.

vi. In vitro *cell senescence (cell ageing in culture)*

Normal cell lines commonly have a finite lifespan, that is, they do not grow beyond a finite number of cell generations (population doublings). For example, the lifespan of normal diploid fibroblasts is in the range of 50–70 population doublings. The process of cell ageing in culture has been shown to involve progressive alterations in a number of cell characteristics. *Table 2* lists some of the changes exhibited by normal diploid fibroblasts as they age in culture.

vii. In vitro *transformation and the transformed cell*

In vitro transformation is the acquisition of altered (usually persistent) characteristics or properties such as elevated saturation density and anchorage independence that distinguish transformed cells lines from normal cell lines.

Cellular transformation can be induced by a variety of conditions including exposure to certain chemical carcinogens, ionizing radiation, or transfection with retroviruses or DNA tumour viruses (22). Transformation can also occur spontaneously, in which case the cells appear to undergo transient proliferative arrest ('crisis') often accompanied by abnormal morphology and degenerative loss in cell number (23). Characteristically, a rapidly dividing cell population then emerges and overgrows the culture.

Table 2 Changes in culture characteristics as a consequence of *in vitro* ageing[a]

Increased cell cycle time
Decreased number of cells in the cell cycle (i.e. decreased growth fraction)
Decreased saturation density
Increased incidence of polyploidy
Decreased adhesion to the substrate
Alterations in cell surface glycoproteins and proteoglycans
Altered cytoskeletal organization
Reduced DNA synthesis
Decreased rate of protein synthesis, degradation, and turnover
Decreased amino acid transport
Diminished response to growth factors

[a] Profile of changes typical of normal diploid fibroblasts (94).

Transformed cell lines commonly have infinite lifespan and may express numerous altered characteristics (*Table 1*). Some transformed cells form tumours upon injection into an immuno-incompetent host. However, cellular transformation is not always associated with tumorigenicity, and transformed cells in culture are not necessarily equivalent to cancerous cells *in vivo*. Thus, *in vitro* transformation is distinguished from *in vitro* neoplastic transformation where it has been determined that the cells form neoplasms (benign or malignant). Likewise, if it has been determined that the cells produce invasive (metastatic) tumours the term *in vitro* malignant transformation is appropriate.

viii. Immortalization

Cell lines that have unlimited lifespan are termed immortal or, preferably, continuous. Although infinite lifespan is generally considered to be a characteristic of transformed cells, not all continuous cell lines exhibit alterations in growth control attributed to cellular transformation (*Table 1*). Thus, the terms immortalized and transformed are not synonymous. Immortalization, the process by which a cell line becomes continuously culturable, is, instead, understood to be one step in the cell transformation process. Therefore, the terms *in vitro* transformation and transformed cell line should be used to describe cell lines that have altered growth control. The terms immortalization and immortal/continuous cell line should be used when the cell line does not show alterations in growth control.

2 Basic components of the cell culture environment

In order for the culture system to promote cell survival and proliferation, the *in vitro* environment must meet the fundamental physiological requirements of the cell. Components of the culture environment that one can control include factors associated with the medium such as its nutrient composition, pH, osmolality, and the volume and frequency of media replenishment. In addition, one can regulate incubation conditions including temperature, relative humidity, and gas composition, as well as the form and composition of the physical substrate for cell attachment.

2.1 Culture medium

As discussed in Chapter 3, the culture medium must provide all essential nutrients, vitamins and cofactors, metabolic substrates, amino acids, inorganic ions, trace elements, and growth factors necessary to support cellular functions and the synthesis of new cells. One should realize that different cells have different nutrient requirements (see Chapter 7). Also, most media were designed for very specific applications, such as the optimal growth of a certain cell line. Thus, one cannot expect a given medium to provide optimal support for any cell line other than the one for which it was developed.

It takes a tremendous effort to establish the conditions that support optimal cell proliferation or to maximize the expression of differentiated function. Frankly, many workers do not attempt to optimize media composition (or other factors) but, instead, find conditions that provide 'adequate' growth and function. Still, there are important advantages to be found in the use of fully characterized, fully defined media (see Chapter 7 and ref. 12).

2.2 Physicochemical factors

Several factors of the culture environment, those contributed in part by the medium and in part by incubation conditions, may fluctuate over time and should be monitored. As a general rule (exceptions do exist) optimal culture conditions for most mammalian cells in culture fit the following profile:

- pH: 7.2–7.5
- osmolality: 280–320 mOsmol/kg
- CO_2: 2–5% in air
- temperature: 35–37 °C

However, just as different cell lines have different nutrient requirements, certain cells can be expected to show unique physiochemical requirements for optimal growth or function. Temperature is a good example: whereas most mammalian cells show optimal growth in the range 35–37 °C, cells from cold-blooded vertebrates commonly grow best at reduced temperature. For example, the A6 toad kidney epithelial cell line grows best at 26–28 °C (24). Also, some transformed cells have been shown to be more sensitive to elevated temperature than normal cell lines (25).

2.3 Stationary versus dynamic media supply

Media delivery is also an important factor in a culture system. Cells must be supplied with enough medium to meet physiological demands and the medium must been replenished at an appropriate frequency and volume. If the volume of medium is too low, nutrients may be utilized too rapidly and the accumulation of metabolites may affect cell function or proliferation. If, however, the initial volume of medium is too great or the frequency of replenishment is too rapid, cells may be unable to condition their environment adequately (see Section 7.1). The recommended volume of medium for routine substrate-dependent culture is 0.2–0.3 ml/cm^2. Since the medium overlaying a layer of cells acts as a barrier to gas diffusion, an excessive volume of medium can restrict cell access to CO_2 and O_2.

Automated delivery and replenishment of medium is an integral feature of many suspension and microcarrier culture systems, particularly in commercial-scale operations (10, 19). Methods have also been devised to improve delivery of medium to routine substrate-dependent cultures. One of the simplest techniques is to place the culture vessel on a rocker platform (e.g. Bellco) within the incubator. As the platform tilts to one side and then to the other (three to six cycles/min) the cells are gently washed by the medium. This acts to mix the medium, thereby minimizing the build-up of potentially deleterious metabolites (waste products) at the cell surface. Once in each tilt cycle the cells are covered by only a thin film of medium. This improves O_2 and CO_2 delivery, which may be beneficial to some cell populations. The roller bottle (e.g. Bellco) used for large-scale culture accomplishes the same end. Here, a cylindrical bottle is incubated while positioned on parallel rollers. As the bottle turns (60–80 rev/h), cells growing on its wall are submerged in medium (at the bottom of the turn) and retain a thin film of medium throughout the remainder of the rotation cycle (4).

3 Sterile technique and contamination control

Successful cell culture is absolutely dependent on fastidious aseptic technique. In order to keep cultures free from microbial and cellular contamination one must understand how contamination is apt to occur, must adopt a well thought-out regime for handling cells and reagents, and pay constant attention to even the most routine manipulations. Laboratory errors can be costly in any discipline. Cell culture has the added burden that not all problems show up immediately. Some forms of contamination such as low level mycoplasma infection or the occurrence of cellular cross-contamination can go undetected for years.

It is obvious that the consequences of sloppy cell culture technique can be very disruptive to the laboratory. Occasional isolated contaminations can happen to anyone, but commonly prove to be no more than a nuisance. Recurrent bacterial or fungal contamination of working cultures is more troublesome and can be quite time-consuming and expensive. An even more serious problem is the contamination of cell line stocks. This can wipe out a model system or, in the event that cellular cross-contamination is involved, can invalidate the results of studies performed with the suspect lines (see Section 3.4). Thus, it is essential that the cell culture specialist adopt the best possible working habits and continually be on the lookout for problems and their causes.

(b) Prepare the area to be as free from particulates as possible. A closed glove box or a bench-top dust hood with a partial window can be used to isolate the field further.

(c) Work at arm's length and follow all routine precautions to avoid contaminating critical surfaces.

(d) Some workers use a burner to flame bottles, caps, and instruments.

(e) Replace instruments frequently. Keep all instruments covered and pipette canisters closed when not in use.

For further details concerning many of the aspects covered in Section 3, see ref. 28.

4 General procedures for the cell culture laboratory

Proper preparation of the culture area along with systematic, routine protocols for maintenance of equipment and stock cultures promotes productivity, lessens the risk of contamination, and improves worker safety.

4.1 Maintenance of the laboratory

Establish a start-up and shut-down routine that keeps the laboratory clean and safe and assures that the ongoing cultures and frozen cell stocks remain in the best possible condition.

4.1.1 Laboratory start-up

(a) Check the incubators first. Record temperature, humidity, and concentration of CO_2 (and of other gases if in use). If there is a problem such as power failure, CO_2 drop, or temperature overshoot, deal with it immediately.

(b) Check the water level indicator of any water-jacketed incubator.

(c) Check the humidification system. Be sure the water pan on the floor of the chamber is filled. If the incubator has an exterior reservoir check the water level and be sure the tubing line is open.

(d) Check gas regulators to assure that line pressure is correct for the incubator, and that tank pressure is adequate for the day. Check gas cylinder moorings as a routine precaution.

(e) Keep an inventory that locates and identifies incubator contents. This is particularly valuable when multiple users are involved. Refer to this record and scan each tray on each shelf for grossly contaminated cultures. Discard any contaminated cultures. Contaminated cultures can be immersed (10 min) in 0.525% sodium hypochlorite (10% household bleach) or placed in an autoclavable bag for sterilization.

(f) Wipe down work surfaces in the laboratory with a disinfectant such as 70% ethanol or commercial anti-microbial solution (e.g. Coldspor, Metrex Corp.). Caution: 70% ethanol is flammable, but is a good alternative to caustic or toxic agents such as bleach (0.5–1%) or phenolics (for further discussion, see Chapter 2, Section 7.2). Dispense 70% ethanol from a squeeze bottle in modest amounts.

(g) Turn on the laminar flow hood and allow it to run for 10–15 min before use. These cabinets are designed to run all day.

(h) If the hood is equipped with an ultraviolet (UV) lamp, be sure that it is off. UV exposure at the hood for even a relatively brief period (many minutes to several hours) can result in painful eye injury, often with some visual impairment (usually short-term).

(i) Wipe down the interior of the hood with 70% ethanol. Any apparatus, equipment, media bottles, trays of cultures, etc. that go into the hood throughout the day should receive the same treatment.

(j) If an aspirator is used, hook up a clean vacuum bottle (containing 1% bleach) and line trap. If a peripheral pump and trap system are used, likewise assemble a clean trap and tubing line.

(k) If a microburner is used in the hood (this is not recommended, as the flame disrupts the laminar airflow pattern) check to see that tubing to the gas supply is in good repair.

(l) Check function of the automatic pipetting devices.

(m) Fill and turn on heated water-bath used to warm culture medium.

(n) Purge the water purification system (e.g. ion-exchange) and check to see that resistivity meets specifications.

4.1.2 Laboratory shut-down

When work is completed:

(a) Remove all equipment from the hood. Turn off vacuum and gas supply valves. Wipe down the hood interior with 70% ethanol.

(b) Discard waste medium. Medium decontaminated with 1% bleach can be washed down the sink. Other waste medium should first be autoclaved. Note: medium from 'high risk' cultures, especially those of human or primate origin, must be autoclaved before disposal.

(c) Check to see that the lids of cryogenic freezers are securely seated. If left open a freezer can lose a substantial volume overnight.

(d) Check the liquid nitrogen log to see if level should be measured. Typical 35 litre freezers are fairly efficient and may not need replenishment for up to two weeks, depending on frequency of use. Establish a calendar for liquid nitrogen replenishment (e.g. 1st, 10th, and 20th days of the month). Never take the risk of running out of liquid nitrogen. The frozen cell inventory may be the laboratory's most valuable resource.

(e) Check the incubators and their gas supplies.

excellent optical clarity. Virgin (unmodified) polystyrene is rather hydrophobic. Many cell types will not adhere at all to unmodified plastic, and those that do tend not to spread well. It has been shown that cell shape influences cell proliferation, and that attachment surfaces that increase the extent of cell spreading promote cell proliferation. Thus, treatments that make a surface more hydrophilic (more wettable) improve its performance for cell propagation. Commercial preparation of tissue culture plastic commonly involves oxidation of the growth surface by a method such as the gas-plasma discharge technique (30). Only the growth surface is treated then the vessel is assembled, packaged, and sterilized by gamma-radiation. Oxygen-plasma treatment increases the O:C and COOH:C ratios of the plastic and imparts an increased negative charge. The magnitude of the charge is important, not its polarity, since cells adhere to both positively and negatively charged surfaces (30).

Some additional points regarding plastic culture substrates are listed below.

(a) Culture flasks made from polyester copolymer [PETG, poly(ethylene) terephthalate glycol] (In Vitro Scientific) are easier to cut than polystyrene and offer some advantages in terms of improved access to cells.

(b) Polystyrene is susceptible to organic solvents. This should be kept in mind in designing cell constituent extraction systems or when processing for morphological analysis (31).

(c) Thermanox plastic coverslips (Nalge Nunc International) are resistant to xylene and acetone.

(d) Plastic coverslips tend to float, but can be secured to the floor of a dry dish with a drop of medium or a dab of silicone stopcock grease (sterilized by dry heat or autoclaving).

5.2 Choice of culture vessels

Culture-grade plasticware is available in a variety of sizes (surface area) and configurations (*Table 3*). Selection of a specific type of vessel is largely a matter of convenience, but there are a few practical considerations to keep in mind. Direct access to cells is much easier with dishes than with flasks. For example, in applications where it is necessary to harvest cells by scraping, dishes are more convenient. Long-handled scrapers or rakes are available to reach into flasks, but this can be laborious and inefficient, especially when harvesting multiple flasks. Direct access to the cells also makes clonal cell isolation (see Chapter 8) or processing for ultrastructural study easier.

Knowing the exact surface area of a flask or dish may be important for some applications. It should be noted that a number of manufacturers supply similar products that differ slightly in surface area for cell growth.

For example:

	'60 mm' dish	**'100 mm' dish**
Source X	21 cm^2	55 cm^2
Source Y	21.15 cm^2	59.3 cm^2
(difference)	0.7%	7.8%

Table 3 Specifications of some common plastic culture vessels

	Approximate working surface area (cm^2)	Working volume of medium (ml)	Examples[a]
Dishes			
35 mm	8	2	C 430165
60 mm	21	5	C 430166
100 mm	55	12	C 430167
150 mm	152	30–40	F 353025
Flasks	25	6	C 430168
	75	15	C 430720
	150	30–40	C 430823
	225	45	C 431080
Multiwell plates			
6-well	9.5	2	C 3516
12-well	4	1	C 3513
24-well	2	0.5	C 3524
48-well	1	0.25	C 3548
96-well	0.32	0.1	C 3596

[a] C, Corning Costar; F, Falcon.

5.2.1 Vented versus sealed culture vessels

Most dishes and multiwell plates are vented, that is, the lid and base are constructed with a narrow interposing ridge that prevents them from forming a tight seal. This narrow gap allows gases to equilibrate with the culture medium, and is essential when using a bicarbonate-buffered medium in a CO_2 incubator. Screw-capped culture flasks can be sealed tightly and, thus, are suitable for use with nominal bicarbonate media and organic buffers (e.g. Hepes). For vented culture the cap is simply loosened a half turn or so. Some manufacturers offer culture flasks with different styles of caps. Conventional plug-seal caps can be used for vented culture. These caps provide the most complete seal when fully tightened and so are recommended for use under closed culture conditions. Rim-seal type and multiple position caps are intended for vented culture, but can be closed as well. Some manufacturers supply perforated caps guarded by a sterile, gas-permeable filter to improve gas–medium equilibration (e.g. Corning Costar; Nunc; Greiner).

Depending on the buffering capacity of the medium, one can expect a noticeable pH shift (toward basic) when a vented dish or flask is removed from the incubator, such as for feeding or viewing with the inverted microscope. This problem can be minimized by incorporating an organic buffer (e.g. 10–20 mM Hepes) in the medium. Taking a culture out of the incubator for a short period of time (minutes) during routine maintenance is usually not a problem. Since changes in pH and temperature can affect cell function, cultures should be disturbed as little as possible especially during timed incubations.

Some additional points to consider regarding the use of vented culture vessels are listed below.

(a) Always check the position of the cap before transferring a flask to the incubator.

(b) Avoid splashing medium into the cap of a flask or between the lip and lid of a dish. This can restrict gas exchange and can lead to contamination.

(c) Vented culture vessels allow evaporation from the medium. Evaporation can be substantial if humidity in the incubator is not sufficiently high. Excessive evaporation will increase the osmolality of the medium, which can affect cell function and even cause cell death.

(d) It is wise to test for evaporative loss over time by determining weight change in blank dishes (without cells), and to do this for shelf sites throughout the incubator.

5.2.2 Culture dish inserts

Culture dish inserts are culture chambers in which the floor is made of fluid permeable, microporous filter membrane material. Some examples include Anocell (Whatman), Transwell (Corning Costar), NuCell (Nucleopore), Cyclopore (Falcon), and Millicell (Millipore) culture dish inserts. These units are designed to be placed within a larger dish such as a 6- or 24-well plate (Corning Costar offers 75 mm diameter inserts, codes 3419 and 3420). Inserts are either free-standing, and thus have short feet to hold them off the floor of the outer well, or they hang suspended from the lip of the outer well. They are available in various types of perforated or mesh, translucent or optically clear membrane. Some membrane types, such as the Millicell CM (Millipore), are hydrophobic and must be coated with an extracellular matrix (ECM) component (e.g. collagen I, fibronectin—see Section 5.4) or other attachment factor to promote cell adhesion. ECM-treated culture dish inserts are also available (Transwell-col, Corning Costar; Biocoat, Collaborative Biomedical Products BD Discovery Labware).

Cells growing within a culture dish insert are in contact with the medium that overlies them, as well as medium that diffuses upwards through the porous substrate. Importantly, this environment promotes cell differentiation and the expression of differentiated cell function. Culture in this configuration presents the opportunity to manipulate the cell population in ways that cannot be carried out when cells are grown on a non-permeable substrate such as glass or plastic. For example, when epithelial cells are grown to confluency within a filter insert, the cells form a barrier that separates the culture environment into two compartments, one that bathes the apical cell surface and one that bathes the lateral-basal cell surface. This makes it possible to manipulate selectively the composition of medium in one or other compartment, or to collect media samples selectively. This is ideal for studies of vesicular trafficking and transcytosis, transepithelial ion and fluid transport, the sidedness of cell response to agents or toxicants, and the maintenance of barrier function. For a publication devoted to culture techniques and methods for assessment of cell function for cultures grown on permeable supports, see ref. 32.

5.3 Artefacts of cell attachment and growth

Artefacts in cell growth pattern can occur but are rarely due to defects in the culture substrate (33). Many problems involving apparent irregular growth pattern can, instead, be traced to errors in routine handling or problems with incubation conditions. Some of the more common artefacts encountered in routine cell culture are discussed below.

5.3.1 Effect of volume of medium

Improper media volume may affect cell distribution and survival. If the volume of the seeding inoculum is too low the cell suspension will be pulled to the edge of the dish, creating a meniscus effect in which cell attachment may be reduced near the centre. Likewise, if a growing culture is fed with an insufficient volume, the medium will pool at the edges and can lead to cell injury and death towards the centre of the dish. If the incubator shelves are not level the depth of medium across the dish will not be uniform and this may create irregularity in growth pattern.

5.3.2 Uneven distribution of cells

Insufficient mixing at seeding can influence the pattern of cell attachment. Even at proper seeding volume, cells tend to settle unevenly, often concentrated more toward the centre of the vessel. This is particularly evident with round dishes and can be minimized by gently tilting freshly seeded cultures side to side then forwards and backwards to disperse the cells more evenly prior to static incubation.

Creating bubbles at seeding can cause spotting. If the medium is shallow a bubble may reach the substrate and prevent cells from attaching at that site, leading to vacant patches or spots within an otherwise confluent culture. The tendency for bubbling or frothing increases with the concentration of serum or other proteins in the medium. One should avoid creating bubbles, not only to prevent attachment artefacts, but also to minimize the potential for creating aerosols which can lead to microbial and cellular contamination (see Section 3.2 and 3.4). In addition, excessive bubbling can cause cell injury.

5.3.3 Peeling of the cell sheet

This may be attributed to several factors. Cell lines differ in how tightly they adhere to the culture substrate; some types may be prone to peeling or sloughing, especially if they are handled too roughly. Heavily confluent cultures are often more likely to shed or peel, particularly if the cell sheet is scratched, as with a transfer pipette, or if medium is directed onto the cell sheet during feeding. Some fluid-transporting epithelial cell lines that are avid dome-formers (34) may separate from the substrate over fairly broad areas as fluid accumulates between the base of the cells and the plastic. Such cultures may require gentle handling.

The consequences of peeling or sloughing of cells differs among cell lines. When a culture sloughs in patches it is common to see the adherent cells at the margin of the 'wound' undergo migration and increased proliferative activity to resurface the available unoccupied substrate. Some cell lines exhibit retraction of the cell sheet when injury occurs. That is, the cells may detach from the substrate but not break free from the remainder of the cell sheet. This can lead to the piling-up of cells in focal regions. Thus cells that normally grow as a true monolayer may show regions suggestive of layering. If this sort of retraction and piling-up phenomenon involves a large portion of the cell sheet and if the retracted region becomes too large, the cell mass may limit access to nutrients and gases and the cells may undergo necrosis.

5.3.4 Effect of vibration

Vibration can affect cell attachment. Excessive vibration can disrupt or delay cell attachment and may cause cells to aggregate within specific areas of the dish. Cells plated at clonal density may be particularly sensitive to vibration.

5.3.5 Effect of temperature variation

Temperature gradients may affect cell growth. The attachment or proliferation of some cell lines may be particularly sensitive to temperature. An example is the demonstration that cell growth pattern can reflect the position of perforations in the shelving upon which the dish rests during incubation (33).

A more common temperature-dependent artefact occurs when temperature within the incubator is not uniform, causing dish-to-dish variability in cell growth. This may be a problem in incubators that see heavy traffic or have a poorly insulated door, a leaky door gasket, or a leaky stopper in the rear access port. Condensation on flasks and trays may indicate temperature gradients in the incubator. This should be investigated without delay as condensation also promotes the growth of micro-organisms.

5.4 Use of attachment factors and bio-substrates

Attachment factors are agents that promote cell adhesion to conventional culture substrates (e.g. glass or plastic). Many also enhance cell spreading, substrate coverage, and cell proliferation, and as such are used on a routine basis to increase culture 'efficiency' (defined in Section 8.8). Attachment factors include basic polymers (e.g. polylysines) used to increase the surface charge of the substrate for binding with the cell surface, as well as a variety of purified ECM components (e.g. fibronectin, laminin, or collagens I and IV) and complex ECM preparations to which cells bind via specific transmembrane receptors. ECMs and complex ECM preparations are often referred to as bio-substrates. It is well established that various ECM preparations influence the differentiation, structure, and function of cultured cells. Indeed, *in vitro* studies with ECM components have provided much of our understanding of the biology of cell–ECM interactions.

Many common attachment factors and bio-substrates are commercially available and can be readily reconstituted for use. One must be aware that these agents are not always supplied in sterile form, and it may be necessary to sterilize a preparation before or after coating (35). Also, some ECMs are expensive, and for certain applications may require additional purification. Thus, some laboratories may find it valuable to carry out their own isolations, and well-tested protocols are available for isolation and purification of most ECM components (12). As an alternative, commercial firms (e.g. Collaborative Biomedical Products BD Discovery Labware) are now marketing cultureware pre-coated with attachment factors.

A popular bio-substrate is Matrigel, a complete basement membrane ECM produced by the Englebreth-Holm-Swarm (EHS) tumour. Matrigel is supplied as a sterile liquid (e.g. by Collaborative Biomedical Products BD Discovery Labware) and can be used to form a thin substrate or a three-dimensional gel (36). Native Matrigel contains growth factors. A growth factor-extracted preparation is also available. In addition, Matrigel-coated culture dish inserts are available. Those with a large pore (8.0 μm) support membrane (Matrigel Invasion Chamber, Collaborative Biomedical Products BD Discovery Labware) have been applied to studies of tumour cell migration and basement membrane invasion (37).

Cell-Tak cell and tissue adhesive (Collaborative Biomedical Products BD Discovery Labware) is a bio-substrate derived from the marine mussel, *Mytilus edulis*. This material, used at 1–5 $\mu g/cm^2$, promotes non-specific cell attachment and can also be used to anchor organ explants *in vitro* (36).

Investigators have also pursued native ECMs derived from specific cultured cell populations for use as bio-substrates. For example, protocols have been developed for the isolation of substrate-adherent ECM produced by bovine corneal endothelial cells and by the PF-HR-9 endodermal cell line (38). Cultures are raised to confluency and then the cells are lysed by treatment with 0.5% Triton X-100 or 20 mM NH_4OH. The natural matrix remains adherent to the dish, and following washing is used as a culture substrate for other cell lines.

The task of selecting an appropriate attachment factor (or bio-substrate) and method of coating is not entirely straightforward. This is because there are numerous methods for preparing ECM-coated substrates, and the various purified components are used over a broad range of concentrations (see *Protocol 2*). It will probably be necessary to optimize substrate coating conditions for specific applications.

Protocol 2

Preparation of attachment factor and ECM-coated culture substrates

A. Polylysine

Poly-D- and poly-L-lysine are available over a range of molecular weights and can be purchased in sterile form (e.g. from Sigma). The D and L isomers of polylysine have a

Protocol 2 continued

comparable cell adhesion promoting effect. Since L isomer amino acids are biologically active, poly-D-lysine is usually used.

1. Prepare sterile poly-D-lysine (0.1 mg/ml) in PBS.
2. Flood the substrate for 5 min at room temperature.
3. Remove the excess by aspiration and rinse with sterile culture-grade water.
4. Allow surface to dry before seeding.

B. Collagen I (thin coating)

Collagen I from rat tail tendon is available as a sterile solution (3.5–5.0 mg/ml) in acetic acid (e.g. from Collaborative Biomedical Products BD Discovery Labware). Other collagen types (e.g. types III, IV, and V) from various species are also available. Suppliers provide suggested coating protocols for each.

1. Dilute collagen to 50 μg/ml in sterile 0.02 M acetic acid.
2. Flood the substrate and allow to stand for 1 h at room temperature.
3. Rinse (and neutralize) with PBS (or medium) and use immediately or allow to air dry before seeding.

C. Gelatin

Gelatin from bovine or porcine skin is available as a powder or sterile solution (2%) (e.g. from Sigma).

1. Prepare a 0.2% solution in PBS (gelatin can be autoclaved).
2. Flood the substrate and allow to stand for several hours (or overnight) at 4 °C.
3. Remove the excess by aspiration and rinse the dish with medium prior to seeding.

D. Fibronectin

Plasma and cellular fibronectin are available in sterile form (e.g. from Sigma or Collaborative Biomedical Products BD Discovery Labware).

1. Prepare a fibronectin solution (50–100 μg/ml) in PBS.
2. Flood the substrate and allow to stand for 1 h (or air dry).
3. Aspirate any excess and rinse the dish with medium before seeding.

E. Laminin

Commercial laminin is commonly derived from the EHS mouse tumour and is available in sterile form (e.g. from Sigma or Collaborative Biomedical Products BD Discovery Labware).

1. Prepare a solution (2–5 μg/ml) in PBS.
2. Flood or smear the substrate with a thin coating.
3. Allow the surface to dry, rinse with medium, then seed.

5.5 Alternative culture substrates

Cultured cells have been grown on a variety of unconventional substrates, some of which include thin silicone rubber films (39) or sheets (40), stainless steel (41), titanium (42), and palladium (43). The nature of the substrate can make unique experimental applications possible. For example, investigators have used stretchable silicone rubber sheeting as a substrate with which to place cardiac myocytes under tension and, thereby, model the physical dynamics of the developing heart (44).

5.6 Three-dimensional matrices

A number of three-dimensional matrices including Gelfoam collagen sponge (Upjohn) (45), collagen-coated sponge (46), medium hydrated collagen gel (47), and fibrin or plasma clot (48) have been used as substrates for normal and immortalized cells. Collagen gels are particularly versatile and have been widely applied in the study of cell–cell and cell–matrix interactions, cell motility, and tumour cell invasiveness. Whereas the typical collagen substrate used to coat a coverslip or dish is a monomeric film or denatured, desiccated fibrillar meshwork, collagen gels form a three-dimensional lattice in which medium surrounds the fibrils. Collagen gels are usually prepared by neutralizing acid-solubilized collagen harvested from rat tail tendon (49). Gels can be prepared at various collagen concentrations, according to the desired physical rigidity, and are commonly used at about 1.5–3.0 mg/ml (see *Protocol 3*). In a typical application, the collagen–medium mixture is pipetted to fill the bottom of a dish, to a depth of 1–2 mm. The gel is allowed to solidify and is then overlaid with medium. Such a gel is fairly fragile and must be handled with care to avoid tearing or dislodging it from the dish. Cells can be seeded to the surface of a solidified gel or mixed within the fluid collagen–medium before the gel has set. Cells growing at the surface of a collagen gel can be subcultured by cutting out and replanting fragments of the gel within a freshly poured (unsolidified) gel. Cells migrate out from the cut edges of these micro-explants, thus subculture can be accomplished without exposing the cells to proteolytic enzymes. Motile and invasive cells tend to migrate into collagen gels, while some cell types have been shown to exhibit histotypic growth when cultured suspended within collagen gel (47).

Protocol 3

Preparation of hydrated collagen gel (HCG) substrates

Reagents

- Type I collagen in 0.2 M acetic acid (Collaborative Biomedical Products BD Discovery Labware)
- 10× Ham's F12 nutrient medium, $CaCl_2$-free (F12K)
- 0.1 M Hepes solution
- 0.5% (w/v) $NaHCO_3$ solution
- 1 M NaOH solution

Protocol 3 continued

A. HCG as a substrate for 'monolayer culture'

1 Combine the following components (chilled), in order, within a centrifuge tube and mix by vortex action. (Adjust the volume of H_2O and collagen to give 2.0 mg/ml collagen.)
- 3.4 ml H_2O
- 1.0 ml of 10× F12K
- 1.0 ml of 0.1 M Hepes
- 0.5 ml of 0.5% (w/v) $NaHCO_3$
- 0.1 ml of 1 M NaOH
- 4.0 ml collagen I

2 Pipette 1 ml of collagen mixture into each 35 mm dish. Tilt dish to achieve complete coverage. Transfer to CO_2 incubator and allow gel to solidify (15 min).

3 Add 2 ml of complete medium and incubate for 10 min to saturate the gel fully.

4 Aspirate overlying medium, taking care not to rupture the gel.

5 Seed 2 ml of cell suspension onto the gel surface. Return cultures to the CO_2 incubator.

For additional details see ref. 49.

B. Culture of cells suspended within HCG

1 Combine (in order) and gently mix the following:
- 1.0 ml of 10× F12K
- 1.0 ml of 0.1 M Hepes
- 0.5 ml of 0.5% (w/v) $NaHCO_3$
- 3.5 ml cell suspension (0.5–1.0 × 10^6 cells) in complete medium

2 Add 4.0 ml collagen I, mix by gentle vortex action, and pipette 1 ml into each 35 mm dish.

3 Incubate (15 min) to allow gel to solidify then overlay with 2 ml complete medium.

For additional details see ref. 47.

5.7 Culture of cells on microcarriers

Microcarriers are small (100–250 μm diameter) beads or rods with a surface suitable for attachment and growth of substrate-dependent cells (50). They were developed to increase the surface area to volume ratio for the mass production of cells, viruses, and diffusible cell products. Microcarriers are available in various sizes and can be made from a variety of materials, some of which include derivatized dextran, collagen or gelatin-coated dextran or plastic, gelatin, DEAE-cellulose, polystyrene, plastic-coated glass, and glass-coated plastic. Most of these have a smooth surface that promotes growth to confluency.

Macroporous microcarriers have an irregular surface that allows cells to grow within, as well as at the outer surface of the bead (see also Chapter 3, Section 5.4.1). One example is ImmobaSil (Ashby Scientific) a substrate made from a gas-permeable synthetic polymer that allows media access and good gas diffusion to cells growing within its matrix. ImmobaSil is available in various shapes and sizes such as fine particles (0.8×0.25 mm) for use in spinner flasks, discs (10×1 mm) for use in packed bed reactors, and large sheets which can be custom-ordered for special applications.

For large-scale culture, microcarrier-adherent cells are usually maintained in bioreactors that allow close control over media conditions. For laboratory applications such cultures can be maintained on a non-adherent substrate (see Section 6.1) or in stirred suspension culture flasks (51).

The principal application of microcarrier technology involves the mass culture of cells (19, 50). The fact that many cell types migrate readily onto microcarriers makes them useful for small-scale endeavours as well. For example, microcarrier beads (e.g. Cytodex beads, Pharmacia) can be used as a transfer substrate for the non-enzymatic subculture of adherent cells grown in typical plastic dishes, or as the substrate for the transport of cells (see Section 10.2.2).

An interesting application is the use of microcarrier beads to isolate endothelial cells from microvessels (52). The animal is perfused with a chelating buffer to weaken cell–cell and cell–ECM interactions, then microcarrier beads are infused and allowed to incubate within the vessels. Endothelial cells preferentially attach to the beads which are subsequently flushed from the vasculature and collected for transfer to conventional substrate-dependent culture. Microcarrier beads have also been applied to studies of epithelial permeability and barrier function. One method is to expose confluent epithelial cultures on beads to conditions of experimental injury in the presence of a dye that can gain access to the bead only when the epithelium is damaged. Quantification of dye uptake into the beads under experimental and control conditions gives an indication of the degree of cell injury (53).

5.8 Mass culture systems for adherent cells

A number of methods have been used to increase the surface area for cell growth on rigid substrates. Some early attempts at mass culture included the use of multiple-tier glass plates (54), stacked titanium discs (42), and spiral plastic sheets within rotating bottles (55). For the most part such systems are physically bulky and difficult to regulate. The roller bottle (see Section 2.3) is one method that is still in common use.

Recent advances in mass culture include the introduction of plastic culture flasks that have the same 'footprint' as conventional flasks but an increased surface area for cell growth, such as the TripleFlask from Nunc which has multiple horizontal culture surfaces. Also, larger capacity modular units that use stacked plastic plates to increase surface area are available (e.g. Nunc Cell Factory, Nunc) including a unit that employs automated control and delivery systems to regulate media supply (Cellcube, Corning Costar).

An additional method suitable for mass culture of substrate-adherent (and non-adherent) cells is a system in which cells are grown in the extra-capillary space of a hollow-fibre cartridge (56). Each cartridge contains several thousand hollow-fibre capillaries made from ultrafiltration membrane, typically with a molecular weight cut-off of 10 000 Da (but other porosities are available to suit specific applications). Medium is circulated through the lumen (intra-capillary space) of the hollow fibres. Low molecular weight substances equilibrate between the intra- and extra-capillary compartments, while molecules too large to pass through the membrane pores do not. This means that essential nutrients in the medium can diffuse through the capillary wall and bathe the cells, and waste products can diffuse away from the cells, but high molecular weight cell products are retained in the extra-capillary space. High molecular weight substances are added to or removed from the extra-capillary space via access ports. Hollow-fibre systems are particularly well suited to the production of high molecular weight, cell-secreted products such as monoclonal antibodies. Since direct access to the cells is restricted, these systems are generally less suitable for cell propagation and harvest. As originally designed, nutrient and metabolite gradients were a major problem with these systems. However, recent advances in instrumentation by companies such as Biovest International (formerly Cellex Biosciences) have largely eliminated these problems. For further information on the operation of hollow-fibre systems, see refs 57 and 58, and Chapter 3, Section 5.4.2.

6 The culture of cells in suspension

6.1 Non-adherent substrates for small-scale culture

For some experimental applications such as the formation of histotypic multicellular aggregates (59) and tumour cell spheroids (60), cell adherence to the substrate is undesirable. In this case, a hydrophobic surface is used. Bacteriological-grade polystyrene may be suitable for certain non-adherent applications, but some cell types may nevertheless attach to these dishes. Cell adherence can be discouraged by incubating the dishes on a rocker platform or gyratory shaker. A simple and inexpensive alternative is to coat the dish with a layer of 0.5–1.0% agarose (61). Another way to reduce cell adherence or prevent cell attachment altogether is to coat the dish with poly(2-hydroxy-ethyl methacrylate). This method has been used to inhibit cell attachment and spreading in studies on the role of cell shape as a regulator of cell proliferation (62).

6.2 Mass culture of cells in fluid suspension

Many immortalized cells are anchorage-independent and will proliferate in fluid suspension (63). For large-scale work, cells may be cultured within spinner flasks (64) or bioreactors (65) in which the medium is continually, but gently, agitated to help keep the environment uniform and to prevent the cells from settling out.

Regulation and control of the environment in suspension culture poses some unique technical problems (65). For example, cells in suspension tend to form multicellular aggregates. This is a problem since aggregates settle out and cells are more likely to be damaged, especially if a spin-bar mechanism is being used. Cell aggregation also makes the culture system more heterogeneous, reduces cell yield, and can adversely affect cell function. It may also be difficult to obtain accurate cell counts, due to problems in taking a representative sample and in enumerating the number of cells within aggregates. For these reasons media used for suspension culture often have a reduced Ca^{2+} concentration to inhibit cell–cell adhesion, and may include cell-protective agents such as methylcellulose or agents to reduce the foaming or precipitation of serum (65).

6.3 Micro-encapsulation

This technique was developed for large-scale production of cell products (e.g. monoclonal antibodies) from cell lines that are capable of growth in fluid suspension (66). For culture, cells are immobilized within spheres (*c.* 200 μm diameter) formed from sodium alginate or other gel-type matrix. A semi-permeable membrane is cast at the surface of the sphere and the alginate is then liquefied. The cells are retained within the membrane, which allows medium to diffuse in and cell products to diffuse out. The membrane protects the cells from injury during incubation within a mass culture bioreactor and promotes growth at very high density (e.g. 10^8 cells/ml).

Recently micro-encapsulation has found use in therapeutic applications such as cell transplantation (67) and *in vivo* tumour therapy (68, 69). Encapsulation of cells within a droplet of alginate or other polymer isolates them from the host and protects against immune-mediated destruction.

7 *In vitro* cell growth behaviour

7.1 Adaptation to culture

With the establishment of a primary culture or with the subculture of an ongoing cultured cell population, cells may be subjected to considerable stress. For example, enzymatic dissociation of organ fragments or substrate-dependent cultured cells breaks cell–cell and cell–substrate interactions. Dissociated cells commonly change shape (round up), lose phenotypic polarity, and show alterations in the distribution of plasma membrane proteins. Clearly not all cells survive culture manipulations. Those that do may experience some degree of physical injury such as alterations in their cell surface. To survive and resume growth, or to express its differentiated function, a cell must be able to repair the injury it suffers and adapt to changes in its environment. Regardless of how 'physiological' the environment in primary culture may be, it does not match conditions *in vivo*. Similarly, when cells already in culture are subcultured their cell–cell and cell–matrix interactions are disrupted and they must attach to a new patch of substrate. Adaptation to culture takes time and is influenced by culture conditions. It has been demonstrated that cells 'condition' their environment, that is, they release into the medium substances (conditioning factors: possibly ECMs or growth factors) that promote attachment and growth. Some cell lines may adapt more quickly, proliferate more quickly, or differentiate sooner or more fully if the medium has been 'conditioned' by exposure to actively growing cells. This is particularly true for cells cultured at clonal density. When conditioned medium is used it is usually filter sterilized to remove free-floating cells, then may be stored frozen. Another method to condition the culture environment is to seed cells into dishes that already contain a viable but non-mitotic cell population. So-called 'feeder layers' are commonly prepared from embryonic fibroblasts such as the 3T3 line by exposing the cells to gamma-radiation or by treating them with mitomycin C (see Chapter 6, *Protocol 7*, and ref. 70).

Conditioning and adaptation to culture may be particularly important for cells plated at clonal density, but these processes play a role in routine culture as well. It should be kept in mind that many cell lines require time following attachment to re-establish their differentiated function. A culture seeded at very high density might show superb attachment such that it is near confluency within hours, but may be dysfunctional for the feature of interest. This is, for example, the case for some epithelial cell lines, where the differentiated expression of vectorial ion and fluid transport may not occur until several days after attainment of confluency (24).

7.2 Phases of cell growth

Normal cells in culture typically show a sigmoidal pattern of proliferative activity that reflects adaptation to culture, conditioning of the environment, and the availability of physical substrate and nutrient supply necessary to support the production of new cells (*Figure 2*).

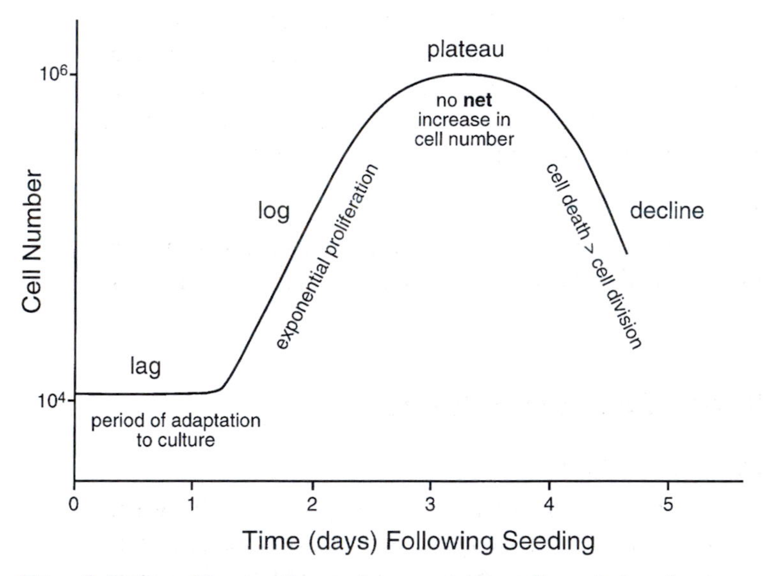

Figure 2 Pattern of the growth curve for normal mammalian cells in culture.

The phases of cell growth in culture are defined below.

7.2.1 Lag phase

Following seeding, cells exhibit a lag phase during which they do not divide. The length of lag phase is dependent on at least two factors, including the growth phase at the time of subculture and the seeding density. Cultures seeded at lower density condition the medium more slowly and have a longer lag phase. Cultures seeded from actively growing stock have a shorter lag phase than cells from quiescent stocks.

7.2.2 Logarithmic (log) growth phase

This is a period of active proliferation during which the number of cells increases exponentially. In log phase the percentage of cells in the cell cycle may be as high as 90–100%, and at a given time cells are randomly distributed throughout the phases of the cell cycle. During log phase the cell population as a whole is generally considered to be at its 'healthiest', so it is common to use cells in log phase when assessing cell function. Cell proliferation kinetics during log phase are characteristic of the cell line and it is during this phase that the population doubling time is determined (see Section 8.2). Factors that influence the length of log phase include seeding density, rate of cell growth, and the saturation density of the cell line.

7.2.3 Plateau (or stationary) phase

The rate of cell proliferation slows down and levels off as the cell population attains confluency. This is the plateau phase. Many cell lines in substrate-dependent culture still show some degree of mitotic activity at confluency. The cells may continue to divide and pack tighter, or daughter cells may be released into the overlying medium. At plateau phase, cell division tends to be balanced by cell death, and the percentage of cells in the cell cycle may drop to 0–10%. For some cell lines the plateau phase can be extended if the medium is replenished. This is not a stable period for most cell lines and plateau phase cells may be more susceptible to injury. Cells in suspension culture also reach plateau phase (saturation density) dependent in part on availability of nutrients.

7.2.4 Decline phase

The plateau phase is followed by a period of decline in cell number (decline phase) due to uncompensated cell death. This is not merely a function of nutrient supply.

8 Determinations of cell growth data

8.1 Calculation of *in vitro* age

It is well established that normal cell lines undergo *in vitro* age-related changes in structure and function and have a finite lifespan. Transformed cell lines may also show changes with successive generations. For example, the distribution of morphologically distinct cell types within phenotypically heterogeneous lines, such as Madin–Darby canine kidney (MDCK) cells, changes with increasing subculture. Since cell lines do not remain stable throughout their lifespan it is important to track the *in vitro* age of the line. One may then restrict studies of a given experimental series to cultures within a selected age range, as a measure to help achieve consistency in the study system.

Two methods are commonly used to track cell age. Many workers simply record the number of times the line has been subcultured (passaged). Passage number documents the number of times a culture is handled, but it is at best a crude estimate of the age of the cell line. A more useful method is to calculate the number of cell generations the line has undergone, to determine the number of cumulative population doublings.

8.1.1 Population doubling level (generation number)

The following explanation follows closely the description of a method used to determine population doubling level (PDL) in studies of cell ageing conducted on normal diploid fibroblasts (71, 72).

The concept of PDL is based on the assumption that cells in culture undergo sequential symmetric divisions such that the population increases in number exponentially (1 to 2 to 4 to 8, etc.). At the end of n generations each cell of the original seeded inoculum produced 2^n cells. Thus, the total number of cells at a given time following inoculation is given by $N_H = N_I 2^n$, where N_H is the number of cells harvested at the end of the growth period, and N_I is the number of cells inoculated. The number of generations is the number of population doublings, and can be expressed using common logarithms (base 10) as:

$$2^n = N_H/N_I \text{ or } n\log 2 = \log\left(\frac{N_H}{N_I}\right) = \log N_H - \log N_I$$

$$\text{since } \log 2 = 0.301, \quad 0.301n = \log N_H - \log N_I$$

$$\text{so } n = 3.32\,(\log N_H - \log N_I)$$

Example. Calculation of population doubling level (PDL)

2.5×10^5 cells seeded	$N_I = 2.5 \times 10^5$	$\log 2.5 \times 10^5 = 5.3979$
6.0×10^6 cells harvested	$N_H = 6.0 \times 10^6$	$\log 6.0 \times 10^6 = 6.7782$

$n = 3.32\,(6.7782 - 5.3979) = 4.58$

Therefore, to yield 6.0×10^6 cells, 2.5×10^5 cells underwent 4.58 population doublings.

- Add the number of generations for growth interval (time between seeding and harvest) to the previous PDL number to give the cumulative PDL. Cumulative PDL is recorded on the flask, and similar calculations are performed at each subsequent subculture.

PDL cannot be calculated without an accurate determination of cell number in the original inoculum and so is usually not attempted with primary cultures. By convention the primary culture at confluency is designated PDL 1.

8.2 Multiplication rate and population doubling time

Two additional calculations give useful information on the log phase growth characteristics of a cell line. **Multiplication rate** (r) is the number of generations that occur per unit time and is usually expressed as population doublings per 24 hours. **Population doubling time** (PDT) is the time, expressed in hours, taken for cell number to double, and is the reciprocal of the multiplication rate (i.e. $1/r$).

$$\text{PDT} = \text{total time elapsed/number of generations}$$

Example. Calculation of multiplication rate and PDT
Supposing that 2.5×10^5 cells were seeded at time zero (t_1), and 6.0×10^6 cells harvested at 96 h (t_2).

$$\text{Multiplication rate } (r) = 3.32 (\log N_H - \log N_I) / (t_2 - t_1)$$
$$r = 3.32\,(6.7782 - 5.3979)/96 = 0.048 \text{ generations per hour}$$
$$\text{or } 1.15 \text{ population doublings/24 hours}$$
$$\text{PDT} = 1/r$$
$$\text{PDT} = 1/(1.15/24)\text{ h} = 24/1.15\text{ h} = 20.9\text{ h per doubling}$$

Population doubling level, multiplication rate, and population doubling time describe the growth characteristics of the cell population as a whole, and strictly speaking do not characterize the division cycle of individual cells. By definition, PDT (commonly 15–25 h) is not the same as **cell generation time** (cell cycle time), which is the interval between successive divisions (mitosis to mitosis) for an individual cell. In practice, however, PDT is used as an estimate of cell cycle time and to determine the length of the phases of the cell cycle (see Section 8.5).

Several points should be considered before calculation of PDL and PDT is adopted:

(a) Accuracy of cumulative PDL, and especially PDT, requires consistent handling of cultures.

(b) Cells must be subcultured at regular intervals and when the culture is in the log phase of growth.

(c) Each subculture should be at the same seeding density, and culture conditions must be consistent, including the same type of substrate, size of flask, volume of medium, and formulation of medium.

(d) The method used to perform cell counts must be consistent and accurate.

(e) When combining (e.g. averaging) data from multiple cultures for the estimation of PDT, it is best to calculate the individual multiplication rates first, average these, and finally calculate the PDT, otherwise a single slow-growing flask (giving a long PDT) can disproportionately skew the result.

8.3 Counting cells in suspension

Many cell culture protocols require an estimate of the number of cells at plating or at harvest. Two common methods used to determine cell number in a cell suspension include manual haemocytometer counting using a light microscope, and semi-automated counting using an electronic particle counter (e.g. Coulter counter).

8.3.1 Haemocytometer counting

This is often the simplest, most direct, and cheapest method of counting cells in suspension. It also allows the percentage of viable (intact) cells in the preparation to be determined using the dye exclusion method. The haemocytometer is a modified microscope slide that bears two polished surfaces each of which displays a precisely ruled, subdivided grid (*Figure 3*). The grid consists of nine primary squares, each measuring 1 mm on a side (area 1 mm^2) and limited by three closely spaced lines (2.5 μm apart). These triple lines are used to determine if cells lie within or outside the grid. Each of the primary squares is further divided to help direct the line of sight during counting. The plane of the grid rests 0.1 mm below two ridges that support a sturdy coverslip. There is a bevelled depression at the leading (outer) edge of each polished surface, where cell suspension is added to be drawn across the grid by capillarity.

Several points should be considered when performing cell counts with a haemocytometer (see *Protocol 4*):

(a) For accuracy and reproducibility, counts must be performed in the same way each time. Standardize the cell dissociation procedure for a given cell line (or

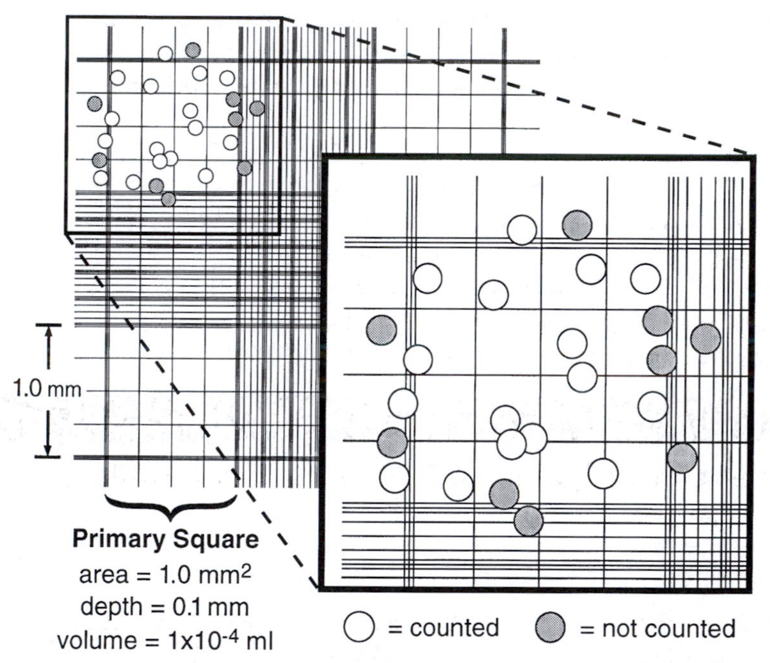

Figure 3 Grid pattern of improved Neubauer ruled haemocytometer. Inset shows cells (enlarged for clarity) distributed over a primary square. Cells that are within, or that touch, the left or top boundary are counted, while those that touch, or are outside, the lower or right boundary are not counted.

tissue for primary culture) and use the same counting protocol and conventions.

(b) When the haemocytometer is properly loaded, the volume of cell suspension that will occupy one primary square is 0.1 mm^3 (1.0 $mm^2 \times 0.1$ mm) or 1.0×10^{-4} ml.

(c) In practice, one counts and totals the cells within ten primary squares (five primary squares per chamber), to give the number of cells within 1.0 mm^3 (10×0.1 mm^3) or 1×10^{-3} ml. Total cell concentration in the original suspension (in cells/ml) is then:

$$\text{total count} \times 1000 \times \text{dilution factor}$$

(d) A dilution factor [(volume of sample + volume of diluent)/volume of sample] is needed if the suspension was diluted with buffer, or with a dye used to perform a viable cell count. Any diluent used must be isotonic.

i. Determining the number of viable cells by dye exclusion

Dye exclusion involves mixing an aliquot of the cell suspension in a volume of buffer or balanced saline containing a water-soluble (membrane lipid-insoluble) dye (e.g. 0.4% Erythrosin B, or Trypan Blue) that is visible when it leaks into cells that have damaged plasma membranes. By counting total cells and stained cells (damaged cells) one can calculate the per cent viability (see *Protocol 4*). Recently, a device has come on the market which automatically counts viable and dead cells as stained by Trypan Blue, using the techniques of microscopy and image analysis (Cedex System, Innovatis GmbH). However, the high price of this instrument is liable to restrict its use to laboratories performing very large numbers of viable cell counts.

One should be cautious in interpreting cell 'viability' by dye exclusion. Dye uptake marks cells that have grossly disrupted membranes and may not detect other forms of injury that affect cell attachment or may progress to cell death. Also, this method does not account for cells that have fully lysed (i.e. are no longer identifiable as cells). In addition, the choice of dye may influence results. Trypan Blue has a greater binding affinity for protein in solution than for injured cells, and should only be used in protein-free solution (73).

Protocol 4

Haemocytometer counting for total and per cent viable cells

Equipment and reagents

- Haemocytometer and coverslip
- Upright bright field microscope, mechanical stage, ×20 objective
- Multi-key manual counter
- Ca^{2+}- and Mg^{2+}-free phosphate-buffered saline (CMF-PBS)
- 0.4% (w/v) Erythrosin B in CMF-PBS

Protocol 4 continued

Method

1. Prepare a dissociated cell suspension in complete medium (e.g. *Protocol 1*).
2. Collect a 0.5 ml aliquot and mix with 0.5 ml Erythrosin B solution in a dilution tube. Place the tube on wet ice (5 min).
3. Seat a clean coverslip squarely on top of the haemocytometer. If the supporting surfaces are polished, lightly moisten them before pressing the coverslip into position, whereupon Newton's rings (alternating light and dark interference fringes) should appear in both the areas of contact—if not, remove the coverslip, clean coverslip and slide, and try again. Note that if the supporting surfaces are ground instead of polished, Newton's rings will not be seen.
4. Use a Pasteur pipette to resuspend the cells gently.
5. Load the haemocytometer so that fluid entirely covers the polished surface of each chamber. To ensure that the chamber is not overloaded, use a wedge of filter paper to blot excess fluid from the filling groove.
6. Using the ×20 objective, locate the upper left primary square of one grid and defocus the condenser to improve visibility of the cells.
7. Count the cells in the centre and four corner primary squares of each grid (ten primary squares). Use separate keys of the tally counter to record unstained (viable) and stained (non-viable) cells. Use the following counting conventions.
   (a) The middle of the triple lines separating each primary square is the boundary line. Cells that touch the upper or left boundaries are included, those that touch the lower or right boundaries are excluded (*Figure* 3, inset).
   (b) If greater than 10% of particles are clusters of cells, attempt to disperse the original cell suspension more completely, and collect another sample. Some workers assign all clusters containing more than five cells a value of five.

   Example. Total and per cent viable cell count

   Aliquot (0.5 ml) collected from 40 ml of cell suspension and mixed with 0.5 ml Erythrosin B solution.

	Stained/unstained cells per primary square					
Grid A	2/28	4/35	1/25	6/32	4/44	= 17/164
Grid B	1/30	3/32	0/26	4/40	4/29	= 12/157
					Total	= 29/321

Total cell count:	350
Per cent viable cells:	$(321/350) \times 100 = 91.7\%$
Dilution factor:	$(0.5 + 0.5)/0.5 = 2$
Cells/ml:	$350 \times 1000 \times 2 = 7.0 \times 10^5$
Total no. of cells:	$(7.0 \times 10^5$ cells/ml$) \times 40$ ml $= 2.8 \times 10^7$
Viable cells/ml:	6.4×10^5

Protocol 4 continued

Notes:

(a) Dilute suspension to give c. 25–50 cells/primary square.

(b) The suspension should be uniform and, ideally, monodisperse (i.e. without cell clusters). Clumping can be minimized by diluting the aliquot in CMF-PBS.

(c) For viable cell counts, allow a set time (e.g. 4–5 min) for dye to diffuse into the cells, as too rapid a count may underestimate the number of damaged cells.

8.3.2 Electronic particle counting

Cells in suspension can be counted accurately and rapidly using a particle counter such as the Coulter counter (Beckman Coulter). Multiple counting runs can be performed in a fraction of the time it takes to do a haemocytometer count. In addition, many instruments are available that allow cell sizing and the determination of size distribution within a cell suspension. Electronic particle counters operate on the following principle. Cells are suspended in an electrolyte and a metered volume of this suspension is pulled through a narrow (20–200 μm) aperture that carries a nominal current. As a cell crosses the orifice it produces an increase in impedance resulting in a voltage pulse. The pulse is amplified, and the number of pulses within the set threshold is counted and displayed on an oscilloscope. Cell sizing can also be performed, because pulse amplitude is directly proportional to cell volume. Cell clusters or debris may also register a pulse, but these extraneous counts can be eliminated by adjusting the threshold of the pulse amplitude. The volume of suspension counted and the aperture size can be changed to suit specific applications. As with counting using a haemocytometer, the quality of the cell suspension is extremely important, and accurate, reproducible counts require a consistent method.

8.4 Counting cells adherent to a substrate

For many applications it may be necessary to obtain an estimate of the number of cells still adherent to the substrate. A number of methods have been used.

8.4.1 Visual counting of cells or nuclei

The number of cells (living cells or fixed and stained specimens) are counted within representative fields of known area under the microscope. This may seem straightforward, but it is very time-consuming and subject to a number of potential errors. Because vital counting commonly takes a long time it is usually necessary to record images on film or electronically. This has the advantage of providing a permanent record, but storage, cataloguing, and expense are added factors to consider. A number of additional points should be considered before attempting visual cell counts:

(a) Sampling within a culture must be representative and fields must be collected by a systematic method to avoid bias.

(b) Criteria for identification of individual cells must be unambiguous and conventions established to account for instances where cell boundaries may not be visible.

(c) Counting nuclei, as with DNA dyes (e.g. propidium iodide) for fluorescence microscopy, is an alternative, but can lead to error if mulinucleate cells are present.

(d) The actual counting of cells must be done using appropriate morphometric conventions to include or exclude cells that straddle the boundaries of the field.

8.4.2 Densitometry of fixed and stained cultures

Densitometry has been used to quantify the density of cell growth within a culture vessel (74). The cells are fixed then stained (e.g. with 0.5% crystal violet) and the light transmittance (or absorbance) of the specimen is determined. The method is most useful in detecting relative differences in culture density or occupancy of the substrate within the same experiment and has been applied to growth factor response studies in primary culture (75) and to assays of colony formation used to assess nutrient requirements for optimal cell growth (76).

8.4.3 DNA assays

Determinations of DNA content are commonly used to estimate changes in population density, for example to establish growth curves or assess culture response to growth factors. Because the DNA content of a cell changes as it progresses though the cell cycle, and because it is very common for a culture to contain binucleate and multinucleate cells as well as mononucleate ones, DNA content may not give an accurate estimate of cell number. In addition, DNA assays do not discriminate between viable and dead cells.

8.4.4 Colorimetric assays

A variety of methods have been developed to estimate cell number based on cellular content of a specific enzyme or substrate, or the uptake and subsequent quantitative extraction of a dye. MTT (3-[4,5-dimethylthiazol-2-yl]-2,5-diphenyltetrazolium bromide) added to the culture medium (at 5 mg/ml) is converted to a coloured formazan by mitochondrial dehydrogenase activity in living cells (77). Since dehydrogenase content is relatively consistent among cells of a specific type, the amount of formazan produced is proportional to cell number. One may expect, however, to find variation among different cell types. An assay for cellular content of the lysosomal enzyme hexoseaminidase has also been used to estimate cell number (78). There is potential for error in both methods. Culture conditions may affect cell enzyme content and activity, and serum in the medium may contain hexoseaminidase.

Dye uptake and extraction can be used to estimate cell number. Methylene Blue (1%) binds to negatively charged groups (e.g. phosphate, nucleic acids, some proteins) in fixed cells and can be extracted by lowering the pH (0.1 M HCl wash)

(79). Janus Green is a supravital dye that, like Trypan Blue, stains only cells that have damaged membranes. If, however, Janus Green (1 mg/ml) is applied to fixed (ethanol-permeablized) cells, all cells stain. The dye can then be extracted with ethanol and quantified spectrophotometrically (80).

8.5 Phases of the cell cycle

The length of each of the successive phases of the cell cycle (G_1, S, G_2, and M) and the total length of the cell cycle can be determined for cells in culture. The most commonly applied methods are based on detection of DNA synthesis or on quantification of DNA content within individual cells.

Cells synthesize DNA only during the S phase (DNA synthesis phase) of the cell cycle, and so incorporation of the radiolabelled nucleotide [^{3}H]thymidine into DNA is limited to this phase. Cells that incorporate label can be identified by autoradiography using light microscopy (81). Labelling index is the fraction of cells labelled, and represents the percentage of cells that were in S phase during the time of exposure to [^{3}H]thymidine. Labelling index is commonly used to quantify the response of a cell population to a mitogenic stimulus. However, labelling index alone does not give information on the duration of any of the phases of the cell cycle.

8.5.1 Labelled mitoses method

The length of the S, G_2, and G_1 phases, and of the entire cell cycle can be obtained by determining the percentage of cells in mitosis that are labelled (82). The method is summarized as follows:

(a) Multiple replicate cultures are pulsed with [^{3}H]thymidine (370 Bq/ml) for 20–30 min, washed free of unincorporated label, and returned to incubation.

(b) Groups of three cultures are then terminated (by fixation) at successive time points (0, 2, 4, . . . 30 h) and processed for autoradiography.

(c) The percentage of labelled mitoses is determined and plotted versus time after [^{3}H]thymidine pulse.

The length of the complete cell cycle for most cultured animal cells falls in the range 15–25 h. Therefore, if the percentage of labelled mitoses is determined at points up to 30–35 h the data will include some cells that have cycled through two rounds of mitosis. The plot of per cent labelled mitoses versus time (*Figure 4*) shows a rapid rise (first rise) to almost 100%, a fall (first fall) to a few per cent, and then another rise (second rise) as some cells enter their second round of mitosis. The lengths of the phases of the cell cycle are estimated from the points on this curve as follows:

i. S phase

The period of DNA synthesis is the time interval between the 50% labelling points on the first rise and the first fall (usually 7–8 h).

ii. G_2 phase (post-DNA synthesis, pre-mitosis gap phase)

This is the time between the end of the [^{3}H]thymidine pulse and the 50% labelling point on the first rise (usually 2–4 h).

iii. Length of mitosis

This is not evident from the curve, but is approximately the same for all normal cells (usually 0.5–1 h).

iv. G_1 phase

The post-mitotic gap which precedes DNA synthesis is estimated as the time between the 50% point on the first fall and the 50% point on the second rise. G_1 is the most variable phase of the cell cycle (for example, it may not exist for some rapidly dividing cells), but is usually 6–15 h.

v. Total cell cycle time ($S + G_2 + M + G_1$)

Some workers estimate the interval between successive mitoses using the time between any two corresponding points for per cent labelled mitosis on the first rise (first mitosis) and the second rise (second mitosis).

8.5.2 Cell cycle analysis by DNA content (flow cytometry)

The amount of DNA within a cell changes as the cell progresses through the phases of the cell cycle. DNA content is greatest during the G_2 and M phases and lowest during post-mitotic G_1 gap, and increases throughout S phase. Certain dyes [e.g. ethidium bromide or bisbenzimide (Hoechst 33258)] bind to DNA. Flow cytometry of a dye-treated cell suspension can then be used to obtain a

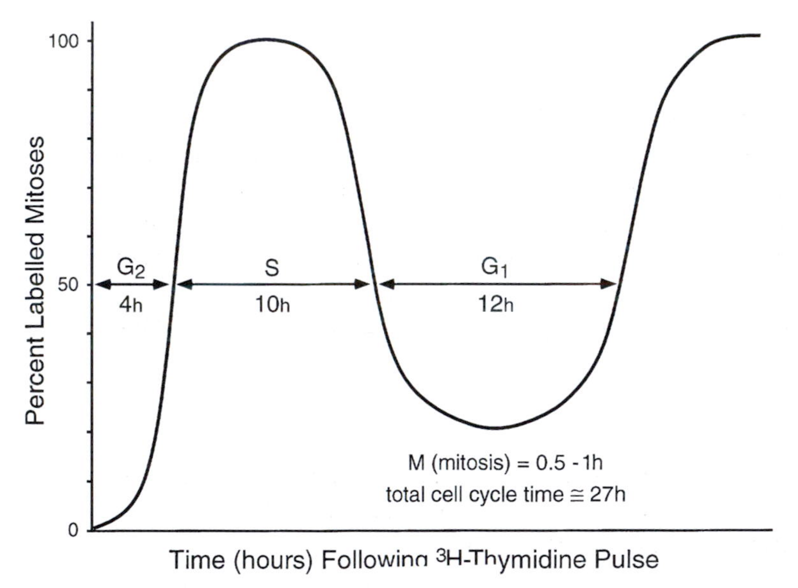

Figure 4 Plot of per cent labelled mitoses used to determine length of phases of the cell cycle. See text for details.

population distribution based on relative DNA content. The percentage of cells in each phase is directly proportional to the length of that phase (83).

8.6 Cell generation time by time-lapse photomicrography

Cell cycle time (also termed cell generation time), and the length of mitosis can be accurately determined for individual cells by direct visual inspection using time-lapse microscopy (84). This methodology requires sophisticated microscopy equipment and necessitates rigorous control of incubation conditions (e.g. temperatures and pH) with the culture vessel positioned on the microscope stage (see also Chapter 4, Sections 5 and 6.1).

8.7 Growth fraction

The growth fraction is an estimate of the percentage of cells active in the cell cycle. This value is commonly high (90–100%) for cultures in exponential growth (log phase) and may be very low (< 10%) when the culture reaches saturation density (plateau phase). One method to determine the growth fraction is to expose the culture to [^{3}H]thymidine for the length of the entire cell cycle (in practice, usually 24 h), and then determine the percentage of labelled cells. A single short pulse with [^{3}H]thymidine as used in the labelled mitoses method (see Section 8.5.1) would underestimate the growth fraction since the labelling index only accounts for cells passing through S phase during the exposure period. Determination of the growth fraction is most useful in assessing the growth-stimulatory effect of an exogenous agent (e.g. growth factor) on a quiescent cell population, that is, in determining the conditions that cause cells to re-enter the cell cycle.

8.8 Expressions of culture 'efficiency'

Cell attachment to a substrate and the capacity of attached cells to form colonies can be quantified. Such determinations are useful in assessing a variety of aspects of cell behaviour such as cell survival following cryopreservation, cell–matrix interactions, and the effects of cell ageing. In addition, assays of cell plating and colony formation are used in toxicity testing and in cell culture quality control to screen sera and reagents. Two commonly used expressions of culture 'efficiency' are discussed below.

8.8.1 Attachment efficiency

This is the percentage of cells that attach to the culture substrate within a given period of time. A known number of cells is seeded. At the end of the attachment period the unattached cells are collected and counted.

Attachment efficiency (%) = [(number of cells seeded − number of cells recovered)/(number of cells seeded)] × 100

8.8.2 Colony-forming efficiency

This is the percentage of cells seeded that form colonies within a prescribed time. Cultures are seeded at known low density and following a growth period (e.g. 144 h) replicates are fixed and the cells are stained (e.g. with Crystal Violet). With the aid of a dissecting microscope, fields of known area within the dish are selected systematically (to avoid bias) and the number of discrete colonies (clusters of 16–50 cells) is recorded. The number of colonies per unit area is determined and a value calculated for the area of the entire substrate.

$$\text{Colony-forming efficiency (\%)} = (\text{number of colonies formed/number of cells seeded}) \times 100$$

A slight variation of this method of assessing colony-forming efficiency is described in detail in Chapter 8, *Protocol 1*.

9 Cryopreservation and retrieval of cells from frozen storage

9.1 Purpose of cell banking

Cultured cells can be perplexingly unstable and exhibit alterations in morphology, growth pattern, function, and karyotype during continued or extended culture. Cell lines show age-related changes *in vitro*, and some cultures may undergo spontaneous transformation to exhibit altered growth or changes in functional characteristics. Fortunately, cells that are properly frozen (and properly stored) can be kept for extended periods (many years, and perhaps indefinitely) without substantial alterations in viability or other characteristics. Thus cell banking, that is, freezing and storage of cultured cell stocks, makes it possible to maintain cultures as a renewable resource so that experimentation over a period of time can be performed on cells of equivalent characteristics (see Chapter 9, Section 3.4). Such a resource also helps soften the impact of incidents of microbial or cellular contamination, or equipment (e.g. incubator) failure.

9.2 Mechanism of cell freezing, and factors that affect viability

In order to survive freezing and thawing, cells must be treated with a cryoprotective agent. The mechanism by which cryoprotectants act has yet to be fully established. One view is as follows: glycerol or dimethyl sulfoxide (DMSO) added to the resuspension medium (5–10%, v/v) acts to permeabilize the plasma membrane and allows water to flow out of the cell as cooling occurs. Cryoprotectants depress the freezing point so that ice crystals begin to form at about −5 °C. If, at −5 °C to −15 °C, the cell is sufficiently dehydrated (accompanied by osmotic shrinkage and concentration of solutes), ice crystals will form in the surrounding medium but not in the cell interior (85). Efflux of water is the key to the process and is affected by several factors. The rate at which cooling occurs is critical, but the type and concentration of cryoprotectant are also important. The aim is to select conditions that minimize or eliminate intracellular ice crystal formation. Ice crystals can form during freezing or thawing and may result in cell lysis (86).

Most laboratories store vials immersed in liquid nitrogen. This makes best use of canister space. However, both glass ampoules and screw-capped plastic cryovials can leak (if improperly sealed or tightened). This can result in cell death or contamination. Also, if a leaky vial accumulates liquid nitrogen it may explode during thawing. Therefore, storage in the vapour phase of liquid nitrogen is recommended. This also overcomes the potential for the transfer through the liquid nitrogen of infectious agents between improperly sealed vials (87).

In practice, the warming rate is more difficult to control than the cooling rate. Some automated controlled-rate freezing units have a thawing cycle, but these expensive instruments are not readily available to most laboratories. General practice (see *Protocol* 6) is to immerse frozen vials (with agitation) in a warm water-bath (37–40 °C) so that the cells thaw in about 1.5 min. Some cells are very susceptible to osmotic stock (swelling and rupture) during the process of warming and transfer to culture medium. Routine protocols commonly describe a one-step dilution (1:10 to 1:20). However, it may be necessary to combine the thawed cell suspension with medium gradually (dropwise) or initially suspend the cells in a non-permeant solute (e.g. sucrose equimolar to cryoprotectant) to allow the cryoprotectant to diffuse out of the cell before transfer to isotonic medium.

9.3 Supplies and equipment for cell freezing

9.3.1 Cryoprotectants

Reagent grade glycerol or DMSO used at 5–10% in complete culture medium are the most common cryoprotectants. Filter-sterilized DMSO is available (e.g. Sigma). Glycerol and DMSO should be replaced yearly and discarded if they become turbid or change colour.

9.3.2 Freezing vials

Freezing vials of various types are available. Commercial cell repositories and laboratories that can afford to purchase an ampoule sealer use borosilicate glass ampoules. When properly sealed (which takes some practice), glass ampoules give the best protection against leakage. Plastic screw-capped vials (1–2 ml autoclavable or supplied sterile) with or without gasket are safer and easier to use. In many designs of vial the gasket may deform and leak if the cap is over-tightened. Therefore, vials without gaskets (e.g. Nalgene Cryovials, Nalge Nunc International) or those where the gasket is supported to prevent deformation on over-tightening (e.g. Cryovial LW3432, Alpha Laboratories) are recommended for the storage of cells in the liquid phase.

When glass ampoules are used and sealed with a gas-O_2 torch if is important to test for leakage. This is done by immersing the vials in 0.05% Methylene Blue (aqueous) at 4 °C for 30 min. Any vial that picks up the blue dye must be discarded. If a glass ampoule or plastic cryovial leaks and accumulates liquid nitrogen it may explode (with considerable force) upon warming. Flying glass or shards of plastic are an obvious hazard and protective eyewear (safety glasses, full face shield) is recommended. Protective (insulated) gloves should be worn and frozen vials should be handled with tongs. Workers should be aware of the potential biohazard if high risk pathogens are involved.

9.3.3 Freezing devices

It is possible to achieve adequate cell freezing on a tight budget. Vials can be wrapped in insulating material (e.g. styrofoam) then placed within a −70 °C mechanical freezer for 3–5 h. Thus frozen, the specimens must then be transferred to liquid nitrogen for long-term storage. The disadvantage is that conditions for freezing are difficult to control, and it may take considerable trial and error to establish a reliable protocol. A similar, but more consistent method is the use of a freezing container (e.g. Cryo 1 °C Freezing Container, Nalge Nunc International) designed to freeze cells at approximately 1 °C/min. The unit is first filled with isopropyl alcohol. Vials of cell suspension are loaded into a holder and the lid is secured. The unit is then placed within a −70 °C freezer and after 4–5 h the vials are transferred to storage in liquid nitrogen.

Biological freezers are available for use with narrow-necked, Dewar-type liquid nitrogen refrigerators (e.g. 35 VHC, Taylor-Wharton) common to most cell culture laboratories. The freezer unit is loaded with vials and inserted in the neck of the refrigerator, placing the vials in the vapour phase of liquid nitrogen (see *Protocol 5*). The depth that the vials extend into the neck can be adjusted (increasing the depth increases the rate of cooling). The manufacturer provides guidelines to achieve a desired cooling rate dependent on the number of vials to be frozen.

The state of the art in cryogenics is the use of a programmable controlled-rate freezing unit (e.g. Cryo-Med; Minnesota Valley Engineering) in which cooling rate is regulated via temperature probe feedback to the liquid nitrogen supply. Some units have a warming programme for use in retrieval of cells from storage. Once standardized, such automated units give repeatable results and commonly provide a print-out of the cycle (temperature versus time) for documentation.

9.3.4 Cell storage refrigerators

As already mentioned, for long-term storage cells are usually kept in Dewar vessels containing liquid nitrogen, and there is an extensive range of vessels available from manufacturers such as Taylor-Wharton. These range from small vessels with a maximum capacity of ninety 1 ml vials, through the popular under-bench units which can hold up to 4000 vials, to huge vessels capable of containing over 33 000 vials. There are two main methods of storing vials within these vessels. They can either be attached to aluminium canes which are placed in canisters, or stored in divisions in the drawers of multi-drawer tessellating racks. Smaller vessels only use the former system and larger vessels only the latter, but medium-sized vessels are available in both configurations. The choice is largely one of user preference, as both systems have advantages and disadvantages. In general, for a given storage capacity, racked systems tend to use more liquid nitrogen due to the larger opening required in the top of the vessel to remove the racks. Freezer units such as those described in the previous section are only available for use with narrow-necked (i.e. cane and canister) cell stores.

As part of the laboratory routine, all cell stores must be topped up with liquid nitrogen on a regular basis, e.g. once or twice per week, but for added protection low level alarms can be fitted to most vessels. Larger vessels can also be equipped with an auto-fill facility, such that the store is refilled with liquid nitrogen from a dedicated supply when the level drops below a pre-set point.

9.4 Additional comments regarding cryopreservation

(a) The condition of the culture will influence how well it survives cryopreservation. Cells to be frozen should be actively growing and free from contamination.

(b) Conditions for optimal cell survival are different for different cell lines.

(c) The main factors that influence cell viability include the cryoprotectant, cooling rate, storage temperature, and warming rate.

(d) The number of cells for suspension in cryoprotectant medium to give optimal survival must be determined empirically, but usually falls within the range 1×10^6 to 1×10^7 cells/ml.

(e) Slow cooling, at approximately 1 °C/min, gives good results for most cultured animal cells. Cooling at too slow a rate can result in substantially decreased cell survival.

(f) Freezing at a continuous rate tends to be more successful than interrupted or stepwise freezing.

(g) For permanent storage, cells must be kept below −130 °C. Normally, this is achieved by storage in liquid nitrogen, either immersed (−196 °C) or in the overlying vapour phase (−140 °C to −180 °C).

(h) Storage at −70 °C is not sufficient to prevent progressive deterioration of the specimen over the long-term, but can be used for short-term storage (e.g. for a few weeks).

(i) A benchmark assessment for good cryopreservation is for viability (dye exclusion) of the thawed cell suspension to be no more than 5–10% below the pre-freeze value.

(j) Viability does not necessarily predict attachment efficiency and potential for cell growth. A 24 h cell or colony count may give a better indication of cell response to freezing.

(k) DMSO used in cryopreservation is readily absorbed through the skin and can penetrate some rubber gloves. It can serve as a vehicle to introduce potentially harmful agents into the body.

(l) Liquid nitrogen is a potential hazard and can cause severe burns especially if caught by loose clothing, spilled down the sleeve of an insulated glove, or caught within shoes. Suffocation or asphyxiation from oxygen deprivation due to displacement of the atmosphere by nitrogen gas is a possibility, and work with liquid nitrogen should be performed in a well-ventilated area. It may be appropriate to fit an oxygen depletion warning device in areas where large volumes of liquid nitrogen are stored or used.

Protocol 5

Cryopreservation of cultured cells

Equipment and reagents

- 1.0 ml plastic cryovials (e.g. from Nalge Nunc International or Alpha)
- Liquid nitrogen refrigerator (e.g. 35VHC from Taylor-Wharton) with canisters
- Complete culture medium
- Refrigerator-neck-style cell freezer (e.g. BF-5 or Handi-freeze freezing tray, Linde)
- DMSO (reagent grade), filter-sterilized (e.g. from Sigma)

Method

1 Prepare a monodisperse cell suspension (e.g. *Protocol* 1) and obtain a viable cell count by the dye exclusion method (*Protocol* 4). Resuspend the cell pellet to give 6×10^6 viable cells/ml in complete medium (chilled) containing 10% DMSO.

2 Load plastic cryovials (labelled with date and cell identity) with 1 ml of cell suspension. Tighten the cap securely but avoid dislodging the rubber gasket.

3 Place the vials on wet ice for 15–30 min to allow equilibration of the cells with DMSO-containing medium.

4 Transfer the vials to the tray of the cell freezer and adjust the height of the tray, according to the manufacturer's instructions, to freeze at 1 °C/min (height to give a desired freezing rate depends on number of vials). Insert freezing unit in the neck of the liquid nitrogen refrigerator. Freeze for approximately 3 h.

5 Transfer frozen vials to a storage canister in liquid nitrogen.

Notes:

(a) Replenish the culture medium 24 h before freezing.

(b) Cell viability may be improved using gas equilibrated (optimal pH) medium to prepare DMSO-containing medium.

(c) To prepare 11 inch (28 cm) canisters for storage of vials in vapour phase, stuff (6–7 inches (15–18 cm) deep) with cotton towelling. Be certain that the towelling has reached vapour phase temperature before transferring frozen vials. Alternatively, use empty canisters and clip frozen vials onto aluminium canes so that they are held above the level of the liquid nitrogen.

Protocol 6

Retrieval of cells from frozen storage

Equipment and reagents

- Protective eyewear, insulated gloves, and tongs
- 40 °C water-bath
- 70% ethanol in squeeze bottle
- Sterile alcohol swabs

Method

1 Pre-equilibrate 15 ml of complete medium within a 75 cm^2 culture flask for 2 h in a CO_2 incubator.

2 Wearing protective eyewear and insulated gloves, remove a cryovial of cells from frozen storage and, using tongs, immerse immediately in warm water (40 °C) with agitation. Thawing should be complete within 1–1.5 min and the cells must not overheat.

3 Dry the vial, then rinse it with 70% ethanol. Wrap an alcohol swab around the cap and gasket and loosen the cap. Protective eyewear must not be removed until this step is completed.

4 Transfer the cell suspension to equilibration medium and incubate.

Notes:

(a) Special care should be taken to avoid sustaining cuts when opening glass cryopreservation ampoules. Score the ampoule with a file, grasp the neck through several layers of sterile cotton gauze, then break.

(b) For many cells it is recommended that the cryoprotectant be diluted out by dropwise addition of medium, and the cells pelleted by gentle centrifugation, before resuspension in the equilibrated medium for incubation.

10 Transportation of cells

10.1 Transporting frozen cells

Frozen cells in vials or ampoules can be packaged in dry ice and shipped by overnight courier. Special refrigerator-type containers (dry shippers) are also available (e.g. CryoPak, Taylor-Wharton; Arctic Express, Barnstead-Thermolyne) for transport of multiple vials in liquid nitrogen vapour. These units contain a liquid nitrogen-absorbent material surrounding the specimen canister. The absorbent material is saturated with liquid nitrogen and then vials are placed within the canister for shipment. Dry shippers are well insulated and have a long holding time (i.e. many days), and there is no danger of spilling liquid nitrogen during transit.

10.2 Transporting growing cells

Actively growing cultures can be transported without the need for elaborate equipment.

10.2.1 Suspension-cultured cells

Transfer an aliquot of cell suspension to a sterile, screw-capped tube. Fill the tube with medium, cap it, and wrap the cap with Parafilm. Place the tube in an insulated container for shipment by overnight courier.

10.2.2 Use of microcarrier beads during transport of substrate-dependent cells

(a) Suspend hydrated beads (e.g. Cytodex, Pharmacia) prepared according to the manufacturer's protocol, in equilibrated culture medium and transfer to a flask containing the actively growing culture for transport (use 10 mg dry weight of beads per cm^2 of flask area). Incubate in stationary culture for 24–48 h to allow the cells to migrate on to the beads (as confirmed by phase-contrast microscopy).

(b) Collect the beads, concentrate them by centrifugation, resuspend them in complete medium, and load into a sterile screw-capped tube for packaging and transport.

10.2.3 Transporting cultures growing in screw-capped flasks

(a) Select healthy cultures in logarithmic growth phase.

(b) Fill the flask to the neck with gas-equilibrated (optimal pH) complete medium.

(c) Securely tighten the cap and wrap it with Parafilm.

(d) Cushion the flask with protective wrap (e.g. bubble-wrapping) and pack it upright in an insulated box for overnight delivery. When shipping to a warm climate destination it is a good idea to include a cold pack (e.g. Blue Ice, Rubbermaid) in the shipping box to reduce the chance that the cells may get too warm.

Further advice on the transportation of cells, including useful insights into the relevant regulations, can be found in ref. 88.

10.2.4 Re-initiating incubation following transportation

Whenever a new cell line is brought into the laboratory, steps should be taken to separate it from ongoing cultures (i.e. quarantine it), as a measure to prevent introduction of microbial contaminants to the laboratory (see Chapter 9).

(a) Unpack the flask or tube and inspect for breaks or leakage. If it is broken, discard it into a receptacle appropriate for biohazardous waste. A small amount of leakage at the cap may not be a problem, but calls for extra care in opening the vessel. In this case, carefully remove any Parafilm, then use the edge of a sterile absorbent paper towel to blot medium from the interface between the cap and neck. Remove the cap and further blot any additional medium from the outside surface of the neck.

(b) Cells may have dislodged from the substrate during transportation but can be sedimented and plated in the same medium used for shipment. Use a pipette to transfer medium from the flask to sterile centrifuge tubes. As in the routine handling of culture flasks, assume that the lip is contaminated and avoid touching it with the pipette. Centrifuge the medium (150 g, 10 min), resuspend any cells (or cells on microcarriers) in fresh medium, and transfer to a new flask (or the original flask) in an appropriate volume.

(c) Examine the cells in the original flask by phase-contrast microscopy. If a substantial number of cells remain adherent to the substrate, retain the flask; if not, discard it. If the flask is to be retained and there was any evidence of the leakage of medium at the cap, wipe the outside of the neck with an alcohol swab and replace the cap. Return the flask to the incubator.

11 Safety in the cell culture laboratory

11.1 Potential risks in routine cell culture

The general cell culture laboratory, one in which activities are limited to work involving well established cell lines or cells derived from pathogen-free subprimate species, is a relatively safe place to work. In this pathogen-free setting the principal safety concern is probably the potential for injury associated with handling liquid nitrogen during cryopreservation of cells and the retrieval of vials from frozen storage (see Section 9). In laboratories using glass pipettes, breakage of these can also pose a significant threat (see Chapter 1, Section 4.3).

When the work of the laboratory involves pathogens carried by cell lines or infecting the animals used to establish primary cultures, the main safety concern is the potential for worker infection. Viruses probably present the greatest risk of infection, but various bacteria, fungi, mycoplasma, and parasites are potential pathogens as well. Regarding the potential for virus infection, Caputo (89) indicates that there are no known incidents of laboratory-acquired infection among workers handling cell lines considered to be free from infectious virus. Still, the potential exists for continuous cell lines to carry latent viruses and for transformed lines spontaneously to produce viruses with oncogenic potential in man (90).

Assessment of risk requires knowledge of the history/status of the cell population. Unfortunately, not all cell lines have been tested to determine if they harbour an infectious agent. Thus, appropriate precautions must be taken to minimize the potential for direct contact with any 'untested' cell line brought into the laboratory. Likewise, it is important to know the pathogen status of animal tissues from which primary and subcultured populations are derived.

Several points are worthy of note:

(a) Risk is low when the work involves cells derived from pathogen-free animals or cell lines that have been determined to be free of adventitious agents.

(b) The potential for infection is increased when the work involves the use or the production of pathogenic agents.

(c) There have been reports of worker infection associated with the primary culture of cells from virus-infected animals (91).

11.2 Awareness of increased risk associated with human cells

Because human cells and bodily fluids can carry infectious agents such as, but not limited to, HIV and hepatitis viruses, cells of human origin must be considered a potential risk. It is important to remember that:

(a) HIV has been isolated from human cells and tissues, cell extracts, whole blood (and blood products), and body fluids including semen, vaginal secretions, cerebrospinal fluid, tears, breast milk, and urine. Other body secretions and fluids may also have the potential to harbour HIV (92).

(b) All work with human cells should be carried out under the assumption that the specimen may carry an infectious agent. There is also the potential risk that certain human or primate cell lines, if introduced to the body, may have oncogenic potential.

11.3 Classification of cell lines as aetiological agents

The American Type Culture Collection has classified cell lines on the basis of their potential to transmit pathogenic agents (7). This classification (*Table 4*) is consistent with the Biosafety Level classification of hazard assessment and laboratory safety procedures adopted by the US Public Health Service and the Centers for Disease Control (92).

Table 4 Classification of cultured cells as hazardous agents[a]

Class 1:	**low risk, do not present a recognized hazard**
	Sub-primate cell lines and primary cultures not contaminated with bacteria, mycoplasma, fungi, or viruses
	Normal primate cell lines that are primate virus-free
Class 2:	**carry the risk of contamination**
	Cell lines that harbour recognized pathogens in man
	All human tissues, cells, and fluids
	All primate tissue
	Primate cell lines that harbour virus
	Primate cell lines from lymphoid or tumour tissue
	Cell lines transformed by primate oncogenic virus
	Cell lines exposed to primate oncogenic virus
	Mycoplasma-containing cell lines
	Any animal cell line new to the laboratory, until proven to be free of adventitious agents

[a] Abstracted from ref. 89.

11.4 Precautions in handling pathogenic organisms and human cells

11.4.1 Education and instruction

- All workers in the cell culture laboratory must be instructed in proper technique for handling pathogens.

Two concise, methodologically oriented reports are particularly well focused on the issue of safety in the cell culture laboratory and provide valuable practical information important to anyone who handles cultured cells, especially workers who handle known pathogens and cells of human origin. These are entitled 'Biosafety procedures in cell culture' (89) and 'Guidelines to avoid personnel contamination by infective agents in research laboratories that use human tissues' (93).

11.4.2 Good laboratory procedure for work involving potential pathogens

Protocols in the cell culture laboratory should be conducted with the understanding that infection takes only one exposure and, depending on the pathogen involved, that one exposure might result in very serious illness, or even death. Where the potential exists that high risk pathogens may be present in the laboratory, the practice of **universal precaution** should be adopted, that is, all specimens should be handled as if they present a real risk of infection (92).

Some practical points of technique that promote safety and lower the risk of infection include the following:

(a) All work involving the handling of potential pathogens should be performed in a Class II or other microbiological safety cabinet appropriate for the organism involved (see Chapter 2, Section 8.3).

(b) Mouth pipetting is **never** permitted.

(c) Hands should be washed before and after handling cells.

(d) Gloves should be worn and should be replaced if torn or punctured.

(e) When handling cultures, workers must avoid touching unprotected body surfaces (e.g. eyes and mouth) with gloves or unwashed hands.

(f) Decontaminate all surfaces and equipment that come in contact with a pathogen.

(g) Laboratory coats or gowns should be worn and must be removed before leaving the laboratory.

(h) Use of sharps such as needles and scalpel blades should be avoided when working with infected cells and tissues. When sharps have to be used they must be disposed of properly in a leak-proof, rigid container.

(i) Pathogen-contaminated pipettes and instruments should be discarded into a stainless steel pan (with lid) containing distilled water, within the MSC. This container is to be covered when it is removed from the hood to be autoclaved.

(j) All reusable items must be autoclaved before they are cleaned for reuse. All disposable contaminated items must be autoclaved or incinerated. Careful consideration must be given to the transportation of materials from the cell culture laboratory to the autoclave or incinerator. In many cases autoclave bags may be inadequate to contain such materials, as the bags may be punctured by pipettes or other items. In such cases, the use of suitable rigid containers (e.g. disposable 'Biohazard' bins) may be more appropriate.

(k) Contaminated medium must be autoclaved.

(l) Cultures that may harbour pathogens must be clearly labelled. A separate incubator should be designated specifically for such cultures.

References

1. Schaeffer, W. I. (ed.) *Methods in cell science* (formerly *Journal of tissue culture methods*). Kluwer Academic Publishers, Dordrecht, The Netherlands. (This journal publishes protocols in cell culture, tissue and organ culture for application to biotechnology, cellular and molecular toxicology, cell biology, cellular pathology, developmental biology, growth-differentiation-senescence, genetics, immunology, infectious disease, neurobiology, plant biology and virology.)
2. Kruse, P. F. Jr. and Patterson, M. K. Jr. (ed.) (1973). *Tissue culture methods and applications.* Academic Press, New York.
3. Jakoby, W. B. and Pastan, I. H. (ed.) (1979). *Methods in enzymology*, Vol. 58. Academic Press, New York.
4. Freshney, R. I. (1994). *Culture of animal cells: a manual of basic technique* (3rd edn). Wiley-Liss, New York.
5. Freshney, R. I. (ed.) (1992). *Culture of epithelial cells, culture of specialized cells.* Wiley-Liss, New York.
6. Butler, M. and Dawson, M. (ed.) (1992). *Cell culture labfax.* BIOS Scientific Publishers, Oxford.
7. Hay, R. J., Caputo, J., and Macy, M. L. (ed.) (1992). *ATCC quality control methods for cell lines* (2nd edn). American Type Culture Collection, Rockville, MD.
8. Pollard, J. W. and Walker, J. M. (ed.) (1990). *Animal cell culture. Methods in molecular biology*, Vol. 5. Wiley, Chichester.
9. Crowe, R., Ozer, H., and Rifkin, D. (1978). *Experiments with normal and transformed cells: a laboratory manual for working with cells in culture.* Cold Spring Harbor Laboratory Press, Cold Spring Harbor, NY.
10. Butler, M. (ed.) (1991). *Mammalian cell biotechnology: a practical approach.* IRL Press, Oxford.
11. Watson, R. R. (ed.) (1992). *In vitro methods of toxicology.* CRC Press, Boca Raton, FL.
12. Barnes, D. W., Sirbasku, D. A., and Sato, G. H. (ed.) (1984). *Methods for preparation of media supplements, and substrata for serum-free animal cell culture. Cell culture methods for molecular and cell biology*, Vol. 1. Alan R. Liss, New York.
13. Spier, R. L. (ed.) (2000). *Encyclopedia of cell technology.* John Wiley & Sons, New York.
14. Celis, J. E. (ed.) (1998). *Cell biology: a laboratory manual* (2nd edn). Academic Press, London.
15. Spector, D. L., Goldman, R. D., and Leinwand, L. A. (1997). *Cells: a laboratory manual.* Cold Spring Harbor Laboratory Press, Plainview, NY, USA.
16. Harrison, M. A. and Rae, I. F. (1997). *General techniques of cell culture.* Cambridge University Press, Cambridge, UK.
17. Mather, J. P. and Barnes, D. (1998). *Methods in cell biology*, Vol. 57. *Animal cell culture methods.* Academic Press, London.

18. Pollard, J. W. and Walker, J. M. (ed.) (1997). *Methods in molecular biology*, Vol. 75. *Basic cell culture protocols.* The Humana Press Inc., Totowa, New Jersey, USA.
19. Lubiniecki, A. S. (ed.) (1990). *Large-scale mammalian cell culture technology. Bioprocess technology*, Vol. 10. Marcel Dekker, New York.
20. Schaeffer, W. I. (1990). *In Vitro Cell. Dev. Biol.*, **26**, 97.
21. McPherson, I. (1969). In *Fundamental technique in virology* (ed. I. McPherson, K. Habel, and N. P. Salzeman), p. 214. Academic Press, New York.
22. Milo, G. E., Casto, B. C., and Shuler, C. F. (ed.) (1992). *Transformation of human epithelial cells: molecular and oncogenic mechanisms*. CRC Press, Boca Raton, FL.
23. Freshney, R. I. (ed.) (1992). In *Culture of epithelial cells*, p. 1. Wiley-Liss, New York.
24. Perkins, F. M. and Handler, J. S. (1981). *Am. J. Physiol.*, **241**, C154.
25. Kase, K. and Hahn, G. M. (1975). *Nature*, **255**, 228.
26. Nelson-Rees, W. A. and Flandermeyer, R. R. (1976). *Science*, **191**, 96.
27. Nelson-Rees, W. A., Daniels, D. W., and Flandermeyer, R. R. (1981). *Science*, **212**, 446.
28. Davis, J. M. and Shade, K. L. (2000). In *Encyclopedia of cell technology* (ed. R. L. Spier), p. 108. John Wiley & Sons, New York.
29. Delly, J. G. (1988). *Photography through the microscope*. Eastman Kodak, Rochester, NY.
30. Ramsey, W. S., Hertl, W., Nowlan, E. D., and Binkowski, N. J. (1984). *In Vitro*, **20**, 802.
31. Dougherty, G. S., McAteer, J. A., and Evan, A. P. (1986). *J. Tissue Cult. Methods*, **10**, 239.
32. Steele, R. and Lane, H. (ed.) (1992). *J. Tissue Cult. Methods*, Vol. 14.
33. Ryan, J. A. (1989). *Am. Biotech. Lab.*, **7**, 8.
34. Lever, J. E. (1985). In *Tissue culture of epithelial cells* (ed. M. Taub), p. 3. Plenum Press, New York.
35. Reid, L. M. and Rojkind, M. (1979). In *Methods in enzymology* (ed. W. B. Jakoby and I. H. Pastan), Vol. 58, p. 263. Academic Press, New York.
36. Elliget, K. A. and Trump, B. F. (1991). *In Vitro Cell. Dev. Biol.*, **27A**, 739.
37. Stearns, M. E. and Wang, M. (1992). *Cancer Res.*, **52**, 3776.
38. Gospodarowicz, D. (1984). In *Methods for preparation of media, supplements, and substrata for serum-free animal cell culture* (ed. D. W. Barnes, D. A. Sirbasku, and G. H. Sato). *Cell culture methods for molecular and cell biology*, Vol. 1, p. 275. Alan R. Liss, New York.
39. Harris, A. K. Jr. (1984). *J. Biomech. Eng.*, **106**, 19.
40. Terracio, L., Miller, B. J., and Borg, T. K. (1988). *In Vitro*, **24**, 53.
41. Birnie, G. D. and Simmons, P. J. (1967). *Exp. Cell Res.*, **46**, 355.
42. Litwin, J. (1973). In *Tissue culture methods and applications* (ed. P. F. Kruse, Jr. and M. K. Patterson, Jr.), p. 383. Academic Press, New York.
43. Westermark, B. (1978). *Exp. Cell Res.*, **111**, 295.
44. Terracio, L. and Borg, T. K. (1986). *J. Mol. Cell. Cardiol.*, **18**, 329.
45. Douglas, W. H. J., McAteer, J. A., and Cavanagh, T. (1977). *Tissue Cult. Assn. Manual*, **4**, 749.
46. Leighton, J., Mark, R., and Justh, G. (1968). *Cancer Res.*, **28**, 286.
47. McAteer, J. A., Evan, A. P., and Gardner, K. D. (1987). *Anat. Rec.*, **217**, 229.
48. Nicosia, R. F. and Ottinetti, A. (1990). *In Vitro*, **26**, 119.
49. McAteer, J. A. and Cavanagh, T. J. (1982). *J. Tissue Cult. Methods*, **7**, 117.
50. Reuveny, S. (1990). In *Large scale mammalian cell culture technology* (ed. A. S. Lubiniecki). *Bioprocess Technology*, Vol. 10, p. 271. Marcel Dekker, New York.
51. Levine, D. W., Wang, D. I. C., and Thilly, W. G. (1979). *Biotech. Bioeng.*, **21**, 821.
52. Ryan, U. S. and White, L. (1986). *J. Tissue Cult. Methods*, **10**, 9.
53. Killackey, J. J. F. and Killackey, B. A. (1990). *Can. J. Physiol. Pharmacol.*, **68**, 836.
54. Schleicher, J. B. (1973). In *Tissue culture methods and applications* (ed. P. F. Kruse, Jr. and M. K. Patterson, Jr.), p. 333. Academic Press, New York.

55. House, W., Shearer, M., and Maroudas, N. G. (1972). *Exp. Cell Res.*, **71**, 293.
56. Gullino, P. M. and Knazek, R. A. (1979). In *Methods in enzymology* (ed. W. B. Jakoby and I. H. Pastan), Vol. 58, p. 178. Academic Press, New York.
57. Hanak, J. A. J. and Davis, J. M. (1995) In *Cell and tissue culture: laboratory procedures* (ed. A. Doyle, J. B. Griffiths, and D. G. Newell), Section 28D:3. John Wiley & Sons, Chichester, UK.
58. Davis, J. M. and Hanak, J. A. J. (1997) In *Methods in molecular biology*, Vol. 75. *Basic cell culture protocols*, (ed. J. W. Pollard and J. M. Walker), p. 77. The Humana Press Inc., Totowa, New Jersey, USA.
59. Moscona, A. A. (1961). *Exp. Cell Res.*, **22**, 455.
60. Bjerkvig, R. (ed.) (1992). *Spheroid culture in cancer research*. CRC Press, Boca Raton, FL.
61. Nitsch, L. and Wollman, S. H. (1980). *Proc. Natl. Acad. Sci. USA*, **77**, 472.
62. Folkman, J. and Moscona, A. (1978). *Nature*, **273**, 345.
63. Earle, W. R., Schilling, E. L., Bryant, J. C., and Evans, V. J. (1954). *J. Natl. Cancer Inst.*, **14**, 1159.
64. McLimans, W. F., Davis, E. V., Glover, F. L., and Rake, G. W. (1957). *J. Immunol.*, **79**, 428.
65. Birch, J. R. and Arathoon, R. (1990). In *Large-scale mammalian cell culture technology. Bioprocess technology* (ed. A. S. Lubiniecki), Vol. 10, p. 251. Marcel Dekker, New York.
66. Goosen, M. F. A. (ed.) (1992). *Fundaments of animal cell encapsulation and immobilization*. CRC Press, Boca Raton, FL.
67. Charles, K., Harland, R. C., Ching, D., and Opara, E. C. (2000). *Cell Transplant.*, **9**, 33.
68. Read, T.-A., Sorenson, D. R., Mahesparan, R., Enger, P. O., Timpl, R., Olsen, B. R., *et al.* (2001). *Nature Biotechnol.*, **19**, 29.
69. Joki, T., Machluf, M., Atala, A., Zhu, J., Seyfried, N. T., Dunn, I. F., *et al.* (2001). *Nature Biotechnol.*, **19**, 35.
70. Rheinwald, J. G. (1989). In *Cell growth and division: a practical approach* (ed. R. Baserga), p. 81. IRL Press, Oxford.
71. Hayflick, L. (1973). In *Tissue culture methods and applications* (ed. P. K. Kruse, Jr. and M. K. Patterson, Jr.), p. 220. Academic Press, New York.
72. Cristofalo, V. J. and Phillips, P. D. (1989). In *Cell growth and division: a practical approach* (ed. R. Baserga), p. 121. IRL Press, Oxford.
73. Philips, H. J. (1973). In *Tissue culture methods and applications* (ed. P. F. Kruse, Jr. and M. K. Patterson, Jr.), p. 406. Academic Press, New York.
74. Terracio, L. and Douglas, W. H. J. (1982). *J. Tissue Cult. Methods*, **7**, 5.
75. Terracio, L. and Douglas, W. H. J. (1982). *Prostate*, **3**, 183.
76. Ham, R. G. (1963). *Exp. Cell Res.*, **29**, 515.
77. Mosmann, J. (1983). *J. Immunol. Methods*, **65**, 55.
78. Landegren, U. (1984). *J. Immunol. Methods*, **67**, 379.
79. Oliver, M. H., Harrison, N. K., Bishop, J. E., Cole, D. J., and Laurent, G. J. (1989). *J. Cell Sci.*, **92**, 513.
80. Rieck, P., Peters, D., Hartman, C., and Courtois, Y. (1993). *J. Tissue Cult. Methods*, **15**, 37.
81. Baserga, R. (1989). In *Cell growth and division: a practical approach* (ed. R. Baserga), p. 1. IRL Press, Oxford.
82. Baserga, R. (1985). *The biology of cell reproduction*. Harvard University Press, Cambridge, MA.
83. Poot, M., Hoehn, H., Kubbies, M., Grossmann, A., Chen, Y., and Rabinovitch, P. S. (1990). In *Flow cytometry* (ed. Z. Darzynkiewicz and H. A. Crissman), p. 185. Academic Press, New York.
84. Angello, J. C. (1992). *Mech. Ageing Dev.*, **62**, 1.
85. Mazur, P. (1988). *Ann. NY Acad. Sci.*, **541**, 514.
86. Karow, A. M. Jr. (1969). *J. Pharm. Pharmacol.*, **21**, 209.

87. Tedder, R. S., Zuckerman, M. A., Goldstone, A. H., Hawkins, A. E., Fielding, A., Briggs, E. M., *et al.* (1995). *Lancet*, **346**, 137.
88. Masters, J. R. W., Twentyman, P., Arlett, C., Daley, R., Davis, J., Doyle, A., *et al.* (1999). *UKCCCR guidelines for the use of cell lines in cancer research*. United Kingdom Coordinating Committee on Cancer Research, London WC2A 3PX, UK.
89. Caputo, J. L. (1988). *J. Tissue Cult. Methods*, **11**, 223.
90. Weiss, R. A. (1978). *Natl. Cancer Inst. Monogr.*, **48**, 183.
91. Barkley, W. E. (1979). In *Methods in enzymology* (ed. W. B. Jakoby and I. H. Pastan), Vol. 58, p. 36. Academic Press, New York.
92. Richardson, J. H. and Barkley, W. E. (ed.) (1988). *Biosafety in microbiological and biomedical laboratories*. HHS Publication (NIH) 88-8395. US Government Printing Office, Washington DC.
93. Grizzle, W. E. and Polt, S. S. (1988). *J. Tissue Cult. Methods*, **11**, 191.
94. Stanulis-Praeger, B. M. (1987). *Mech. Ageing Dev.*, **38**, 1.

Chapter 6
Primary culture and the establishment of cell lines

Caroline MacDonald
University of Paisley, Paisley PA1 2BE, UK.

1 Introduction

A wide range of cultured animal cell lines has been described in the literature and is available from tissue culture collections such as the European Collection of Cell Cultures (ECACC) in the UK and the American Type Culture Collection (ATCC) in the USA (see Appendix 2). These lines are from different animal species, including humans, and increasingly insect cell lines have become available as interest grows in the baculovirus expression system for the production of proteins. Such lines, while providing a valuable resource and being appropriate for many uses, are not suitable for all purposes. In this situation the scientist may consider establishing his or her own primary culture and, if feeling particularly adventurous and optimistic, trying to isolate a new cell line. The aim of this chapter is to describe some of the approaches which are available and to indicate, wherever possible, how to go about deciding which method to use.

2 Establishment of primary cultures from various sources

2.1 Source of material

In theory, and in most cases in practice, primary cell cultures can be obtained from any source of tissue. However, the condition of the cells and their behaviour in culture will be affected by the starting material chosen. There are three main decisions to be made:

- whether to use normal or tumour-derived tissue
- whether to obtain the tissue from an adult or embryo
- which species to choose

You should also be aware of any legislation covering the work you propose. In the United Kingdom, if you are working with human tissue you are likely to need permission from the local ethical committee and you should ensure that any work with fetal material is permitted. Work with animals is covered by the Animals (Scientific Procedures) Act, 1986.

2.1.1 Normal 'versus' tumour

Many of the existing cell lines have been derived from tumour tissue, and this is particularly true of human cell lines where tumour tissue is often easier to obtain and easier to culture than normal material. The choice of tumour tissue may give important benefits if the cell type of interest is rare in the normal tissue but present in large quantities in an individual with a cancer of that cell type, e.g. a particular stage of lymphocyte differentiation. The advantage of normal tissue is, of course, that there is less concern about pathological changes present when the culture is established.

2.1.2 Adult 'versus' embryo

It is generally accepted that, in the case of tissue which is capable of some division in culture, the potential number of cell doublings *in vitro* correlates well with the age of the tissue at isolation (1). Thus one can expect the maximum number of cell divisions to be obtained from embryonic tissue. The disadvantage of embryonic material is that in some tissues, e.g. liver, there are substantial phenotypic differences between fetal and adult material. Also, in the case of human tissue, there are ethical objections to the use of fetal tissue for experimentation, and any spontaneously aborted fetus must be presumed to be abnormal.

2.1.3 Human 'versus' animal

For most scientists animal tissue is more readily available than human material. This is particularly true in the case of mouse and rat, and to a lesser extent for guinea pig and rabbit. Certain isolation techniques, e.g. organ perfusion, can be carried out on animals but not humans. The amount of material available may be larger or smaller in humans than animals depending on whether the whole organ (or a large part of it) can be taken, or whether one is restricted to a small biopsy sample. Finally, the biochemical characteristics of the tissue may be different in different species, and if one is ultimately interested in the behaviour of human cells then human primary cell culture will be more appropriate.

2.1.4 Source of material

The best procedure to follow to obtain human material is to approach the consultant surgeon at a local hospital and, if appropriate, obtain approval from the local ethical committee. If co-operation and approval are forthcoming then the next step is to liase with the theatre sister and whoever is performing the operation, since these are the people who will be providing the tissue samples. Surgical specimens are a convenient source of material and can be acquired under (virtually) aseptic conditions. Once the tissue has been removed, it should be placed in a sterile container of serum-free medium as soon as possible. It is important to provide a number of specimen jars, preferably with a wide neck to facilitate their use, and to make sure that the surgical team do not use jars of formalin or similar preservatives favoured by pathologists!

The removal of organs from small animals should be done after swabbing the outside of the skin with 70% isopropanol and using instruments which have been sterilized in alcohol and flamed. If possible the process should be carried out in a laminar flow cabinet; however, the use of one situated in a cell culture facility is not recommended because of the risk of contaminating cell lines with organisms associated with the animal. A vertical laminar flow cabinet should be suitable, but a Class II microbiological safety cabinet offers a higher level of protection for the operator. Once again the tissue should be placed into culture medium in a suitable sterile container. The media used both to collect and culture tissue samples should be supplemented with antibiotics (see Section 2.4.1).

2.2 Isolation of cells

Cells which are present in the circulation can easily be obtained from blood. There is no need to release individual cells – the problem is one of separating the cell type of interest from associated cells (*Protocol 1*). If, however, the cells of interest are organized into solid tissue, e.g. skin or liver, then it is necessary to release the cells and purify the cell type of interest. The choice of method for disaggregating the cells depends both on the nature of the tissue and the amount of material available. Best results are obtained with fresh tissue, but storage for short periods, e.g. less than 24 hours at 4 °C, is acceptable for some tissues. The tissue should be washed free of blood and the fat and damaged tissue removed prior to isolation of the cells.

Protocol 1

Separation of mononuclear cells from human peripheral blood

Reagents

- Histopaque 1077 (Sigma) or LSM (lymphocyte separation medium, ICN) or similar
- Human peripheral blood

Method

1. Collect blood in a heparinized syringe or obtain as buffy coat from the local Blood Transfusion Service.
2. Carefully layer 10 ml blood onto 10 ml of Histopaque 1077 or other lymphocyte separation medium in 50 ml polypropylene centrifuge tubes or universals (Sterilin). Histopaque should be pre-warmed to room temperature.[a]
3. Centrifuge at 400 g for exactly 30 min at room temperature.

Protocol 1 continued

4. After centrifugation use a Pasteur pipette to remove the upper layer to within 5 mm of the opaque interface containing the mononuclear cells. Discard the upper layer.
5. Carefully transfer the opaque interface to a clean (sterile) conical centrifuge tube using a Pasteur pipette.
6. Wash the cells three times with 30 ml phosphate-buffered saline (PBS) or medium without serum and centrifuge at 250–300 g for 5–10 min.
7. Wash the cells one more time, then resuspend in an appropriate culture medium, e.g. RPMI 1640, with 10% FCS.

Note:

This method works very effectively for human cells, but due to differences in the density of cells it cannot always be used for other animal species. For example, peripheral blood mononuclear cells (PBMC) from macaques are more readily separated on Percoll gradients (density 1.080 g/ml) (2), whereas PBMC from mice can be separated on metrizamide gradients (3).

[a] It is also possible to dilute the blood in twice its volume of RPMI 1640 medium or Hank's balanced salt solution, and layer the diluted blood on the separation medium in the ratio of four volumes of diluted blood to one volume of separation medium.

2.2.1 Enzyme digestion

Digestion by incubation of tissue in proteolytic enzymes is effective and can be varied to suit the tissue. Trypsin is probably the most commonly used enzyme, but for fibrous tissue collagenase is often more effective. Other enzymes which are used include elastase, hyaluronidase, pronase, dispase, or combinations of these. Digestion can be carried out slowly, at 4 °C over a long period, or rapidly, at 37 °C for a short period (*Protocol 2*), or a combination of the two. Cell damage is likely to be less at low temperature, resulting in higher yields. Collagenase and dispase digest tissue less aggressively than trypsin and pronase and consequently cause less damage to the cells. It has been reported (4) that if mouse embryonic tissue is digested at low temperatures a wider variety of different cell types can be isolated than if warmer temperatures are used.

Protocol 2

Enzyme digestion

Equipment and reagents

- Erlenmeyer flask (250 ml capacity)
- Trypsin (Difco crude 1:250)
- Bovine pancreatic DNase (Sigma) (optional)

Protocol 2 continued

Method

1. Chop a minimum of 1 g of tissue into pieces about 3 mm in diameter using sterile scalpels in a sterile Petri dish.
2. Transfer the chopped pieces to an Erlenmeyer flask[a] and add sufficient 0.25% trypsin to cover completely. If 10 g of tissue is used, a 250 ml flask plus 100 ml trypsin should be suitable.
3. Stir the pieces with a sterilized magnetic follower at 37 °C for 15–30 min at about 200 r.p.m.
4. Leave to settle then remove the supernatant and centrifuge at 500 g for 5 min. Resuspend the pelleted cells in 10 ml of medium containing serum and store on ice.[b]
5. Add fresh trypsin to the tissue pieces and stir for another 15–30 min.
6. Repeat the process until the tissue is completely disaggregated or sufficient cells have been obtained.
7. Plate out the cells in medium containing serum.

[a] If necessary the tissue pieces can be washed to remove cell debris and blood cells by incubating in serum-free medium or PBS for 5–10 min at 37 °C in a shaking water-bath.

[b] If the supernatant is difficult to remove because of the formation of a viscous gel, add a few drops of 4 mg/ml bovine pancreatic DNase solution and incubate at 37 °C for a further 2–3 min.

2.2.2 Perfusion

Perfusion is a specific type of enzymatic digestion which involves pumping a proteolytic solution through the tissue and collecting the cells in the enzyme solution as they are stripped off. It is commonly used for isolating cells from organs such as liver. Alternatively, tissue such as a vein may be clamped at either end and digested with a solution which is left in contact with the inner surface for some time, as described in *Protocol 3* (see also Chapter 7, *Protocol 4*).

Protocol 3

The isolation of human umbilical cord vein cells by collagenase perfusion

Equipment and reagents

- Venisystems butterfly 21 needle (Abbot Laboratories)
- Surgical clamps
- Fungizone (Life Technologies)
- Worthington Type 1 collagenase (Cambridge Bioscience)

Method

1. Collect the cord in sterile PBS containing 2.5 μg/ml Fungizone and store for up to 24 h at 4 °C until required.

Protocol 3 continued

2 Remove regions of damaged tissue, for example where it has been clamped during delivery, and rinse the ends of the cord with 70% alcohol.
3 Insert a butterfly needle into the vein and secure with a surgical clamp.
4 Clamp the other end of the cord and add 50–100 ml serum-free Dulbecco's modified Eagle's medium (DMEM) or sterile PBS until the vessel is slightly distended, in order to clear excess blood.
5 Remove the wash medium, add 5 ml of 0.2% Worthington Type 1 collagenase, and incubate the cord in PBS pre-warmed to 37 °C for 10 min to digest selectively the single layer of endothelial cells.
6 Release the lower clamp and collect the collagenase solution containing the cells.
7 Rinse the cord with 10 ml DMEM plus 20% FCS and add to the collagenase solution.
8 Pellet the cells by centrifugation at 500 g for 5 min.
9 Discard the supernatant, resuspend the cells in 5 ml DMEM containing 20% FCS, 100 IU/ml penicillin/100 μg/ml streptomycin, and 100 μg/ml gentamicin.

2.2.3 Mechanical disaggregation

Cells can be released from some tissues by mechanical means, for example in the case of spleen cells (*Protocol* 4) simply by squeezing the cells through a wire mesh, or for soft tissue such as brain by forcing the tissue through a series of sieves of gradually reducing mesh size (*Protocol* 5). Syringe needles can also be used. Mechanical disaggregation is faster than enzymatic digestion but may result in more damage to the cells and consequently lower recovery rates. It has the advantage, however, of not requiring large quantities of expensive enzymes such as collagenase and not suffering the problems of variation in the activity of enzymes from different batches.

Protocol 4

Preparation of murine spleen cells

Equipment

- Sterile scissors, forceps, and wire mesh

Method

1 Kill mice by cervical dislocation and immerse in 70% ethanol. Using sterile scissors and forceps, lift and cut the skin on the left side of the abdomen and pull back the skin. Spray the exposed peritoneum with 70% ethanol and, using fresh sterile scissors and forceps, make an incision over the spleen. Remove spleen with forceps, carefully cutting away the connective and vascular tissue. Transfer spleen to sterile PBS or balanced salt solution (BSS), supplemented with 5% FCS.

Protocol 4 continued

2 Transfer spleen and PBS/BSS on to a sterile wire mesh (typically 5 cm × 3 cm) in a Petri dish and, using the plunger of a 5 ml syringe, grind spleen until a fine suspension is obtained.

3 Transfer spleen cell suspension to 15 ml centrifuge tubes and allow debris to sediment at 1 g, or centrifuge at 100 g for 30 sec.

4 Pipette off the cells in the supernatant and add to a fresh 15 ml tube.

5 Centrifuge at 300 g for 5 min and resuspend the cell pellet in RPMI 1640 + 10% FCS (or other suitable medium).

6 Perform cell count (Chapter 5, *Protocol* 4).

Protocol 5

Mechanical disaggregation of soft tissue

Equipment

- Sterile stainless steel sieves (mesh sizes 1 mm, 100 μm, 20 μm)

Method

1 Chop the tissue into pieces of about 5 mm in diameter and place into a sterile stainless steel sieve with a mesh size of around 1 mm.

2 Place the sieve in a 90 mm diameter Petri dish containing medium and use the piston of a disposable plastic syringe to force the tissue through the mesh into the medium.

3 Add more medium to the sieve to wash the cells through.

4 Pipette the partially disaggregated tissue into a sieve with a smaller mesh, e.g. 100 μm and repeat steps 2 and 3.

5 Repeat with a 20 μm mesh sieve if a single cell suspension is required, then plate out cells.

2.2.4 Explant cultures

Explant cultures are usually established from punch biopsies containing a plug of material from a tissue such as skin. The procedure (described in *Protocol* 6) involves fine chopping of the tissues, usually with sterile scalpels, then placing in a tissue culture grade Petri dish with a small amount of medium for a few hours to allow attachment of the small pieces of tissue to the substrate. Once the pieces of the biopsy have attached, more medium is added and, if all goes well, cells will grow out from the piece of tissue and gradually spread out over the surface of the dish to form a monolayer. This is the method of choice if only very small amounts of material are available, but it is the slowest of the methods described here.

Protocol 6

Explant culture

Equipment

- Sterile scalpels

Method

1. Rinse and trim the tissue, then chop it finely into pieces of about 1 mm in diameter using sterile scalpels in a sterile Petri dish.
2. Wash the pieces in a universal containing medium without serum.
3. Seed 25–30 pieces on to a 60 mm diameter culture dish in a very small quantity of medium (< 1 ml) containing 50% serum.
4. Tilt the dish to spread the pieces evenly, then leave the tissue in a humidified incubator at 37 °C to attach to the dish.
5. Once the pieces have attached (probably after overnight incubation) gradually increase the volume of the culture medium to the customary 5 ml.
6. Change the medium regularly (every five to seven days) until cells can be seen growing out from the tissue pieces.
7. If desired, pick the explanted tissue pieces away from the outgrowing cells and re-attach to a fresh flask.
8. When the outgrowing cells cover about 50% of the surface of the culture vessel, trypsinize, dilute, and re-seed them in a fresh flask.

2.3 Substrate for attachment

Although some cells, notably those derived from blood, are able to grow in suspension, most cells require a surface for attachment. The first substrate used routinely was glass because it was cheap and easily washed and the cells could be visualized through it. Most laboratories use disposable plastic material now, in order to reduce labour costs and problems associated with breakages and incomplete cleaning. Tissue culture grade plasticware is made from polystyrene which has been specially treated to provide a hydrophilic surface (see Chapter 5, Section 5.1.2) and is supplied after treatment to sterilize it (usually γ-irradiation or ethylene oxide treatment).

A number of specialized substrates have been developed. These may be for specific applications, e.g. microcarrier beads or fibres for large-scale culture of adherent cells, or for specific cell types, e.g. gelatin, collagen, laminin, chondronectin, or fibronectin (see Chapter 5, Section 5.4), to promote the growth and/or differentiation of cells such as mammary epithelial cells, muscle cells, nerve cells, and hepatocytes (5–8). In the past, preparation of coated flasks had to be carried out by the individual scientists, but they are now available commercially, e.g. the Primaria range from Falcon which has been specially developed for primary cell culture. Good results have been claimed for a variety of different types of cells including hepatocytes (9).

2.4 Culture conditions

The whole process of setting up a culture involves selection of a subpopulation of cells. The choice of culture conditions will, however, decide how rigorous this selection process is. Inevitably the conditions will be 'abnormal' in that they will vary considerably from the *in vivo* situation from which the cells were removed. Some cells will be lost during isolation, dissociation, and purification, others as the culture is established - they will fail to attach or to divide in the synthetic medium - and others will be lost because they are outgrown by more robust cells.

One of the major criticisms of work performed on cultured cells is that we cannot be sure to what extent they represent the *in vivo* situation, and this limitation should always be borne in mind.

2.4.1 Type of medium

Numerous recipes have been developed to support the growth of different types of cells. Some media, e.g. Eagle's media, were designed for high density cell culture, while others, such as Ham's F12, are rich media which were originally developed for growing cells at low density. In general, many adherent lines can be grown in either one of the modifications of Eagle's basal medium [DMEM or Glasgow modified Eagle's medium (GMEM)], or Ham's F12, or a mixture of DMEM and F12. Some recipes have been developed specifically for particular types of cells, e.g. Williams' E for hepatocytes (10) and MCDB 152 for keratinocytes (11). Most suspension cells, e.g. lymphoid cells, myeloma, and hybridoma cell lines, can be grown in RPMI 1640, Iscove's modified Dulbecco's medium (IMDM), or DMEM/F12.

In general, the routine use of antibiotics in cell culture should be avoided if possible. The presence of antibiotics in medium can mask a low-level infection and encourage poor aseptic technique. However, the tissue used as a starting material in primary culture is often not sterile and the addition of antibiotics and antimycotics may be essential in order to establish the culture. Surgical tissue from internal organs, or tissue taken aseptically from an animal should not cause too many problems although it is wise to add penicillin (100 IU/ml) and streptomycin (100 μg/ml) for the first few days. Some tissue is very likely to contain micro-organisms, for example gut, tonsil, or nasal tissue, and the addition of Fungizone (Life Technologies) and gentamicin is recommended. These are normally used at 1–2.5 μg/ml and 50 μg/ml respectively, but higher concentrations may be required and can be used provided the cells can survive.

2.4.2 Selection against some cell types

When cells are isolated from tissue, even if a purification step is included in the procedure used, a mixture of cell types is usually obtained. The mixture will contain cells which have different growth rates, and the fastest growing can take over the culture quite rapidly. One such cell type is the stromal fibroblast which is found in connective tissue and is frequently present in a primary culture. It is difficult (sometimes impossible) to rid a culture of fibroblasts, but the addition of compounds such as hydroxyproline or phenobarbitone has been reported to suppress fibroblast growth (12, 13). The use of D-valine as an alternative to L-valine selects for epithelial cells which possess D-amino acid oxidase (14). Serum-free media are also sometimes used to select against one cell type in favour of another (see Chapter 7).

Another approach to select for or against a particular cell type is to exploit differences in the adhesion or detachment of cells from a substrate. If a suspension containing a mixture of cells is seeded into a flask and left for 30 minutes, then transferred to a second flask with subsequent transfers every hour, the cells which adhere most rapidly will be found in the first flask and the slowest to adhere in the last. In general, fibroblasts adhere more quickly to tissue culture flasks than epithelial cells. Fibroblasts can also be detached by a briefer trypsin treatment than is required for epithelial cells.

2.4.3 Conditioned medium

Conditioned medium is medium which has already been used to support the growth of cells. Although the original cells will have depleted it of components such as glucose and glutamine, it will contain growth factors and hormones secreted by these cells. It can, therefore, be used to stimulate the growth of some cells. Conditioned medium is generally prepared by removing medium which has been in contact with cells in logarithmic growth for about three days and centrifuging it to remove cells, followed by filtration through a 0.22 μm filter to remove micro-organisms and cells. The medium is stored at −20 °C and used as about 10–25% of the new medium.

2.4.4 Feeder cells

Feeder cells are often used to condition the culture medium and stimulate cell growth, in particular at low (e.g. cloning) density. Cells, such as irradiated or mitomycin C-treated fibroblasts, peripheral blood lymphocytes, or peritoneal exudate cells, provide an undefined mixture of substances which promote cell growth, and may be the only way of obtaining growth at low cell density. Peritoneal or peripheral blood cells can be used directly since the cells do not divide in culture, but fibroblast feeder cells must be pre-treated to prevent cell division. *Protocol 7* describes the procedure for preparing mitomycin C-treated mouse fibroblasts.

Protocol 7

Preparation of mouse fibroblast feeders

Reagents

- Mitomycin C (Sigma)
- Mouse fibroblast cells

Protocol 7 continued

Method

1. Seed out mouse fibroblast cells at just below confluence, e.g. 2×10^4 cells/cm^2.
2. Add mitomycin C at 10 μg/ml for 2 h.
3. Remove, wash cells twice with medium, and replace with fresh medium.[a]
4. Seed primary cells on top of the feeder layer.
5. Retain one dish of feeder cells and observe for a period of four weeks to ensure that there is no outgrowth of colonies from the feeder cells.

[a] Treat waste medium and all washings with care. Mitomycin C is a suspected mutagen and should only be used with protective clothing and extreme care.

3 Evolution of primary cultures

3.1 Nomenclature

The term **primary culture** should only refer to the original cell culture prior to passage or subculture. Primary cultures which have been passaged give rise to a **secondary culture** and so on. This terminology, however, is cumbersome and not frequently used, but it is important to define the terms used and to agree on a common understanding of what is meant. It is particularly important to distinguish between 'normal' cells with a limited lifespan in culture, and those continuous, established cell lines which can be subcultured indefinitely.

3.2 Subculture

All cell cultures reach a stage where cell division ceases. This happens with suspension cells when they reach a certain density (which varies according to cell type, medium, and culture conditions, but is usually between 10^6 and 10^7 cells/ml of culture medium). Most adherent cell lines divide until the surface they are growing on is covered, a stage known as confluence, and then stop. Some cells, often tumour cells, do not suffer from contact inhibition or density-dependent growth inhibition and continue to grow to produce colonies or foci on top of the cell monolayer. It is good practice, however, to maintain all cultures in a state of logarithmic growth and to subculture when a certain density is reached. Maximum growth is achieved if the cells are subcultured ('subbed', 'split', or 'passaged') just before stationary phase or confluence is reached. This process involves diluting the cells, resulting in either more culture vessels containing all of the cells, or the continuation of growth of only a proportion of cells in the same size vessel. It is standard practice to maintain a stock of cells in an appropriately sized flask and to use the remaining cells for experimentation.

Suspension cells can be subcultured simply by dilution (after either counting to give a fixed number of cells, or dilution, e.g. 1:10) and transfer of a proportion of cells into a new vessel with fresh medium. Adherent cells must first be removed from the substrate before they are diluted (see Chapter 5, Section 5).

Because 'normal' cells have only a limited lifespan in culture, approximately 50 generations for normal human fibroblasts (15), it is important to keep accurate records of the number of subcultures (passages) which the cell line has undergone. If the dilution ratio (or split ratio) is 1:2 then the passage number corresponds to the generation number. However, if a greater ratio is used then the approximate generation number must be calculated (see Chapter 5, Section 8.1.1). Most cells are capable of between 10 and 75 generations before reaching senescence.

3.3 Growth phases

Cell growth typically follows the pattern consisting of an initial lag phase, a period of logarithmic growth, and then a stationary phase (see Chapter 5, *Figure 2*). This pattern repeats every time the cells are subcultured, until, in the case of cells with a limited lifespan, senescence is reached. At senescence the lag phase becomes longer with each subculture, and the cells slow down and ultimately stop dividing completely. Some cells can be rescued from this crisis, and a small proportion of the cells in the population will eventually outgrow and form a permanent cell line. This process occurs occasionally with rodent cells, but is not known to occur spontaneously with human cells. However, treatment of cells with oncogenes (Section 8) can enhance the process. Treatment with chemical carcinogens and infection with tumour viruses are alternative methods, and were used to isolate many of the original transformed cell lines.

4 Characteristics of limited lifespan cultures

4.1 Lifespan and senescence

Limited lifespan cultures, as the name indicates, are capable of growth for a certain period only. The length of this period depends on the type of cell and the age of the donor, but also varies for reasons which are not clear. Eventually, however, the cells enter senescence and the culture is lost. There has been considerable discussion about what happens during senescence, and Holliday has proposed a 'commitment theory' of fibroblast ageing (16). He speculates that there is a subpopulation of cells with the potential to undergo an unlimited number of doublings. For any given population there is a probability (P) that such cells will become committed to finite growth. The rate of division remains the same for both the committed and the uncommitted cells but after a number of cell divisions (M) all of the descendants of each of the committed cells die. The lifespan of a particular culture is then dictated by its values of P and M and these are determined by many factors including the conditions of culture and the tissue of origin. Provided that P and M are large, the cultures will eventually die because the number of uncommitted cells will decrease with time and eventually will become so low that the cells will be diluted out when subcultured. However, the exact time of loss of the last uncommitted cell will vary between populations, thus accounting for differences which have been observed in the lifespan of different cultures from the same line. Permanent cell lines may contain cells with a limited replicative potential, but there are sufficient numbers of uncommitted cells to divide and increase at each subculture. Thus the difference between mortal and immortal populations of cells may be quantitative rather than qualitative.

4.2 Phenotype

The advantage of limited lifespan cell lines over continuous lines is the more 'normal' phenotype of the former. For example, transformed cells often have altered growth characteristics, become tumorigenic, secrete increased levels of proteases such as plasminogen activator, and have altered cell surface markers, e.g. decreased levels of fibronectin (17). Transformed cells also stop expressing many differentiated, or tissue-specific, enzymes – most liver cell lines have stopped expressing the drug-metabolizing enzymes which function to detoxify xenobiotic compounds (18).

4.3 Karyotype

In addition to the difference in phenotype, limited lifespan cell lines are karyotypically different from permanent cell lines. The latter have chromosomal abnormalities, and there is considerable heterogeneity within a culture. Limited lifespan cultures, in contrast, have the chromosome complement of the donor, and this had led to the use of the name diploid cell lines to describe such cells. Lines from individuals with cytogenetic abnormalities have been used extensively for mapping genes not merely to individual chromosomes but to specific regions of chromosomes (19).

Protocol 8 describes how to prepare a chromosome preparation for karyotyping and *Protocol 9* the procedure for Giemsa staining.

Protocol 8

Karyotyping

Equipment and reagents

- Acid washed slides
- Colcemid solution (Life Technologies) at 100 μg/ml
- 0.075 M KCl
- Freshly prepared methanol/acetic acid (3:1)
- Giemsa stain (Merck)

Method

1 Set up cultures at 25% confluence in a 75 cm^2 flask in 20 ml of culture medium.

2 After 24 h of growth add Colcemid solution for 1–2 h. For a cell line with a doubling time of 20 h use 10 μl of a 100 μg/ml stock. If the cells grow more slowly increase the exposure time.

3 Tap cells to release the loosely attached mitotic cells and pour off old medium into a

Protocol 8 continued

universal (Sterilin). It is important to retain these cells and the spent medium in order to maximize the yield of cells at metaphase.

4. Trypsinize the monolayer and use old medium to stop the trypsin.
5. Centrifuge to pellet the cells and resuspend pellet in 10 ml of 0.075 M KCl to swell for 10–12 min.
6. Centrifuge cells at 500 g for 2–3 min (including this in the 10–12 min swelling time). Remove all but 0.25–0.5 ml of the KCl by aspiration and add 5 ml freshly prepared methanol/acetic acid (3:1) dropwise with agitation to fix the cells. Add a further 5 ml. It is important that the cells do not clump at this stage.
7. Fix for a minimum period of 30 min at 4 °C.
8. Centrifuge and resuspend pellet in fixative, taking care not to lose the pellet which is rather soft.
9. Centrifuge again and resuspend in fixative to give a pale milky solution.
10. Drop cells on to wet, acid washed slides which have been rinsing in running water for at least 30 min. Hold the slide at an angle of 45° and drop four to six drops on to each slide from a distance of 50–70 cm above the slide.
11. Leave flat to dry.
12. The slides can be stained with Giemsa to visualize the chromosomes, banded by digestion with trypsin and then stained with Giemsa to identify individual chromosomes, or hybridized *in situ* with a fluorescently labelled probe.

Protocol 9

Giemsa staining

Equipment and reagents

- Staining jar
- Coverslips
- Giemsa stain (Merck)
- Chromosome spreads on microscope slides (*Protocol 8*)
- DPX mountant (Merck)

Method

1. Immerse slides in Giemsa stain in a staining jar for 2 min.
2. Place jar in sink and add a five- to tenfold excess of cold tap-water to the jar allowing the surplus to overflow from the jar.
3. Leave for a further 2 min.
4. Wash the slides in cold running water to remove excess stain, rinse in distilled or deionized water, and leave to dry.
5. Mount coverslips over the slide using DPX.

Fluorescence *in situ* hybridization (FISH) involves the hybridization of appropriate DNA probes with the DNA in the host cell chromosomes (20). The probe is labelled non-isotopically by incorporating nucleotides which have been modified with molecules such as biotin or digoxigenin, which are then detected with a fluorochrome-labelled antibody against the biotin or digoxigenin. The location and extent of amplification of both recombinant and non-recombinant genes can be detected in metaphase spreads even when the sequences are only present in a single copy.

Chromosome painting (or spectral karyotyping, SKY) involves the use of a collection of probes which are chromosome-specific. These paints are available commercially from a number of suppliers, e.g. Biovation, Cambio, and Oncor, and are produced such that each chromosome can be labelled with a different colour. This allows complex translocation events to be detected and analysed even when only small fragments of a chromosome are rearranged.

5 Establishment of continuous cell lines

5.1 Spontaneous

A large number of established cell lines are listed in the catalogues of the culture collections. Many of these were isolated decades ago, and it is not always clear what events led to their establishment. Some cell lines have been obtained in the absence of any known exposure to a transforming agent. Such lines include the RPM promegakaryocyte line derived from rat bone marrow which retains several differentiated functions including the ability to synthesize factor VIII:antigen and fibrinogen (21). The pre-adipocyte line 3T3-L1 is derived from mouse embryo and is capable of lipid accumulation (22). However, in general it is unusual to isolate cell lines spontaneously, and those that have been obtained are predominantly from fetal rodent tissue.

5.2 Chemical transformation

Methylcholanthrene has been widely used to establish a variety of cell lines including the mouse L-cell line and the rat muscle line L6 (23). Another carcinogen, azoxymethane, has been used to transform normal human colon mucosal cells to give malignant lines with altered morphology, culture longevity, growth in soft agar, substrate adherence, and peanut agglutinin binding (24). Human mammary epithelial cells exposed to benzo[α]pyrene develop an extended lifespan and apparently immortal cell lines can be isolated (25). These lines do not appear to be malignantly transformed as they do not form tumours in nude mice and they show little or no anchorage-independent growth. However, they resemble tumour-derived mammary epithelial cells more closely than their normal progenitors. The mutagen EMS (ethyl methanesulfonate) has also been used in conjunction with simian virus 40 (SV40) infection to isolate the β cell line HIT from Syrian hamster pancreatic islets (26).

5.3 Viral transformation

The provision of protocols for the propagation, titration, and infection methods for different viruses is beyond the scope of this chapter and is described elsewhere (4). However, some examples of cell lines isolated as a result of viral infection are given below.

5.3.1 Simian virus 40

The monkey virus SV40 has been used to isolate transformed mouse and human fibroblast lines. In the case of the human cells, however, these lines are not truly immortal, but merely display a delay in the onset of senescence (27). Temperature-sensitive mutants with a defective large T gene have been used in order to isolate cells which can grow at the permissive temperature when T is expressed, but at the high (non-permissive) temperature stop growing and express differentiated characteristics. This approach has led to the isolation of lines such as the TPA30-1 human placental line (28) and the RLA fetal rat hepatocyte line (29).

5.3.2 Epstein–Barr virus

The Epstein–Barr virus (EBV) has been widely used to isolate transformed cell lines from human B lymphocytes. Such lines have important applications in the study of genetic disease in individuals with chromosome translocations or inherited disorders and in the production of human monoclonal antibodies (30). Recent work suggests that some lymphoblastoid cell lines generated in this way are not, in fact, immortal as previously believed; however, they appear to have a greatly extended precrisis lifespan of up to 160 population doublings (31).

5.3.3 Other viruses

Cell lines have been isolated following infection with other viruses, e.g. murine lymphoid cell lines with the Abelson murine leukaemia virus (A-MuLV) (32). Infection with the Rous sarcoma virus (33) was used to isolate a rat cerebellar cell line, WC5, which expresses glial fibrillary acidic protein (GFAP). A bipotential haematopoietic cell line has been isolated using the Friend leukaemia virus (34), and continuous acute promyelocytic leukaemia cell lines by infection with the Friend and Abelson murine leukaemia viruses (35). Polyoma virus will transform hamster cells and has been used to isolate the Syrian hamster line PyY (36).

6 Properties of continuous cell lines

6.1 Aneuploidy

Continuous cell lines are usually aneuploid, and often have a chromosome number between the diploid ($2n$) and tetraploid ($4n$) values. The cells within the line are also heteroploid, that is there is considerable variation in both the number of chromosomes and the specific chromosomes present among the cells in the population.

6.2 Heterogeneity and instability

Continuous cell lines show phenotypic heterogeneity as well as genotypic variation. Even after cloning, the cells quickly regain this heterogeneity, and this is readily observed by chromosome analysis. Many lines exhibit morphological variation, and again reappearance of this variation can be observed soon after cloning (see Chapter 8, Section 1.3).

6.3 Differentiated status

Most established cell lines have stopped expressing tissue-specific genes, retaining only the so-called housekeeping enzymes required for continued growth in culture. As a result, most established lines are phenotypically more like each other than like the original tissue of origin. Examples of the properties commonly lost include synthesis of serum proteins and clotting factors in liver cell lines (37) and production of milk-specific products and enzymes in mammary epithelial cell lines (38).

6.4 Tumorigenicity

Some, though by no means all, continuous cell lines are tumorigenic *in vivo*. The test of malignancy is whether or not the cell line is capable of forming transplantable tumours when 10^7 cells are injected into a genetically identical (isogeneic) animal. Tumorigenicity in human cells is detected by injection into an immunodeficient animal, e.g. the athymic 'nude' mouse (39), but this method is less satisfactory due to the difficulties associated with handling such animals.

7 Cell fusion

7.1 Methods

Cell fusion can occur spontaneously in culture at low levels but its incidence can be increased by treatment of cells with a fusogenic agent. Hybrid cells can be isolated from parental cells as varied in origin as plant and animal cells, and between different species of animal cells.

7.1.1 Sendai virus

The first cell fusion experiments were performed with Sendai virus which had been inactivated either by ultraviolet (UV) irradiation or treatment with β-propiolactone. This technique was rapidly superseded by polyethylene glycol-mediated fusion, largely due to difficulties in obtaining supplies of Sendai virus.

7.1.2 Polyethylene glycol

Polyethylene glycol (PEG) is still widely used to fuse cells because of its ease of use and reliability. PEG is an efficient fusing agent, but is toxic to cells. The cells are normally only exposed to PEG briefly and the concentration of the solution is critical. Most laboratories use it at a concentration near to 50% (w/v), but there is considerable variation in the molecular weight used (from 6000 to 1000 Da). The lower molecular weight forms are more efficient fusogens, but are also more toxic to the cells. Fusion is normally carried out in suspension (*Protocol 10*) but can be done in monolayer if the parental cells are seeded together.

Protocol 10

Polyethylene glycol fusion

Equipment and reagents

- Glass universal container
- Polyethylene glycol, mol. wt. 1500

Method

1. Prepare 46% PEG solution by weighing 2.3 g PEG 1500 in a glass universal which has been marked at the 5 ml level.
2. Autoclave PEG and while still molten (at about 75 °C) add RPMI, or other medium (without serum) which has been warmed to 37 °C, to the 5 ml mark.
3. Prepare suspensions of cells to be fused and mix together 10^6 of each cell type. Retain some of each cell type for unfused control cultures.
4. Centrifuge cells at 500 g for 5 min and resuspend in 10 ml of medium without serum to wash.
5. Recentrifuge at 500 g for a further 5 min, discard the supernatant, and tap the pellet to resuspend the cells in the very small volume of medium left in the tube (less than 100 μl).
6. Add 1 ml of PEG solution to the cell pellet, slowly over 3–5 min, running it down the sides of the tube and tapping gently.
7. Add 1 ml of serum-free medium gradually over a period of 2 min. Top up with 15 ml of medium (without serum) and leave for 5 min.
8. Centrifuge for 5 min, discard medium, and resuspend the cells in serum-containing medium. Plate out 10^5 cells/5 cm dish and incubate at 37 °C. As controls, plate out two dishes of each parental cell line at the same density.
9. On the next day add selective medium (see Section 7.2.1).

7.1.3 Electrofusion

Cells can also be fused as a result of the application of an electric field to a mixture of cells. As with PEG fusion this is normally done in suspension, but successful electrofusion in monolayer has also been reported by the application of repetitive square-wave pulses of a few microseconds duration (40). Electrofusion in suspension (*Protocol 11*) involves cell alignment in response to an alternating field followed by fusion as a result of the application of a single field pulse of high intensity (41).

Protocol 11

Electrofusion

Equipment and reagents

- Electrofusion equipment, e.g. Kruss CFA 400 apparatus
- 0.28 M inositol solution
- Phenol red-free culture medium

Method

1. Mix together 10^6 of each of the two cell types to be fused.
2. Centrifuge and resuspend in an iso-osmotic, non-conducting medium such as 0.28 M inositol solution.
3. Add cells to a fusion chamber, such as the Kruss polysulfone pipetting chamber, in a volume of 0.25–0.5 ml.
4. Determine pulse conditions by trial and error using the minimum AC field voltage required to produce observable cell alignment within 30 sec, i.e. when the cells can be seen to come together to form a line. Suggested conditions on the Kruss CFA 400 electrofusion apparatus are:
   - AC field: 800 Khz, 250 V
   - gate: 20 msec
   - number of pulses: 3
   - pulse duration: 10 μsec
   - pulse interval: 2.9 sec
   - pulse voltage: 3000 V/cm
   - ramp: 60 sec

   However, these parameters will vary with the type of cell depending on its fragility, and conditions should be optimized for each cell type.
5. Remove cells from the chamber and plate out into culture medium lacking phenol red. Cells remain porous for several hours after pulsing and are consequently sensitive to the toxic effects of chemicals such as phenol red which are normally excluded by the cell membrane.

7.2 Properties of hybrids

The first outcome of a fusion event is the formation of a binucleate cell as a result of membrane fusion. If two different types of cells have fused, the binucleate cell is termed a heterokaryon. After around 24 h the nuclei fuse to form a hybrid cell. Nuclear fusion is a rare event, and it is necessary to use selection techniques to identify the hybrid cells. This usually involves the use of genetically marked cells as either one or both parental cells.

7.2.1 Selection of hybrids

Cell fusion occurs at a low frequency, 10^{-3} or less, which means that a procedure for selecting hybrid cells from unfused parental cells and like-with-like parental fusions is necessary. A number of different strategies for selecting hybrid cells have been adopted.

i. Complementation of conditional lethal mutants

Cell lines exist (or can be selected, e.g. as in *Protocol 12*) which contain mutations in non-essential biochemical pathways such as the purine and pyrimidine salvage pathways. Normal cells synthesize purines and pyrimidines by endogenous pathways supplemented by these salvage pathways, so that when the endogenous pathway is blocked by a folic acid analogue such as aminopterin, the cells can survive by utilizing hypoxanthine and thymidine via the salvage pathway. Cells which are deficient in enzymes involved in the salvage pathway, such as thymidine kinase (TK) or hypoxanthine-guanine phosphoribosyltransferase (HGPRT), can grow in normal medium, but cannot use thymidine or hypoxanthine, respectively, as an alternative to the endogenous pathway. Thus, medium containing hypoxanthine, aminopterin, and thymidine (HAT medium), selects for cells which contain both TK and HGPRT. If a cell lacking HGPRT ($HGPRT^-$) is fused to one lacking TK (TK^-) then, although neither parental cell line will be able to grow in HAT the hybrid cell line will, since it will have TK from the $HGPRT^-$ cell and HGPRT from the TK^- cell. Thus, HAT medium can be used to select hybrids from two conditional-lethal parental mutant cell lines. Other variations on this procedure have also been described (42).

Protocol 12

Selection for variant cells deficient in hypoxanthine-guanine phosphoribosyltransferase (HGPRT)

Equipment and reagents

- Disposable Pasteur pipettes
- Ethyl methanesulfonate (EMS) at 100–300 μg/ml, or *N*-methyl-*N′*-nitro-*N*-nitrosoguanidine (MNNG), or other mutagen
- 6-thioguanine
- Giemsa stain (Merck)
- Hypoxanthine, aminopterin, thymidine (HAT) (Life Technologies)

A. Treatment of cells with mutagenic agent[a]

1. Set up ten 75 cm^2 flasks, seeding the cells at around 10^6 cells per flask.
2. After attachment, treat the cells with a mutagen such as EMS or MNNG. EMS is prepared by dilution to a concentration of 100–300 μg/ml in medium without serum. It is rapidly hydrolysed and must be used immediately.[b]

Protocol 12 continued

3 Expose the cells for 18 h (approximately one cell division), remove the EMS, and wash the cells thoroughly with medium without serum.

NB. Mutagenic chemicals are, by definition, extremely hazardous and they must be treated with extreme caution.

4 Grow the cells for a further four days, subculturing if necessary, then start the selection process.

B. Isolation of cells resistant to 6-thioguanine (6-TG)[c]

1 Set up cells at an appropriate density: for most cell lines this will be around 8×10^3 cells/cm^2 if adherent, or 10^5 cells/ml if non-adherent. However, very rapidly growing cells may require to be seeded at a lower density and it may be possible to seed very small cells at a higher density. Incubate cultures overnight or until cell attachment has occurred.[d]

2 Add 6-TG to the culture medium at concentrations ranging from 10–100 μg/ml. Incubate for three days.[e]

3 Remove spent medium and add fresh medium containing 6-TG. Incubate for a further three days.

4 Repeat step 3.

5 Scan the culture for the presence of living cells by inverted phase microscopy.

6 Continue to culture until colonies of cells are visible and can be picked. Transfer colonies of cells into individual wells of a 24-well plate by gently touching the cells with the end of a disposable Pasteur pipette.[f]

7 Expand the cells in the absence of selective agent until the population is large enough for freezing and for testing for drug resistance and reversion frequency.

C. Testing cells for drug resistance

1 Set up cells at low density (around 10^3 cells/5 cm dish) and incubate cultures overnight to allow the cells to attach.

2 Add 6-TG to the culture medium at concentrations ranging from 10–250 μg/ml, including a control without 6-TG. Incubate for three days.

3 Remove spent medium and add fresh medium containing 6-TG. Incubate for a further three days.

4 Repeat step 3 until colonies are visible to the eye.

5 Stain the colonies with Giemsa to visualize and count colonies. Compare colony-forming efficiency (proportion of cells seeded giving rise to a colony, see Chapter 8, Section 3.1) in the presence of drug with the drug-free control.

D. Testing cells for reversion back to the wild-type phenotype

1 Plate the cells into ten 75 cm^2 flasks at 5×10^5 cells/flask.

2 24 h later remove the medium and add medium containing HAT.

Protocol 12 continued

3. Renew HAT medium every three days.
4. After two to three weeks, examine flasks for surviving cells. These can be visualized most easily by Giemsa staining and examination by eye.
5. If colonies are present calculate the reversion frequency. For example, two colonies from 5×10^6 cells seeded means a reversion frequency of $2/(5 \times 10^6)$ or 4×10^{-7}. If this is lower than the expected frequency of fusion then the cell line can be used as a fusion partner. Ideally there should be no reversion.

[a] Pre-treatment of cells with mutagen will increase the frequency of drug-resistant cells. In the case of an enzyme such as HGPRT which is encoded on the X chromosome (and consequently only one active copy of the gene is present in cells) it should not be necessary. For enzymes whose genes are located on autosomes (e.g. TK) both copies of the gene need to be altered to isolate a TK^- cell line and mutagen treatment will certainly be necessary.

[b] MNNG is used at 5×10^{-5} M and cells are exposed for 2 h at 37 °C.

[c] 6-thioguanine is a purine analogue which is toxic to cells when metabolized to the nucleotide and incorporated into nucleic acid. Cells which lack HGPRT are resistant to 6-TG because they cannot convert it to the nucleotide.

[d] The density must be such that the cells can undergo a further three or four divisions, because 6-TG is only toxic to dividing cells.

[e] 8-azaguanine can be used as an alternative to select $HGPRT^-$ cell lines.

[f] Disposable Pasteur pipettes give superior results to glass pipettes because the cells cling better to the rough edges of the plastic. Alternatively, cloning rings may be used if preferred (see Chapter 8, *Protocol* 5). It is important to ensure that each well only contains cells from a single colony to ensure the clonal nature of the mutants.

ii. Dominant drug selection

Rodent cells are naturally resistant to levels of ouabain about 1000-fold higher than the level tolerated by human cells. This difference can be exploited to remove the human parent after a rodent × human fusion, because the hybrid cells show resistance to ouabain at levels comparable to that of the rodent parent (43). Differential sensitivity to polyene antibiotics and diptheria toxin has also been used to select for hybrid cells (42).

iii. Fluorescence-activated cell sorting

The fluorescence-activated cell sorter (FACS) detects light scatter and fluorescence emission from cells as they pass through a laser beam in a stream of droplets. Cells producing a signal within a defined range can be deflected and sorted from the remainder of the cells. The signal used can be cell size (detected as low-angle light scatter), or a parameter such as expression of a cell surface marker, DNA content, or presence of a damaged membrane and consequent permeability, detected by the use of appropriate fluorescent markers. Fused cells can therefore be sorted on the basis of their increased size, and hybrid cells selected on the basis of the presence of two specific markers, only one of which is present on either of the parental cells. Cell sorting is unlikely to result in the isolation of the hybrid cells directly, but two or three rounds of sorting should result in their enrichment to levels at which they can be isolated by cloning. Methods for isolating hybrids in this way are described in detail elsewhere (44).

7.3 Applications

7.3.1 Isolation of differentiated lines

The phenotype of the hybrid cell might reasonably be predicted to be the sum of the phenotypes of the parental cells. Thus, if one parent expressed activities A and B, and the other activities C and D, one could expect the hybrid to express A, B, C, and D. This situation often occurs so that a common consequence of fusion in gene expression terms is that the hybrid retains the characteristics of both parents. Thus, for example, if both parents express their form of an enzyme such as malate dehydrogenase, the hybrid cell would express both forms. However, although this pattern of gene expression often occurs, other more complex phenotypes can also be found: the hybrid may stop expressing a differentiated function expressed by one parent (extinction); subsequent re-expression can occur, for example due to loss of a repressor (see below); or there may be activation of a gene not expressed in one of the parents. Gene regulation can be investigated in such situations (Section 7.3.2) and information has been obtained about the likelihood of groups of genes being expressed simultaneously (45).

i. Hybridomas

Hybridoma cell lines are formed by fusing B lymphocytes to cells from an established myeloma cell line. Usually the B cells are taken from an immunized animal, e.g. mouse, and fused with a myeloma line derived from B cells, but which no longer secretes antibody. The hybrid cells continue to synthesize antibody like the parental B cell, but acquire the ability to grow in culture from the myeloma cell line, resulting in the production of permanent cell lines capable of secreting antibody. For more details see Chapter 7, Section 3.6.1.

ii. Other cell types

Cell fusion is a powerful technique which has resulted in the isolation of thousands of murine hybridoma cell lines producing a specific antibody. Unfortunately, the technique does not work so well for human lymphocytes, nor has it proved to be as successful for other cell types. Differentiated hybrids have been reported from the fusion of rat liver cells and a mouse hepatoma line (46) and dorsal root ganglia neurons with a neuroblastoma cell line (47). However, neither of these groups of hybrids is as differentiated as hybridoma cells and the method has not been shown to have general applicability.

7.3.2 Gene regulation

Hybrid cells have been used extensively to investigate aspects of gene regulation. For example, it has been shown that hybrids between hepatocytes producing aldolase B and hepatic alcohol dehydrogenase (ADH) and mouse fibroblasts which do not produce these liver-specific functions no longer produce the enzymes. One simple explanation is that the hybrid cells have lost the gene which codes for the particular enzyme. However, after a period in culture it is possible to obtain hybrids which have regained the ability to produce aldolase B and ADH (48, 49). Presumably, synthesis is repressed in the hybrid by a negative control element present in mouse fibroblasts and it is only as a result of loss of the repressor that synthesis is regained.

7.3.3 Gene mapping

Hybrid cells are genetically unstable and prone to lose chromosomes. This loss is essentially random if the parental cells are from the same species. If different species of parental cells are used, however, the chromosome loss is more extensive and is specific to one species. Thus, in hybrids between mouse and human cells, the human chromosomes are preferentially lost. This chromosome loss, or segregation, has been used to correlate expression of a particular gene with the presence of a particular chromosome. This is possible provided the product of the gene can be distinguished from that produced by the other parental cell. This may be done by gel electrophoresis to distinguish species differences in electrophoretic mobility (50) or by using molecular probes to bind to the chromosome *in situ* (FISH).

8 Genetic engineering techniques

Another approach to the isolation of new cell lines, which is gaining widespread use, is to introduce viral oncogenes into cells. These genes are cloned into plasmid vectors and introduced either singly or in combinations, into primary cells. This section will outline both the methods used for gene transfer and the types of genes which have been used most successfully, but more detailed descriptions and protocols for gene transfer techniques are available in companion volumes (51, 52).

8.1 Introduction of genes

Although some gene transfer techniques are efficient in terms of uptake of DNA (resulting in about 50% of cells expressing the foreign gene transiently) most of the DNA taken up initially by the cell is subsequently degraded. For stable expression the recombinant gene must be integrated into the host chromosome and this is a rare event which occurs in less than 1% of the transfected cells. In order to isolate these rare stable integrants, most gene transfer protocols include a selection for those cells which have taken up and are expressing the foreign gene(s). A variety of selectable markers is available: the kanamycin–neomycin resistance genes (*neo* or *aph*) from the bacterial transposons Tn*601* and Tn*5* encode aminoglycoside phosphotransferases (APH) which confer resistance to the antibiotic G418 (Geneticin), and the *Escherichia coli* xanthine guanine phosphoribosyltransferase gene (*gpt*) encodes resistance to mycophenolic acid. These selectable markers do not even need to be present on the same vector if co-precipitation techniques are used, although for electroporation the genes need to be physically linked. The sensitivity of different cell lines to antibiotic must be predetermined, then the transfectants selected on the basis of their resistance.

An alternative approach is to transfect cells with a fluorescent marker, such as the chromophore green fluorescent protein (GFP) (53) then to separate the labelled cells using a fluorescence-activated cell sorter (FACS). Following interrogation by a laser light beam the droplets can be separated into those which contain fluorescent cells and those which do not (see Section 7.2.1.iii). It takes only milliseconds to sort each droplet so up to ten million cells per hour can be analysed.

8.1.1 Co-precipitation

DNA can be readily taken up by cells if it is formed into a co-precipitate with calcium or strontium phosphate and Hepes buffer. The precipitate, which is believed to protect the DNA from degradation, sediments onto the cells, becomes adsorbed onto the cell membrane, and appears to be taken up by the cells through a calcium-dependent process. After an incubation period which can range from 4 h to overnight, the cells are washed to remove the precipitate. The general method is described in detail elsewhere (52, 54).

i. Calcium phosphate

Calcium phosphate precipitation (55) is probably the most commonly used gene transfer technique. It is cheap and simple, though somewhat variable, and although it is inefficient, large numbers of cells can be treated at the same time. There are two major disadvantages; first, it is toxic to some cells, particularly those cells which are sensitive to calcium; secondly, it is unsuitable for cells which grow in suspension.

ii. Strontium phosphate

Strontium phosphate precipitation was developed for cell types which are particularly sensitive to calcium. The technique is virtually identical, using strontium chloride rather than calcium chloride, and it has been used successfully for bronchial epithelial cells amongst others (56).

8.1.2 Electroporation

Electroporation is most commonly used for cells which are not suitable for calcium phosphate precipitation, such as suspension cells, but has the disadvantage that the method needs to be optimized for each cell type. The cells are immersed in a DNA solution and subjected to an electrical field. Pores form transiently in the cell membrane and the DNA enters (57). However, in contrast to co-precipitation, it is unusual for more than one or two molecules of DNA to enter the cell and so co-transformation with unlinked genes is rare. This makes the method particularly appropriate for the generation of cells with a low or single copy number. The disadvantages of this method are the time required for optimization and the need for comparatively large quantities of DNA.

8.1.3 Lipofection

The introduction of DNA into cells with a commercially available cationic liposome preparation (e.g. Lipofectin Reagent, Life Technologies) is probably the most straightforward technique for the beginner to attempt, and a wide variety of reagents and protocols are available. The method works well for a variety of cell types (58) and is efficient; its main drawback is the cost of the reagent.

8.1.4 Spheroplast/protoplast fusion

DNA which has been cloned into bacterial plasmids can be introduced into mammalian cells by direct fusion with bacterial spheroplasts (59). Bacterial spheroplasts (also known as protoplasts) are produced by removing the cell wall with lysozyme but care must be taken to ensure that this digestion is complete otherwise the mammalian cells will rapidly be overgrown with bacteria. The protoplasts are then fused to the mammalian cells with PEG in a method similar to that used for mammalian cell fusion.

8.1.5 Microinjection

DNA can be directly injected into mammalian cells using glass capillary micropipettes. Injection can be into either the nucleus or cytoplasm, but the procedure requires highly skilled personnel and specialized equipment such as micromanipulators and high specification inverted microscopes. Microinjection is extremely efficient (when carried out by experts!) and is generally used in situations where only a limited number of recipient cells are available, e.g. for introducing DNA into embryos or in situations where there is the likelihood that another cell type present will take up DNA more efficiently (60).

8.1.6 Biolistic particle bombardment

Gold or tungsten microbeads can be coated with the gene of interest and delivered to the cell by high voltage electronic discharge, spark discharge, or helium pressure discharge (61). However, these methods are laborious, require costly equipment, and are therefore normally restricted to specialized applications.

8.1.7 Viral vectors

Infection with viral vectors (e.g. retroviral vectors) is another efficient method for introducing genes into cells (62). Viral infection is initiated when the virus binds to a cell surface receptor. This is followed by internalization and either autonomous replication or, in the case of the retroviruses, integration into the host cell genome.

Retroviral vectors are developed from wild-type virus by deletion of the region containing the three structural genes *gag*, *pol*, and *env*, which code for the viral core proteins, reverse transcriptase, and the envelope proteins, respectively. This deleted region is used for cloning the gene of interest, thus generating replication-defective vectors which require complementation of viral functions *in trans*. The vector retains the long terminal repeat (LTR) sequences present at the ends of the viral genome which are required for integration and transcription; the primer binding sites for reverse transcription of viral RNA adjacent to the LTRs; the packaging sequence Ψ, which is necessary for efficient packaging of the viral RNA into virions; and the viral splice donor and acceptor sequences which are used for the production of a sub-genomic mRNA. The missing viral genes are provided either by replication in the presence of helper virus, or more frequently, by propagating the vector in a packaging cell line which provides the missing functions. The packaging cell line has been genetically engineered to contain a defective retroviral genome integrated within the cellular DNA in order to provide the viral functions *in trans*. The integrated genome lacks a packaging signal and is not, therefore, packaged to give virus particles. The retroviral vector DNA is transfected into the packaging cell line and fully infectious but replication-defective virus containing the gene of interest is produced. This virus can infect target cells at high efficiency with the host range being determined by the envelope gene contained within the packaging cell line. Thus to produce vector derived from Moloney murine leukaemia virus which will infect mouse cells the vector should be propagated in an ecotropic packaging cell line such as Ψ_2 (63) or Ψ_{cre} (64), whereas if the target cells are human, feline, or canine an amphotropic cell line such as Ψ_{crip} (64) or PA317 (65) must be used. After infecting the target cell the vector undergoes a single round of reverse transcription and the viral DNA is integrated into the cellular genome as a provirus.

One of the major concerns when using retroviral vectors is the need to ensure that the viruses really are propagated in a replication-defective form and that no replication-competent retrovirus (RCR) is produced. The first retroviral packaging lines such as Ψ_2 were particularly prone to generate RCRs because the viral protein coding sequences were all present on a single, contiguous piece of DNA. This meant that a single recombination event with *cis*-acting sequences from the vector could result in the production of replication-competent helper virus. This risk has been reduced by producing packaging lines in which the viral genes are deleted for these *cis*-acting sequences at the 3′ end. Another approach involves introducing the *gag-pol* and *env* genes on separate plasmids and transfecting them sequentially and not simultaneously.

The main stages involved in constructing a usable retroviral expression system are shown in *Table 1*. Detailed protocols are not provided because they fall outside the scope of this chapter, and the interested reader is referred to a contribution in a companion volume (66).

One of the major disadvantages of retroviral vectors is the fact that they only infect dividing cells. Not all primary cells are capable of cell division, so for some applications, e.g. gene transfer *in vivo* for gene therapy, alternative vectors must be found. There have been significant developments in the use of adenoviral vectors and also, more recently, lentiviruses. A system for producing high titre lentivirus vectors which can efficiently infect brain, liver, muscle, and retinal tissues *in vivo* has been described (69) and is likely to have important applications.

Table 1 The steps involved in gene transfer using a retroviral vector

Step	Process
1	The gene of interest is cloned into a plasmid which contains the relevant viral sequences, i.e. the LTRs and the packaging signal.
2	The vector is introduced into a packaging cell line by transfection or electroporation.[a]
3	Clones of transfected cells are selected on the basis of expression of an antibiotic resistance gene.
4	The packaging cell lines are grown up, frozen stocks prepared, and the secreted virus titrated and assayed for the presence of wild-type recombinants.
5	Viral stocks are harvested from appropriate packaging lines and used to infect target cells.
6	The stocks of packaging cell lines and target cells must be assayed for the presence of wild-type virus at regular intervals in order to ensure that recombination between vector and integrated virus, or vector and naturally occurring endogenous retroviruses, has not occurred.

[a] The efficiency of retroviral infection can be increased by the addition of cationic polymers or liposomes (e.g. Lipofectamine) to the virus for 30 minutes immediately prior to addition to the cells (67, 68).

8.2 Oncogene immortalization

Experiments with polyoma virus have shown that the transformation of rat embryo fibroblasts occurs as two successive steps: the first step is immortalization and is due to the expression of the large transforming protein coded by *plt*; the second step of transformation results from expression of the middle transforming gene *pmt* (70). Oncogenes have been divided into either immortalizing or transforming genes on the basis of whether they can either extend the lifespan of primary embryo fibroblasts (immortalization) or induce NIH 3T3 cells to form transformed foci (transformation). Representatives of the two groups can act together, or co-operate, in a two-step process to transform primary rodent cells in which immortalization is a single genetic event (71).

The situation is more complex in human cells, however, and there is evidence, from work with SV40, that immortalization involves at least two separate genetic events in human cells, with SV40 large T antigen being necessary (in this experimental system), but not sufficient, for immortalization (72). This role of large T is dependent on its ability to interact with the growth suppressors p53 and retinoblastoma protein (pRB) as evidenced by the fact that mutants of large T which do not bind cannot extend the lifespan of cells. This initial extension of lifespan is limited, however, and the culture ultimately enters crisis phase as the cells undergo apoptosis (73).

It is known that human fibroblasts undergo progressive shortening of their telomeres and it has been postulated that telomere length can be used to predict lifespan and may be involved in the regulation of replicative capacity (74). Strong support for this hypothesis has come from the observation that the introduction of the catalytic subunit of telomerase into primary human cells can result in telomere lengthening and extend the lifespan of the cells (75).

The model which emerges from this work is as follows: normal human cells undergo progressive shortening of their telomeres with each round of cell division and this, coupled with their low levels of telomerase activity, results in the cells undergoing senescence. Introduction of the catalytic subunit of telomerase in conjunction with SV40 large T (which inactivates both the p53 and pRB pathways) results in immortalization with transformation to a cancerous state being induced when the *ras* oncogene is added (76). Rodent cells, in contrast, have telomerase activity and also longer telomeres which could explain the different genetic mechanism required to transform rodent cells (77).

8.2.1 Choice of oncogene

Many immortalization experiments were carried out prior to this increased understanding of the process involved, and the experience of most laboratories was that the early region of the DNA tumour virus SV40 was the coding sequence whose expression was most likely to result in the successful immortalization of a primary cell. This region codes for two transforming genes, large T antigen (mentioned above) and small t, with the former being believed to have immortalizing activity. Cell lines have been isolated from many species - human, mouse, rat, rabbit, bovine, and hamster; and from many tissues such as heart, liver, kidney, adrenocortex, bone marrow, and trachea. The procedures used and characteristics of the lines have been reviewed (78, 79). The immortalized clones vary considerably in the extent to which they retain the characteristics of the original tissue, and considerable screening may be required to find the most useful cell lines.

Other viral oncogenes which have been found to possess immortalizing activity include polyoma large T, the E1A gene of human adenoviruses, the E6 and E7 genes from human papillomaviruses (HPV), v-*myc*, Ha-*ras*, and v-*raf*/*myc*. All of these genes have activity but in some cases it is limited to only a narrow range of cells, e.g. human papillomaviruses 16 and 18 only immortalize keratinocytes. In many cases the basis of the activity has been found to be similar to SV40 large T, i.e. the ability of the oncogene to interact with and inactivate either the p53 or pRB pathway. Mutant p53, which binds to the normal allele has also been shown to be capable of immortalization. The role of telomere maintenance in cell immortalization has recently been reviewed (80, 81).

8.2.2 Regulation of oncogene expression

Although immortalized cell lines can be isolated even if the oncogene is expressed constitutively, there is evidence to suggest that a more differentiated phenotype can be obtained if oncogene expression can be switched off. This can be done either with an inducible promoter regulating oncogene expression, or using a mutant encoding an altered form of the protein.

i. Promoter regulation

Heavy metal induction

The mouse and human metallothionein genes are expressed in response to the presence of heavy metals such as cadmium and zinc. The promoter region can be used to regulate the expression of an oncogene such that it is only produced if heavy metal is present. This approach has been used to isolate rat Schwann cell lines in which SV40 T antigen is expressed under the control of a modified mouse metallothionein promoter (82). Reduction in the level of T expressed in the cells (following withdrawal of zinc induction) resulted in increased expression of Po (a protein specific to myelin-forming Schwann cells) and a decreased expression of glial fibrillary acidic protein, a protein expressed only in non-myelin-forming Schwann cells.

Hormone-responsive promoters

Hormone-responsive promoters include the mouse mammary tumour virus LTR which is active in the presence of dexamethasone. This system has been used to regulate expression of polyoma large T antigen in human fibroblasts (83), and the immortalized phenotype was shown to be strictly dependent on the induction of T antigen expression.

Tetracycline-inducible promoters

The promoters described above have the disadvantage that they are based on endogenous cellular elements which respond to exogenous signals or stresses in a pleiotropic fashion by affecting more than one gene and consequently producing multiple, apparently unrelated, phenotypic effects. Furthermore, although they can be controlled in culture systems they are much harder to work with in experimental animals or in man. Systems have been developed which are based on non-mammalian elements or mutated elements which no longer respond to endogenous inducers and these have been reviewed recently (84). The tetracycline-inducible system which was developed by Bujard and colleagues (85) is entirely prokaryotic and comprises a chimeric protein which represses transcription in the presence of tetracycline. More recently, a variant form has been described which activates transcription in the presence of the drug (86).

ii. Protein stability

Mutant viruses containing variant forms of oncogenes which produce proteins which are temperature-sensitive (ts), for example, can be used to study the consequences of regulating protein production. A number of ts SV40 large T mutants exist, e.g. tsA58, which express T at the low or permissive temperature, but produce little or no T at the non-permissive (high) temperature (28, 29, 87).

9 Safety considerations

Safety aspects of the experimental procedures described here should be considered before the work is commenced. A risk assessment should be prepared for all work, and particular care taken in procedures adopted for handling human tissue. All biopsy material carries a risk of infection, in particular from hepatitis B and human immunodeficiency viruses. All workers should be immunized against the former, and care should be taken when handling all material. Manipulations should be carried out in a Class II microbiological safety cabinet (which gives protection to both operator and specimen) and **never** in the type of laminar flow cabinet which directs the air towards the operator (horizontal-flow cabinet) – see Chapter 1, Section 3.1.1 and Chapter 2, Section 8.3. All waste materials should be autoclaved or soaked overnight in a disinfectant such as Chloros or a similar hypochlorite bleach (see Chapter 2, Section 7.2.2). The removal of tissue for biopsy usually requires the consent of the hospital ethical committee, the clinician, and the patient; helpful information can be found in ref. 88. For this type of work, tissue samples, including blood, should **never** be taken from an individual who works in the same laboratory.

Experiments with animal tissue do not carry the same health hazards, but it should be remembered nevertheless that animals may harbour agents potentially pathogenic to humans, and at least one fatality has been reported due to a virus carried by a cell culture (89). The use of animal tissue may require consent from the Home Office or a similar body. In the UK a written risk assessment (COSHH, Control of Substances Hazardous to Health) must be prepared before any experimental procedures begin, and similar regulations exist in other countries. Finally, genetic modification experiments are controlled by European Union legislation, and must be notified to the appropriate body, e.g. the UK Health and Safety Executive (HSE). The procedures require that both the construction or modification of a cell by genetic manipulation **and its subsequent use and/or storage** are reported either in advance or in an annual retrospective return depending on the type of activity involved. Similar guidelines apply in the USA and Japan. Experiments involving the use of oncogenic DNA have been subjected to particular scrutiny in the light of experiments which showed that tumours could be produced in mice in which the skin had been damaged and oncogenic DNA rubbed in (90). The use of retroviral vectors capable of efficiently infecting human cells (and consequently laboratory workers) and containing immortalizing oncogenes is governed by particularly stringent regulations and requires notification and approval in advance of work commencing.

10 Concluding remarks

One of the aims of this chapter is to give an overview of the techniques available for isolating primary cells and attempting to establish permanent lines from those cells. As usual in biological systems, there does not appear to be one ideal method, but rather different approaches are appropriate in different circumstances. I have tried to discuss the advantages of particular approaches and their applicability in specific situations. Some of the observations are the result of hard-won experience, others are more in the nature of prejudices. Inevitably, different laboratories have different rates of success with techniques such as cell immortalization, and although we are increasingly able to offer rational explanations for the differences, it is still not always possible.

10.1 Advantages of cell culture over *in vivo* experimentation

In the UK, and to a lesser extent perhaps in other countries, public opinion has moved strongly in favour of the use of *in vitro* techniques wherever possible. Cell lines provide a relatively homogeneous source of material, can be maintained in a controlled environment, and can be experimented on directly. It is difficult to introduce reagents directly to a particular cell type *in vivo*, and even more difficult to do so at a defined concentration. However, even good systems have their limitations, and it is important to be aware of them.

10.2 Limitations

All cell culture systems share a number of disadvantages - they provide an over-simplification of the natural situation because the cells are isolated from other types of cells with which they may interact. Cell cultures must be maintained according to rigorous procedures in order to minimize the likelihood of contamination by micro-organisms. It is difficult to produce large quantities of material: 10^9 to 10^{10} cells is a significant number for any laboratory other than the pilot plants found in pharmaceutical and biotechnology companies. Thus, while it is possible to obtain gram quantities of most tissues (by a judicious choice of species), the production of 10 g of cultured cells would not be a trivial undertaking.

10.2.1 Limitation of primary cells

Primary cells provide the closest approximation to the situation *in vivo*, and as a consequence suffer from some of the disadvantages of fresh tissue. They are often heterogeneous - even if prepared from inbred animals, procedures and levels of success vary from day to day and cell preparations are not always identical. This problem is exacerbated when material from outbred species (e.g. humans) is used. All primary cell cultures are expensive and time-consuming to prepare and the cells often have only a very short lifespan.

10.2.2 Limitations of diploid cell lines

Diploid, or finite-lifespan, cell lines often suffer from the loss of a differentiated phenotype. Furthermore, the number of generations before senescence varies with the cell type and although some 'normal' cell lines, e.g. WI38, are available in large enough quantities to permit their use for vaccine production, more often insufficient generations are available to enable large numbers of cells to be accumulated.

10.2.3 Limitations of established lines

Established cell lines are capable of indefinite growth but frequently suffer from the drawback of not expressing interesting characteristics. In general, they grow well, but are highly aneuploid and the cultures rapidly become heterogeneous. As in all situations, however, the final choice of material depends on the particular application - one great advantage of working with cultured mammalian cells is that the choice is extensive.

Acknowledgements

I am grateful to the members of my research group who have developed many of the techniques described here: Chris Darnbrough, Una FitzGerald, Alastair Grierson, Rhys Jagger, Helena Kelly, Ute Kreuzburg-Duffy, Fiona MacKenzie, Ian McPhee, Carol Murray, Susan Robertson, Shirley Slater, Maureen Vass, Pat Watts, and Brian Willett. Funding for work on cell immortalization has been provided by the Humane Research Trust, The EC Science Stimulation Programme, SmithKline Beecham, the Science and Engineering Research Council Biotechnology Directorate, and the Biotechnology and Biological Sciences Research Council, and their support is gratefully acknowledged.

References

1. Martin, G. M. (1977). *Am. J. Pathol.*, **89**, 484.
2. Mills, K. H. G., Barnard, A. L., Williams, M., Page, M., Ling, C., Stott, J., *et al.* (1991). *J. Immunol.*, **147**, 3560.
3. Mills, K. H. G., Barnard, A. L., Watkins, J., and Redhead, K. (1993). *Infect. Immun.*, **61**, 399.
4. Freshney, R. I. (1994). *Culture of animal cells* (3rd edn). Wiley-Liss, New York.
5. Michalopoulos, G. and Pitot, H. C. (1975). *Exp. Cell Res.*, **94**, 70.
6. Yang, J., Elias, J. J., Petrakis, N. L., Wellings, S. R., and Nandi, S. (1981). *Cancer Res.*, **41**, 1021.
7. Kleinman, H. K., McGoodwin, E. B., Rennard, S. I., and Martin, G. R. (1981). *Anal. Biochem.*, **94**, 308.
8. Reid, L. M. and Rojkind, M. (1979). In *Methods in enzymology* (ed. W. B. Jakoby and I. H. Pastan), Vol. 58, pp. 263–78. Academic Press, London.
9. Pittner, R. A., Fears, R., and Brindley, D. N. (1985). *Biochem. J.*, **225**, 445.
10. Williams, G. M., Bermundez, E., and Scaramuzzino, D. (1977). *In Vitro*, **13**, 809.
11. Tsao, M. C., Walthall, B. J., and Ham, R. G. (1982). *J. Cell. Physiol.*, **110**, 219.
12. Kao, W. W.-Y. and Prockop, D. J. (1977). *Nature*, **266**, 63.
13. Fry, J. and Bridges, J. W. (1979). *Toxicol. Lett.*, **4**, 295.
14. Gilbert, S. F. and Migeon, B. R. (1975). *Cell*, **5**, 11.
15. Hayflick, L. and Moorhead, P. S. (1961). *Exp. Cell Res.*, **25**, 585.
16. Holliday, R., Huschtscha, L. I., Tarrant, G. M., and Kirkwood, T. B. L. (1977). *Science*, **198**, 366.
17. Risser, R. and Pollack, R. (1974). *Virology*, **59**, 477.
18. Hammond, A. H. and Fry, J. R. (1990). *Biochem. Pharmacol.*, **40**, 637.
19. Kamarck, M. E., Barker, P. E., Miller, R. L., and Ruddle, F. H. (1984). *Exp. Cell Res.*, **152**, 1.
20. Pinkel, D., Straume, T., and Gray, J. W. (1986). *Proc. Natl. Acad. Sci. USA*, **83**, 2934.
21. Weinstein, R., Stemerman, M. B., MacIntyre, D. E., Steinberg, H. N., and Maciag, T. (1981). *Blood*, **58**, 110.
22. Green, H. and Kehinde, O. (1974). *Cell*, **1**, 113.
23. Yaffe, D. (1968). *Proc. Natl. Acad. Sci. USA*, **61**, 477.
24. Moyer, M. P. and Aust, J. P. (1984). *Science*, **224**, 1445.
25. Stampfer, M. R. and Bartley, J. C. (1985). *Proc. Natl. Acad. Sci. USA*, **82**, 2394.
26. Santerre, R. F., Cook, R. A., Crisel, R. M. D., Sharp, J. D., Schmidt, R. J., Williams, D. C., *et al.* (1981). *Proc. Natl. Acad. Sci. USA*, **78**, 4339.
27. Huschtscha, L. I. and Holliday, R. (1983). *J. Cell Sci.*, **63**, 77.

28. Chou, J. Y. (1978). *Proc. Natl. Acad. Sci. USA*, **75**, 1409.
29. Chou, J. Y. and Schlegel-Haueter, S. E. (1981). *J. Cell Biol.*, **89**, 216.
30. Winger, L., Winger, C., Shastry, P., Russell, A., and Longenecker, M. (1983). *Proc. Natl. Acad. Sci. USA*, **80**, 4484.
31. Tahara, H., Tokutake, Y., Maeda, S., Kataoka, H., Watanabe, T., Satoh, M., *et al.* (1997). *Oncogene*, **15**, 1911.
32. Rosenberg, N., Baltimore, D., and Scher, C. D. (1975). *Proc. Natl. Acad. Sci. USA*, **72**, 1932.
33. Giotta, G. J. and Cohn, M. (1981). *J. Cell. Physiol.*, **107**, 219.
34. Greenberger, J. S., Davisson, P. B., Gans, P. J., and Moloney, W. C. (1979). *Blood*, **53**, 987.
35. Dexter, T. M., Allen, T. D., Scott, D., and Teich, N. M. (1979). *Nature*, **277**, 471.
36. Stoker, M. and Macpherson, I. (1964). *Nature*, **203**, 1355.
37. Deschatrette, J., Fougère-Deschatrette, C., Corcos, L., and Schimke, R. T. (1985). *Proc. Natl. Acad. Sci. USA*, **82**, 765.
38. Bissell, M. J. (1981). *Int. Rev. Cytol.*, **70**, 27.
39. Giovanella, B. C., Stehlin, J. S., and Williams, L. J. (1974). *J. Natl. Cancer Inst.*, **52**, 921.
40. Teissie, J., Knutson, V. P., Tsong, T. Y., and Lane, M. D. (1982). *Science*, **216**, 537.
41. Zimmermann, U. (1982). *Biochim. Biophys. Acta*, **694**, 227.
42. Choy, W. N., Gopalakrishnan, T. V., and Littlefield, J. W. (1982). In *Techniques in somatic cell genetics* (ed. J. W. Shay), pp. 11–21. Plenum Press, New York.
43. Jha, K. K. and Ozer, H. (1976). *Somat. Cell Genet.*, **2**, 215.
44. Jongkind, J. F. and Verkerk, A. (1982). In *Techniques in somatic cell genetics* (ed. J. W. Shay), pp. 81–100. Plenum Press, New York.
45. Fougère, C. and Weiss, M. C. (1978). *Cell*, **15**, 843.
46. Szpirer, J., Szpirer, C., and Wanson, J.-C. (1980). *Proc. Natl. Acad. Sci. USA*, **77**, 6616.
47. Platika, D., Boulos, M. H., Baizer, L., and Fishman, M. C. (1985). *Proc. Natl. Acad. Sci. USA*, **82**, 3499.
48. Bertolotti, R. and Weiss, M. C. (1972). *J. Cell. Physiol.*, **79**, 211.
49. Bertolotti, R. and Weiss, M. C. (1972). *Biochimie*, **54**, 195.
50. Nichols, E. A. and Ruddle, F. H. (1973). *J. Histochem. Cytochem.*, **21**, 1066.
51. MacDonald, C. (1991). In *Mammalian cell biotechnology: a practical approach* (ed. M. Butler), pp. 57–83. IRL Press, Oxford.
52. Chisholm, V. (1995). In *DNA cloning: a practical approach* (ed. D. M. Glover and B. D. Hames), Vol. 4, pp. 1–41. Oxford University Press, Oxford.
53. Cody, C. W., Prasher, D. C., Westler, W. M., Prendergast, F. G., and Ward, W. W. (1993). *Biochemistry*, **32**, 1212.
54. Gorman, C. (1985). In *DNA cloning: a practical approach* (ed. D. Glover), Vol. 2, pp. 143–90. IRL Press, Oxford.
55. Graham, F. L. and van der Eb, A. J. (1972). *Virology*, **52**, 456.
56. Brash, D. E., Reddel, R. E., Quanrud, M., Yang, K., Farrell, M. P., and Harris, C. C. (1987). *Mol. Cell. Biol.*, **7**, 2031.
57. Andreason, G. L. and Evans, G. A. (1988). *Biotechniques*, **6**, 650.
58. Felgner, P. L. Gadek, T. R., Holm, M., Roman, R., Chan, H. W., Wenz, M., *et al.* (1987). *Proc. Natl. Acad. Sci. USA*, **84**, 7413.
59. Rassoulzadegan, M., Binetruy, B., and Cuzin, F. (1982). *Nature*, **295**, 257.
60. Garcia, I., Sordat, B., Rauccio-Farinon, E., Dunand, M., Kraehenbuhl, J.-P., and Diggelmann, H. (1986). *Mol. Cell. Biol.*, **6**, 1974.
61. Jiao, S., Cheng, L., Wolff, J. A., and Yang, N.-S. (1993). *Bio/Technology*, **11**, 497.
62. Wiehle, R. D., Helftenbeim, G., Land, H., Neumann, K., and Beato, M. (1990). *Oncogene*, **5**, 787.
63. Mann, R., Mulligan, R. C., and Baltimore, D. (1983). *Cell*, **33**, 153.

64. Danos, O. and Mulligan, R. C. (1988). *Proc. Natl. Acad. Sci. USA*, **85**, 6460.
65. Miller, A. D., Law, M.-F., and Verma, I. M. (1985). *Mol. Cell. Biol.*, **5**, 431.
66. Brown, A. M. C. and Dougherty, J. D. (1995). In *DNA cloning: a practical approach* (ed. D. M. Glover and B. D. Hames), Vol. 4, pp. 113–42. Oxford University Press, Oxford.
67. Hodgson, C. P. and Solaiman, F. (1996). *Nature Biotechnol.*, **14**, 339.
68. Porter, C. D., Lukacs, K. V., Box, G., Takeuchi, Y., and Collins M. K. L. (1998). *J. Virol.*, **72**, 4832.
69. Kafri, T., van Praag, H., Ouyang, L., Gage, F. H., and Verma, I. M. (1999). *J. Virol.*, **73**, 576.
70. Rassoulzadegan, M., Cowie, A., Carr, A., Glaichenhaus, N., Kamen, R., and Cuzin, F. (1982). *Nature*, **300**, 713.
71. Land, H., Parada, L. F., and Weinberg, R. A. (1983). *Nature*, **304**, 596.
72. Shay, J. W. and Wright, W. E. (1989). *Exp. Cell Res.*, **184**, 109.
73. Jha, K. K., Banga, S., Palejwala, V., and Ozer, H. L. (1998). *Exp. Cell Res.*, **245**, 1.
74. Allsopp, R. C., Vaziri, H., Patterson, C., Goldstein, S., Younglai, E. V., Futcher, A. B., *et al.* (1992). *Proc. Natl. Acad. Sci. USA*, **89**, 10114.
75. Bodnar, A. G., Ouellette, M., Frolkis, M., Holt, S. E., Chiu, C., Morin, G. B., *et al.* (1998). *Science*, **279**, 349.
76. Hahn, W. C., Counter, C. M., Lundberg, A. S., Beijersbergen, R. L., Brooks, M. W., and Weinberg, R. A. (1999). *Nature*, **400**, 464.
77. Weitzman, J. B. and Yaniv, M. (1999). *Nature*, **400**, 401.
78. MacDonald, C. (1992). In *Animal cell biotechnology*, Vol. 5 (ed. R. E. Spier and J. B. Griffiths), pp. 47–73. Academic Press, London.
79. MacDonald, C. (1990). *Crit. Rev. Biotechnol.*, **10**, 155.
80. Colgin, L. M. and Reddel, R. R. (1999). *Curr. Opin. Genet. Dev.*, **9**, 97.
81. Yeager, T. R. and Reddel, R. R. (1999). *Curr. Opin. Biotech.*, **10**, 465.
82. Peden, K. W. C., Charles, C., Sanders, L., and Tennekoon, G. I. (1989). *Exp. Cell Res.*, **185**, 60.
83. Strauss, M., Hering, S., Lubbe, L., and Griffen, B. E. (1990). *Oncogene*, **5**, 1223.
84. Rossi, F. M. V. and Blau, H. M. (1998). *Curr. Opin. Biotech.*, **9**, 451.
85. Gossen, M. and Bujard, H. (1992). *Proc. Natl. Acad. Sci. USA*, **89**, 5547.
86. Gossen, M., Freundlieb, S., Bender, G., Muller, G., Hillen, W., and Bujard, H. (1995). *Science*, **268**, 1766.
87. Jat, P. S. and Sharp, P. A. (1986). *J. Virol.*, **59**, 746.
88. Masters, J. R. W., Twentyman, P., Arlett, C., Daley, R., Davis, J., Doyle, A., *et al.* (1999). *UKCCCR guidelines for the use of cell lines in cancer research.* UKCCCR, PO Box 123, Lincoln's Inn Fields, London WC2A 3PX, UK.
89. Hummeler, K., Davidson, W. L., Henle, W., LaBoccetta, A. C., and Ruch, H. G. (1959). *N. Eng. J. Med.*, **261**, 64.
90. Burns, P. A., Jack, A., Neilson, F., Haddow, S., and Balmain, A. (1991). *Oncogene*, **6**, 1973.

Chapter 7
Specific cell types and their requirements

J. Denry Sato and David W. Barnes
National Stem Cell Resource, American Type Culture Collection, 10801 University Boulevard, Manassas, VA 20110, USA.

Izumi Hayashi
Department of Molecular Genetics, Beckman Research Institute of the City of Hope, 1450 E. Duarte Road, Duarte, CA 91010, USA.

Jun Hayashi, Ginette Serrero, and Daniel J. Sussman
Department of Pharmaceutical Sciences, University of Maryland School of Pharmacy, 20 North Pine Street, Baltimore, MD 21201-1180, USA.

Hiroyoshi Hoshi
Research Institute for Functional Peptides, Yamagata 990, Japan.

Tomoyuki Kawamoto
Department of Biochemistry, Okayama University School of Dentistry, Okayama, Japan.

Wallace L. McKeehan
Center for Cancer Biology and Nutrition, Albert B. Alkek Institute for Biosciences and Technology, Texas A&M University, 2121 Holcombe Blvd., Houston, TX 77030-3303, USA.

Ryoichi Matsuda
Department of Biology, Tokyo University, Tokyo 153, Japan.

Koichi Matsuzaki
Third Department of Internal Medicine, Kansai Medical University, 10-15 Fumizono-cho, Moriguchi City, Osaka 570-8507, Japan.

Tetsuji Okamoto
Department of Molecular Oral Medicine and Maxillofacial Surgery I, Hiroshima University School of Dentistry, Hiroshima 734, Japan.

Mikio Kan
Zeria Pharmaceutical Co. Ltd., Central Research Laboratories, 2512-1 Oshikiri, Kohnan-machi, Ohsato-gun, Saitama 360-01, Japan.

1 Introduction

Although attempts at maintaining tissue explants *in vitro* have been reported since 1907, the beginning of the modern era of cell culture was marked by the establishment of the first cell lines, the mouse L-cell fibroblast line and the human HeLa cervical adenocarcinoma line, and by initial studies to determine the nutritional requirements of cells in culture (1–3). However, it was not until the 1960s that functional differentiated mammalian cell lines were first established from transplantable tumours (4, 5) following the realization that measures had to be taken to prevent the routine overgrowth of differentiated cells by fibroblasts in culture (6). A wide range of normal, immortalized, and transformed cell types can now be maintained or propagated *in vitro* (7–13). In general, normal cells, which may divide slowly or intermittently *in vivo*, exhibit a reduced capacity to proliferate *in vitro* in comparison with their partially or fully transformed counterparts. Furthermore, cells that cease dividing *in vivo* after undergoing terminal differentiation, such as neurons, skeletal muscle cells, and adipocytes, do not proliferate *in vitro*. However, as the growth requirements of individual cell types become more explicitly defined and as the interactions between cells and their environments are better understood, cells are able to survive longer or proliferate to a greater extent in culture. Although cells in culture, particularly continuous cell lines, may not display all the properties of differentiated cells *in vivo*, strides are being made towards the goal of being able to investigate physiologically relevant properties of specialized cells in completely defined culture environments.

In this chapter we will provide an outline of general principles for culturing specialized cell types, and we will then illustrate how these principles have been used to define the requirements *in vitro* of selected types of cells. For the sake of brevity we will not provide an exhaustive review of all the culture conditions or media formulations that have been used to culture each cell type. For additional information the reader is referred to refs 7–13, which include protocols for culturing a variety of cells *in vitro*.

2 General principles

Culture media for a number of individual types of cells have been improved through the optimization of the compositions of basal nutrient media (8) and the replacement of serum with purified protein and non-protein supplements (9, 10). These complementary strategies have demonstrated that while the compositions of basal media can be improved for individual cell types, they do not usually provide for optimal growth without added hormones and growth factors; conversely, combinations of purified supplements used for individual cell types can often be simplified when added to optimized basal medium. The combination of these two experimental approaches to defining the growth requirements of cells in culture has produced media formulations that minimize or eliminate the use of serum or other undefined supplements, such as conditioned medium or tissue extracts, and has given rise to the following concepts.

(a) Cells require a quantitatively balanced set of nutrients of which some may be cell type-specific.
(b) Cell growth and differentiated functions are regulated by overlapping sets of hormones, polypeptide growth factors, and extracellular matrix proteins, many of which are present in serum.
(c) Polypeptide and non-polypeptide regulators of cellular functions can be produced *in vivo* by the same cells that respond to them (autocrine factors) or by neighbouring cell types (paracrine factors) in addition to the distant tissue sources of classical hormones.
(d) Polypeptide growth factors can be produced by a diversity of cell types and they usually act on a multiplicity of target cells.
(e) Most cells in culture respond to the serum components insulin or insulin-like growth factors, transferrin complexed with iron, and one or more species of plasma lipids.
(f) Immortalized or transformed cells usually exhibit a reduced serum or growth factor requirement when compared with their normal counterparts, but often remain responsive to particular growth factors.

This relatively small number of general principles can be used by the researcher to take a rational approach to defining the survival, growth, and differentiation requirements of specific types of cells or to improve upon culture media described in the literature. Although understanding the requirements of cells *in vitro* is not a research priority for all investigators, any study of cellular physiology that uses cell culture techniques can only be enhanced by maintaining cells *in vitro* under conditions that are as completely defined as possible. In this way the behaviour of isolated cells in culture can be better correlated with the activities of cells in the intact animal. The growing availability of purified reagents and the continuing identification and characterization of novel growth and differentiation factors makes this approach to experimentation increasingly more feasible for a wider range of cell types.

3 Growth requirements of cells *in vitro*

3.1 Epithelial cells

3.1.1 Epidermal keratinocytes

Historically, normal human epithelial cells, including epidermal keratinocytes, were very difficult to maintain *in vitro*, and they could not be passaged. In retrospect, the difficulty in culturing keratinocytes resulted entirely from an insufficient understanding of the nutrient and growth factor requirements of this cell type. The use on keratinocytes of culture media and culture conditions that had been optimized mainly for fibroblasts promoted overgrowth by fibroblasts and inhibited keratinocyte proliferation in part by stimulating dividing cells to differentiate terminally. The first breakthrough in culturing human keratinocytes was the method of Rheinwald and Green (14) in which keratinocytes could be grown and passaged on a feeder layer of irradiated 3T3 fibroblasts; the cells proliferated in Dulbecco's modified Eagle's medium (DMEM) supplemented with fetal bovine serum (FCS) and hydrocortisone, and they were responsive to epidermal growth factor (EGF). Ham and his colleagues went on to refine these culture conditions for the clonal growth of keratinocytes by optimizing nutrient and growth factor concentrations and adding trace elements such that feeder cells were eliminated and serum was replaced by bovine pituitary extract (15, 16). In this nutrient medium, MCDB 153, keratinocyte growth was stimulated by EGF, insulin, transferrin, hydrocortisone, ethanolamine or phosphoethanolamine, and acidic or basic fibroblast growth factor (aFGF/FGF-1; bFGF/FGF-2) (17, 18). Transferrin, an iron transport protein found in serum, could be replaced by ferrous sulfate, bovine pituitary extract could be replaced by ethanolamine and phosphoethanolamine, and EGF could be replaced by FGF (15, 17). In addition to these growth requirements it was found that a low Ca^{2+} concentration (< 0.1 mM) was necessary for keratinocyte proliferation; in the presence of higher concentrations of Ca^{2+} the cells would terminally differentiate (19). Coating of culture surfaces with attachment factors considerably increased keratinocyte plating efficiency (20). In a low calcium nutrient medium supplemented with insulin, transferrin, ethanolamine, hydrocortisone, EGF, and pituitary extract or FGF-1 or FGF-2 (*Table 1*), normal human keratinocytes can be serially passaged on type I collagen-coated plates for approximately 40 population doublings. As MCDB 153 was optimized for clonal growth, media formulations with higher concentrations of nutrients other than calcium may be useful in maintaining higher density cultures. EGF can be replaced in keratinocyte medium by the homologous polypeptide transforming growth factor-α (TGF-α); the finding that TGF-α is synthesized by normal human keratinocytes (21) suggests that it is a naturally occurring autocrine factor for these cells. A novel and specific polypeptide growth factor for keratinocytes, keratinocyte growth factor (KGF or FGF-7), has been cloned and sequenced (22). FGF-7 is produced by stromal fibroblasts and not by epithelial cells, which suggests that it is a paracrine factor involved in interactions between dermis and epidermis *in vivo*. However, FGF-7 has not been shown to be required for keratinocyte growth *in vitro*.

3.1.2 Sub-maxillary gland epithelial cells

Studies of androgen-dependent sub-maxillary gland functions have been possible *in vivo*, but suitable cell culture models for *in vitro* studies were difficult to develop. Long-lived primary cultures of epithelial cells could be established from tissue explants in Waymouth's MB752/1 medium supplemented with calf serum, insulin, and hydrocortisone (23), but those cells lost the ability to proliferate and were overgrown by fibroblasts when passaged. Serial propagation of mouse sub-maxillary gland (MSG) epithelial cells *in vitro* became practical with the finding that low calcium nutrient medium (< 0.1 mM Ca^{2+}) permitted the growth of these cells while reducing terminal differentiation and inhibiting fibroblast proliferation. In the presence of insulin and transferrin under serum-free conditions, EGF or FGF-1 is essential for proliferation and either growth factor can be used to establish MSG epithelial cell lines.

Protocol 1

Primary culture of MSG epithelial cells

Reagents

- Seven-week-old BALB/c male mice
- Ca^{2+}- and Mg^{2+}-free PBS (Life Technologies)
- Rat tail type I collagen (UBI)
- MCDB 152 medium (Kyokuto Pharmaceutical Industrial Co.)
- Kanamycin sulfate (Sigma)
- Bovine insulin (Sigma)
- Human transferrin (Fe^{3+}-free) (Sigma)
- 2-aminoethanol (Sigma)
- 2-mercaptoethanol (Sigma)
- Sodium selenite (Sigma)
- Oleic acid (Sigma)
- Fraction V BSA (fatty acid-free) (Serologicals Proteins, Inc.)
- Mouse EGF (receptor grade) (UBI) or bovine FGF-1 (UBI)
- Porcine heparin (Sigma)
- Trypsin (Difco Laboratories)
- Soybean trypsin inhibitor (Sigma)

Method

1. Excise sub-maxillary glands from mice sacrificed by cervical dislocation. Free the glands of connective tissue and wash them in Ca^{2+}- and Mg^{2+}-free PBS.
2. Mince the glands, and plate them in serum-free medium in a 60 mm culture dish coated with rat tail type I collagen at 10 μg/ml in PBS (see Chapter 5, *Protocol 2*).
   (a) The basal medium consists of MCDB 152 medium (21) containing sodium pyruvate, sodium bicarbonate, Hepes buffer, and 90 μg/ml kanamycin sulfate. The Ca^{2+} concentration of this basal medium is 30 μM.
   (b) Complete medium consists of basal medium supplemented with the following components:
   - bovine insulin, 10 μg/ml
   - human transferrin (Fe^{3+}-free), 5 μg/ml
   - 2-aminoethanol, 10 μM
   - 2-mercaptoethanol, 10 μM
   - sodium selenite, 10 nM
   - oleic acid, 4 μg/ml complexed with Fraction V BSA (fatty-acid-free), 1 mg/ml
   - mouse EGF, 10 ng/ml, or bovine FGF-1, 1 ng/ml, with porcine heparin, 10 μg/ml

   These supplements are made as sterile 100× stock solutions and stored at 4 °C.
3. Incubate the cells at 37 °C in a humidified atmosphere of 5% CO_2.
4. Subconfluent primary cultures of MSG epithelial cells can be passaged after 14–21

Protocol 1 continued

days. The cells are trypsinized with cold 0.05% trypsin in 0.04% EDTA for 10 min. The trypsin is inactivated with 3 ml of 0.1% soybean trypsin inhibitor in Ca^{2+}- and Mg^{2+}-free PBS for 30 min at 4 °C. The cells are recovered by centrifugation, resuspended in complete serum-free medium, and subcultured at a density of 1×10^5 cells per 60 mm collagen-coated dish. Fibroblasts initially present in the primary cultures do not survive to the first passage. Cells explanted into medium containing 1 mM Ca^{2+} become highly keratinized and cannot be successfully subcultured.

5. Passage continuous cultures of MSG epithelial cells every six days in the medium used for their initial isolation.

Three cell lines established in low calcium serum-free medium containing EGF or FGF-1 or both growth factors have undergone more than 45 passages without exhibiting a reduced proliferative capacity (T. Okamoto *et al.*, unpublished results). The cells proliferate best in medium containing both EGF and FGF-1, but their rate of growth is only slightly reduced in medium with either EGF or FGF-1. The cells isolated in the presence of EGF have become aneuploid while those isolated in the presence of FGF-1 have remained diploid. This cell line, designated MSG-β2, is therefore similar to the EGF-dependent SFME mouse embryo cell line (24) isolated in serum-free medium containing insulin, transferrin, fibronectin, sodium selenite, high density lipoprotein, and EGF, which retained a diploid karyotype for over 200 population doublings. Further characterization of SFME cells has shown that they are neural stem cells, which undergo apoptosis in the absence of EGF (25). The MSG-β2 and SFME cell lines are both growth-inhibited by serum and have remained non-tumorigenic.

FGF-1 has been found to be a normal product of the mouse sub-maxillary gland (26). However, like EGF and nerve growth factor (NGF), sub-maxillary gland FGF-1 is androgen-dependent, and it is therefore expressed in a sexually dimorphic manner. FGF-1 levels in sub-maxillary glands of normal males increased with age from three to eight weeks, while it was low and did not increase with age in normal female mice (26). FGF-1 expression did not increase in sub-maxillary glands of castrated males, and it could be induced in female glands by daily administrations of testosterone.

These results combined with those from the *in vitro* culture experiments described above suggest that FGF-1 has important roles in the growth and differentiation of the sub-maxillary gland.

3.1.3 Prostate epithelial cells

The function and proliferation of normal prostate epithelial cells *in vivo* are dependent on the pituitary gland (27). Ablation experiments with whole animals demonstrated that the effects of the pituitary on the prostate were mediated mainly by androgen produced by the testes in response to gonadotropins. In addition, other pituitary factors such as prolactin, growth hormone, and adrenocorticotropic hormone (ACTH) have been suggested to be involved in the maintenance and function of the prostate. A role for androgen in regulating gene expression in the prostate has been established (28), but a direct mitogenic role of androgen in prostate epithelial cell growth has been difficult to define. Although androgens cause epithelial cell hyperplasia of prostate tissue *in vitro*, they are not unambiguously mitogenic for isolated normal prostate epithelial cells in culture (29). Furthermore, prostatic adenocarcinomas, which are often initially androgen-dependent, invariably give rise to androgen-independent tumours (30).

With the development of serum-free culture conditions for normal rat prostate epithelial cells, it was possible to clarify the hormonal growth requirements of these cells and to create a base with which to compare the growth requirements of prostatic tumour cells. This process allowed prostate epithelial cells to proliferate in primary cultures while suppressing the growth of fibroblasts (29). An optimized nutrient medium for rat prostate epithelial cells, WAJC 404 (29), was developed from MCDB 151 medium which had been created for the clonal growth of human keratinocytes in small amounts of dialysed serum protein (31). Two differences between the media reflecting different growth requirements of the two cell types were the higher Ca^{2+} concentration of WAJC 404 (130 μM) and the replacement of cysteine with cystine. Normal prostate epithelial cells proliferated equally well in WAJC 404 medium supplemented with insulin, EGF, cholera toxin (an adenylate cyclase activator), prolactin, dexamethasone, and bovine pituitary extract (*Table 1*) as they did in basal medium containing horse serum. Subsequently, the active component of bovine pituitary extract was found to be a protein factor designated prostatropin (32), which was later shown to be identical to FGF-1. A similar complete culture medium, based on PFMR-4A nutrient medium (33), was shown to support the clonal growth of normal human prostate epithelial cells except that a higher concentration of bovine pituitary extract was required (100 μg/ml), and it could not be replaced by FGF-1. FGF-1 also could not completely replace pituitary extract in cultures of rat prostate tumour cell lines; at low density these cells required either FGF-1 or EGF in addition to lipoproteins or BSA–oleic acid, in place of pituitary extract (32).

A number of growth factors, hypothalamic releasing factors, and hormones derived from the pituitary and other endocrine organs were tested for mitogenic activity on rat prostate epithelial cells (29). The effect of prolactin on cell growth was variable, and the following hormones had no effect on cell proliferation: epinephrine, follicle-stimulating hormone, luteinizing hormone, luteinizing hormone releasing factor, multiplication stimulating activity, oxytocin, platelet-derived growth factor, progesterone, prostaglandins $F_{2\alpha}$, E_1, or E_2, growth hormone, thyrotropin, thyrotropin-releasing factor, thyroxine, and vasopressin. In addition, under serum-free culture conditions, androgen was not directly mitogenic for primary cultures of rat prostate epithelial cells. Rat prostate epithelial cells express receptors for and respond to FGF-7 (KGF), which is produced by prostate stromal cells in response to androgen (34). These results generated *in vitro* suggest that the pituitary regulates prostate function in an indirect manner:

androgen produced in response to gonadotropins in turn induces the synthesis of FGF-7 by prostate stromal cells, which acts directly on prostate epithelial cells. A testable prediction of this hypothesis is that androgen-independent prostatic tumours should exhibit an abnormal response to FGF-7, and further studies on the malignant transformation of rat prostate epithelial cells suggest this is indeed the case (35). As the cells become malignant, they lose the expression of the FGF-7-binding isoform of FGF receptor 2 (FGFR-2IIIb) through splice switching, and the level of FGFR-2 expression is reduced. In addition the malignant epithelial cells acquire the ability to express FGFR-1, a normal product of stromal cells, which mediates mitogenic responses to FGF-1 and FGF-2 but not FGF-7. Thus, through specific alterations in FGF receptor expression, transformed prostate epithelial cells become both refractory to FGF-7 and responsive to FGF-1 and FGF-2. Transfections with FGF-7 receptor (FGFR-2IIIb) or FGF-2 receptor (FGFR-1) expression constructs indicated that FGF-7 receptors were associated with a non-malignant phenotype while FGFR-1 was associated with the malignant progression of prostate epithelial cells (35).

3.1.4 Mammary epithelial cells

Although human mammary carcinoma cell lines can be propagated in serum-containing or serum-free media (36, 37), normal mammary epithelial cells have proved difficult to maintain in culture under any conditions. Stampfer and colleagues were able to passage mammary epithelial cells serially several times on feeder cells in a medium containing FCS, hormones, cholera toxin, and medium conditioned by epithelial cells (38). Through basal medium optimization and modification of the medium supplements used to culture normal human keratinocytes (15), the undefined components of culture medium were reduced and then eliminated (39). As with keratinocytes, whole bovine pituitary extract proved extremely useful. The addition of transferrin and bovine pituitary extract enhanced the plating efficiency and clonal growth of mammary cells in the absence of feeder cells, cholera toxin, and conditioned medium. Adjustments to the concentrations of L-glutamine, sodium pyruvate, zinc sulfate, and L-cysteine in the nutrient medium allowed the concentration of pituitary extract to be reduced and then replaced by prolactin and prostaglandin E_1 (PGE_1). Unlike keratinocytes, mammary epithelial cells required a high Ca^{2+} concentration (2 mM) for optimal growth. Cells isolated in the resulting defined medium (*Table 1*) proliferated for 10–20 passages and their growth was inhibited by serum. Although these epithelial cells were shown to synthesize fibronectin, it is possible that the use of fibronectin- or collagen-coated culture dishes could increase plating efficiency or cell growth with increasing passage number. In addition, FGF may stimulate the growth of mammary epithelial cells as it does the growth of epithelial cells from other tissue sources.

A modification of the method of Tomooka *et al.* (40) for establishing primary cultures of mouse mammary epithelial cells is provided in *Protocol 2*. In this protocol isolated mammary epithelial cells are cultured on collagen-coated tissue culture plates whereas in the original procedure the cells were cultured within a collagen gel. Insulin, BSA, and EGF were found to be essential components of the culture medium (*Table 1*), and LiCl at 5–10 mM was found to enhance the proliferation of collagen-embedded cells (40). The proliferative capacity of normal mammary epithelial cells may be extended by using a low-calcium basal medium such as MCDB 152 or MCDB 153 in place of DMEM/F12.

Protocol 2

Primary culture of mouse mammary epithelial cells (40)

Equipment and reagents

- 150 μm mesh
- Collagenase (Worthington)
- Ca^{2+}- and Mg^{2+}-free PBS (Life Technologies)
- DNase (Life Technologies)
- Pronase (Calbiochem)
- Percoll (Pharmacia)
- DMEM/F12 (Life Technologies)
- EGF (UBI)
- Insulin (Sigma)
- Transferrin (Sigma)
- Cholera toxin (Sigma)
- Bovine pituitary extract (UBI) or FGF-1 (UBI)
- Heparin (Sigma)

Method

1. Sacrifice female mice by cervical dislocation.
2. Sterilize the mice with 70% ethanol.
3. Excise mammary glands with the surrounding adipose tissue from five mice, and mince the glands with scalpels.
4. Digest the minced tissue with 0.1% collagenase in Ca^{2+}- and Mg^{2+}-free PBS (10 ml/g tissue) containing 1 mg/ml BSA at 37 °C for 90 min.
5. Pass the suspension through a 150 μm mesh, and collect the tissue fragments by centrifugation at 80 g for 5 min. Resuspend the pellet in a small volume of PBS, add several drops of 0.04% DNase in PBS, and digest the pellet with 0.1% pronase containing 1 mg/ml BSA for 30 min at 37 °C. Add 0.1 volume of FCS to the cell preparation, collect the cells by centrifugation, and resuspend the cell pellet in 2 ml of culture medium.
6. Mix the cells in 30 ml of 42% Percoll in PBS and centrifuge at 10 000 g for 1 h. Harvest the layer of epithelial cells (1.07–1.08 g/ml), and wash them in PBS.
7. Plate the cells on a collagen-coated substrate (see *Protocol 1*) in culture medium consisting of DMEM/F12 medium with the following supplements:
   - EGF, 10 ng/ml
   - insulin, 10 μg/ml
   - transferrin, 10 μg/ml
   - cholera toxin, 10 ng/ml
   - Fraction V BSA, 1–5 mg/ml

Protocol 2 continued

This medium may be modified by the addition of 10–100 μg/ml bovine pituitary extract or 10 ng/ml FGF-1 and 10 μg/ml heparin.

8 Incubate the cells at 37 °C in a humidified atmosphere of 5% CO_2.

3.1.5 Hepatocytes

Liver is composed of parenchymal hepatocytes and non-hepatocytes such as sinusoidal endothelial cells, Kupffer cells, fat-storing (Ito) cells, and fibroblasts. Unlike most other organs and tissues, the development of a procedure for perfusing liver with collagenase (41) provided highly pure populations of hepatocytes (> 90%) with good viability (> 80%). Despite the fact that hepatocytes are the source of many of the components of serum, they have proved refractory to serial cultivation. Freshly isolated hepatocytes attach well to culture dishes coated with fibronectin (42) or collagen (43), forming monolayers; they express liver-specific proteins and they respond to growth factors. EGF, TGF-α, FGF-1, and hepatocyte growth factor (HGF or scatter factor) stimulate hepatocyte DNA synthesis or cell division (44–47) while transforming growth factor-β (TGF-β) inhibits DNA synthesis induced by these growth factors (48). The hormones insulin, glucagon, and glucocorticoid support increased cell survival. A typical serum-free growth medium for primary rat hepatocyte cultures consists of Williams' E medium supplemented with insulin, dexamethasone, and the protease inhibitor aprotinin on a fibronectin-coated culture surface (49) (*Table 1*).

Although the mechanisms have not been elucidated, dimethyl sulfoxide (DMSO) (50) or phenobarbital (51) support the increased maintenance of liver-specific functions by primary hepatocyte cultures. Co-culturing hepatocytes with non-hepatocytes will also prolong the expression of liver-specific functions and hepatocyte survival (52). A method for isolating sinusoidal endothelial cells, Kupffer cells, and fat-storing cells has been developed (53), and these cells can be maintained in culture. Since hepatocyte growth and differentiation should be regulated at least in part by factors provided by non-hepatocyte cells, the characterization of these cells and their products will be important to understand the paracrine interactions between the various types of liver cells and to decipher fully the growth requirements of hepatocytes. This information should lead to the development of culture conditions that allow hepatocytes to be passaged *in vitro*. The growth factors EGF and HGF may stimulate hepatocyte differentiation as well as proliferation and may therefore be inappropriate for stem cell renewal (54). Isolation of stem cell-rich subpopulations of liver cells may be necessary to identify growth factors involved in the renewal of stem cells that give rise to hepatocytes.

3.1.6 Thymic epithelial cells

Several thymic epithelial cell culture systems have been described in the literature (55–58). Thymic epithelial cells are relatively difficult to culture since in the thymus, epithelial cells are largely outnumbered by thymocytes. Moreover, epithelial cells are not the only non-lymphoid cell component found in the thymus. Fibroblast overgrowth and macrophage contamination are common problems encountered in thymic epithelial cell cultures. In addition, many of the thymic epithelial cell culture systems suffer from a common problem of epithelial cell cultures, that fully differentiated functional epithelial cells will not divide. Thus, it is difficult to obtain continuous functional thymic epithelial cell lines. Furthermore, selection towards undifferentiated (or dedifferentiated) epithelial cell growth often takes place resulting in growth of cells with limited expression of differentiated functions.

A number of cell lines have been established using techniques such as chemical transformation (59) and SV40 transformation (60). However, these transformed thymic epithelial cell lines often lose their functional properties and are often not suitable for physiological studies. To overcome these difficulties, a serum-reduced selective culture system for growing thymic epithelial cells was developed and used to establish cloned functional thymic epithelial cell lines from rat (61). The following strategy was used (*Protocol 3*):

(a) A low Ca^{2+}-containing basal medium was used to maintain thymic epithelial cells in an undifferentiated state.

(b) Low serum concentrations discouraged fibroblast outgrowth.

(c) Supplementation of the medium with hormones and factors that stimulated epithelial cell growth was used to promote selectively the growth of undifferentiated epithelial cells.

Protocol 3

Primary culture of rat thymic epithelial cells (61)

Equipment and reagents

- T25 flasks
- Four- to five-week-old male rats
- Dexamethasone (Sigma)
- Type II collagenase (Sigma)
- 0.125% trypsin/0.05% EDTA
- Iron-supplemented calf serum (HyClone)
- WAJC 404 medium (see Section 3.1.3)
- High-glucose DMEM (Life Technologies)
- Bovine insulin (Sigma)
- Human transferrin (Sigma)
- Cholera toxin (Sigma)
- Mouse EGF (UBI)

Method

1 Inject rats with 0.5 mg dexamethasone per 100 g of body weight. Sacrifice them three days later by CO_2 inhalation.

2 Surgically remove and mince the thymuses. Wash the tissue fragments twice with PBS, and digest them with 400 U/ml Type II collagenase for 4 h at 37 °C. Allow clumps of cells to settle under unit gravity, and discard the cortisone-resistant thymocytes remaining in suspension.

Protocol 3 continued

3. Wash the recovered tissue fragments twice with PBS, and digest them with 0.125% trypsin/0.05% EDTA for 20 min at 37 °C. Add cold 10% iron-supplemented calf serum (sCS) in PBS. Collect the cells by centrifugation, resuspend the cells in culture medium, and plate them in T25 flasks. This isolation procedure gives rise to 1–2 $\times$ 10^7 viable epithelial cells per thymus.
4. The basal culture medium chosen for thymic epithelial cells is WAJC 404A. Mix Ca^{2+}-free WAJC 404 medium with high-glucose DMEM at the ratio of 92.5:7.5 (v/v) to bring the Ca^{2+} concentration to 100 μM, and supplement with 2 mM pyruvate and 15 mM Hepes to give WAJC 404A medium. For complete medium (*Table 1*), add the following supplements to WAJC 404A basal medium:
   - bovine insulin (I), 10 μg/ml
   - human transferrin (T), 10 μg/ml
   - dexamethasone (D), 10 nM
   - cholera toxin (CT), 20 ng/ml
   - mouse EGF (E), 10 ng/ml
   - iron-supplemented calf serum (sCS), 2% by volume
5. Incubate the cells at 37 °C in a humidified atmosphere of 5% CO_2. Selective growth of thymic epithelial cells takes place under these culture conditions free of macrophage contamination, and the epithelial cells can be passaged.

Subconfluent primary cultures of thymic epithelial cells were subjected to mild trypsin treatment (0.05% trypsin, 0.025% EDTA for 10 min at room temperature) to remove cells from the periphery of colonies. A rapid outgrowth of epithelial cells occurred from these colonies over several days. Trypsinization of primary colonies was repeated, and after several such treatments the detached cells were used to establish secondary cultures. These cultures could be continuously passaged for several months.

A cloned cell line was established from passaged thymic epithelial cell cultures. The cells were passaged at a low seeding density in WAJC 404A medium supplemented with I, T, D, E (see *Protocol 3* for abbreviations), 2% sCS, and 20% conditioned medium from the parent culture. Conditioned medium was prepared by culturing primary thymic epithelial cells in complete medium for four days; the medium was collected and stored at 4 °C. When the colonies reached 100–200 cells, they were passaged using cloning rings (see Chapter 8, *Protocol 5*). After three successive subclonings, the TEA3A1 rat thymic epithelial cell line was obtained. Using an identical approach, a mouse thymic epithelial cell line designated BT1B was established. These cell lines are now adapted to grow in WAJC 404A medium supplemented with I, T, D, and 2% sCS. While stock cells are maintained in a low-Ca^{2+} medium, experimental cells are cultured in a high-Ca^{2+} medium consisting of a 1:1 (v/v) mixture of WAJC 404A and DMEM which gives a Ca^{2+} concentration of 1.2 mM. In high-Ca^{2+} medium, the cells differentiate, stop proliferating, and form desmosomes and highly keratinized structures that resemble Hussel's bodies. Both BT1B and TEA3A1 cells express two neuroendocrine cell markers, namely GQ ganglioside and S100 protein. These differentiated cells will produce the thymic hormones thymosin-α1 and thymulin and various other cytokines including interleukin IL-1α, IL-6, and IL-7. The production of thymic hormones and cytokines by these cells is regulated by hormones such as glucocorticoids, triiodothyronine, and prostaglandins. These cell lines provide useful models for studying thymic endocrine physiology and regulation of early T cell development in the thymus.

3.2 Mesenchymal cells

3.2.1 Fibroblasts

Fibroblasts exist widely in mammalian tissues and are a major component of connective tissue. When mechanically or enzymatically disrupted organs are cultured in most serum-supplemented nutrient media, fibroblasts grow out of explants rapidly and become the predominant cell type in the culture. After the cells are passaged, fibroblasts are often the only proliferating cells that survive. One of the reasons for this outcome is that the commonly used nutrient media were optimized to support fibroblast cell lines; as mentioned above, one of the keys to establishing successful epithelial cell cultures was the development of culture conditions that suppressed fibroblast survival while supporting epithelial cell growth. Although normal human fibroblasts have a spindle-shaped morphology in culture, not all spindle-shaped cells are fibroblasts. There are no known fibroblast-specific markers, but fibroblasts produce and release type I collagen, which is characteristic of connective tissue. Mouse embryonic fibroblasts such as 3T3 cells can be induced to differentiate into adipose cells (62), which may indicate that they are undifferentiated or primitive mesodermal cells.

Although they exhibit a limited lifespan *in vitro* (63), fibroblasts are the least difficult cells to propagate in serum-containing medium. Fibroblasts were the first normal human cells to be grown in defined medium. Starting with F12 medium, Ham and his colleagues developed an optimized nutrient medium, MCDB 104 (64), which included selenium as a required trace element and would support clonal cell growth in dialysed serum. The serum requirement of a modified MCDB 104 formulation was subsequently replaced by the hormones insulin, dexamethasone, PGE_1, $PGF_{2\alpha}$, EGF, and an emulsion of defined lipids composed of soya bean lecithin, cholesterol, and sphingomyelin (65) (*Table 1*). Of these additives insulin, EGF, cholesterol, lecithin, and sphingomyelin elicited major growth responses while dexamethasone and the prostaglandins were of lesser importance. Vitamin E, dithiothreitol, glutathione, and phosphoenolpyruvate had marginal effects on clonal cell growth, and dibutyryl cyclic GMP and dibutyryl cyclic AMP had no effect. PDGF, which increased the saturation density of WI-38 fibroblasts (66), and FGFs 1 and 2 (18) were also found to be significant but non-essential mitogens for normal human fibroblasts. Non-delipidated Fraction V BSA was found to increase the longevity of serially passaged lung fibroblasts from 20 to 80 population doublings (67). Transferrin (66), fibronectin (66, 67), and type I collagen may also be useful protein supplements for serum-free culture of normal fibroblasts. Three media formulations developed for normal human fibroblasts are provided in *Table 1*, and the reader is referred to Greenwood *et al.* (68) for a detailed discussion of culture methods for immortalized and oncogenically transformed fibroblasts.

3.2.2 Vascular endothelial cells

Endothelial cells form a single cell layer that lines the inner surface of large and small blood vessels; capillaries are endothelial cell tubes surrounded by a basement membrane. Endothelial cells have been an important focus of research in the past three decades because a number of pathological conditions including cardiovascular diseases and solid tumour progression directly affect the growth and function of these cells. A major limitation that had to be overcome in order to study endothelial cell physiology *in vitro* was the difficulty in establishing culture systems that allowed the survival and expansion of the small number of normally quiescent endothelial cells that could be recovered from vascular tissue samples by perfusion with collagenase (69) or trypsin. Although normal human endothelial cells cannot yet be propagated long-term under defined culture conditions, efforts to prolong the lifespan of endothelial cells *in vitro* illustrate how newly discovered growth factors can lead to advances in culturing cells that were previously difficult to maintain *in vitro*.

Bovine aortic arch endothelial cells could be serially passaged for 35–40 population doublings in medium with 30% serum (70), and they could be grown on extracellular matrix (ECM) in serum-free medium in the presence of high density lipoprotein (HDL) with or without transferrin and FGF-2 (71). Bovine capillary endothelial cells were serially passaged for more than eight months in serum-containing medium supplemented with medium conditioned by mouse sarcoma cells, aortic endothelial cells, or human foreskin fibroblasts (72). By contrast, human vascular endothelial cells did not proliferate well at low density in medium supplemented with serum only (73). However, in serum-containing medium supplemented with bovine brain FGF-2 and thrombin (74) or with bovine brain endothelial cell growth factor (ECGF) (75), human endothelial cells proliferated and could be passaged 20–30 times. EGF, insulin, and transferrin were found to be weak mitogens for human endothelial cells that could not replace serum or ECGF. ECGF was later found to be identical to FGF-1 and distinct from FGF-2. Thus, the discovery and purification of the first two members of the FGF family of growth factors, which remain the most potent endothelial cell mitogens known, were instrumental in the development of culture conditions under which human endothelial cells proliferated at low density (*Protocol 4* and *Table 1*). These cells can proliferate at high density in the absence of exogenous FGFs in culture dishes coated with fibronectin, collagen, or gelatin; however, this growth may result from autocrine activity of FGF-2 and FGF-1 produced by human vascular endothelial cells (76).

Protocol 4

Primary culture of human umbilical vein endothelial cells (HUVECs)

Equipment and reagents

- Locking polypropylene scissor-type forceps (Nalge Nunc International)
- Collagen-coated T25 flask
- Heparin (Sigma)
- 0.25% (w/v) trypsin/5 mM EDTA
- DMEM/F12 (Life Technologies)
- Kanamycin (Sigma)
- EGF (UBI)
- FGF-2 (UBI)
- VEGF (UBI)

Method

1. Use an unplugged sterile 1 ml disposable plastic pipette attached to a 10 cc syringe with a two-way stopcock to flush the blood from the umbilical cord vein with cold sterile PBS containing 50 μg/ml heparin. When disposing of the cord blood, treat it as biohazardous material. If blood has clotted within the vein, discard the cord.
2. Ligate one end of the umbilical cord with locking forceps, insert the 1 ml pipette into the opposite end of the cord vein taking care not to puncture the vein, and secure the pipette in place with a second pair of locking forceps.
3. Slowly inject 10–20 ml of an ice-cold solution of sterile 0.25% trypsin/5 mM EDTA in PBS into the cord vein taking care not to rupture the vein. Close the stopcock, and wait for 10 min.
4. Withdraw the trypsin solution from the vein, transfer it to a sterile 50 ml conical centrifuge tube, and add FCS to 10% (v/v) to inactivate the trypsin. Wash the vein with 10 ml of sterile PBS, and pool it with the trypsin solution.
5. Repeat steps 3 and 4 four times.
6. Collect endothelial cells by centrifugation at 150 g for 5 min. Resuspend the cells in each tube in 5 ml of DMEM/F12 (1:1, v/v) supplemented with 10% (v/v) FCS and 90 μg/ml kanamycin, and plate the cells in each isolate in a collagen-coated T25 flask. Allow the cells to attach to the substratum in a 37 °C incubator with an atmosphere of 5% CO_2.
7. Once the cells have attached and spread, replace the medium with DMEM/F12 containing the following supplements:
   - FCS, 10% (v/v)
   - EGF, 20 ng/ml
   - FGF-2, 10 ng/ml
   - VEGF, 10 ng/ml
   - heparin, 10 μg/ml
   - kanamycin, 90 μg/ml
8. To propagate HUVECs, harvest the cells just prior to confluence by trypsinization and re-plate in collagen-coated culture vessels at a split ratio of 1:4 in the medium described in step 7.

The serum requirement of human endothelial cells, which cannot at present be completely eliminated *in vitro*, may reflect a need for lipoproteins, fatty acids, and proteinase inhibitors (77). Alternatively, it may indicate that there are growth factor requirements of these cells that remain to be determined. For example, at very low concentrations, TGF-β_1 is mitogenic for endothelial cells in the presence of FGF-1 and heparin (78). In addition, a novel endothelial cell-specific growth factor, designated vascular endothelial cell growth factor (VEGF) or vascular permeability factor (VPF), has been isolated and cloned (79, 80). VEGF is a less potent endothelial cell mitogen than FGF-1 or FGF-2, it is active in the absence of FGFs, and its mitogenic effects are additive with those of FGFs (81, and J.-H. Chen and J. D. Sato, unpublished results). VEGF stimulates but is not required by human endothelial cells selected in the presence of FGF. However, it is not clear whether there exist subpopulations of endothelial cells that respond to VEGF but not to FGFs. Recent studies on VEGF-activated signalling pathways in endothelial cells indicate that the ras/raf/MEK (MAP/ERK kinase)/MAP kinase pathway and the phosphatidylinositol 3′-kinase (PI 3′-kinase)/p70 S6 kinase pathway are both required for a mitogenic response (82), while PI 3′-kinase signalling through protein kinase B (PKB/Akt) promotes endothelial cell survival (83). As relatively few direct acting endothelial cell mitogens are currently known, it is quite possible that others remain to be discovered.

3.2.3 Adipocyte precursors

Primary culture of adipocytes and their precursors is very useful to investigate the physiological mechanisms controlling cell proliferation, differentiation, and maintenance of function in adipose tissue. Although the great majority of studies concerning adipose differentiation use adipogenic cell lines, primary culture of adipocyte precursors is becoming more and more frequent. Culture of isolated pre-adipocytes was first described by Ng and colleagues: stromal cells from normal and obese patients were isolated by collagenase digestion and cultured in standard serum-supplemented medium (84). Subsequently, primary culture of pre-adipocytes freshly isolated from human neonatal adipose tissue (85) and from developing rats (86) was reported. In general, the cultures were carried out in serum-supplemented medium containing a lipid source. Although these cultures have allowed studies of the differentiation process, in general they support only limited differentiation owing most likely to the presence of large concentrations of serum which has an inhibitory effect on differentiation and which supports the growth of non-adipocytes derived from the stromal vasculature.

Substantial progress has been achieved by culturing adipocyte precursors isolated from adipose tissue in defined media developed for adipogenic cell lines such as 3T3-L1 (87), 1246 (88), and Ob17 (89). These defined media have served as the basis for the development of optimal conditions supporting the growth and differentiation of adipocyte precursors isolated from fat pads of several animal species at different stages of development. Reports in the literature have described the culture of pre-adipocytes from adipose tissue of sheep (90), man (91), adult rat (92), newborn rat (93), and adult mouse (94). For ovine pre-adipocytes, culture conditions were established for clonal cell growth. In the other cases, conditions were developed for high density monolayer cultures (10^3 to 10^4 cells/cm^2). In general, the basal medium used for these primary cultures was DMEM/F12 (91–93); ovine pre-adipocytes were grown at clonal density in MCDB 202 medium (90). DMEM/F12 could not reproducibly support the survival and growth of pre-adipocytes from 10- to 15-day-old mice (94); however, a 1:1 mixture of DMEM and low calcium WAJC 404 medium used for epithelial cells optimally supported the proliferation and differentiation of the pre-adipocytes (94). Two approaches to cell plating can be used; the pre-adipocytes can be plated in medium with 10% serum, which is replaced by defined medium after the cells have attached (91, 92), or they may be plated in defined medium on dishes coated with fibronectin (93) or collagen (G. Serrero *et al.*, unpublished results). Insulin and transferrin are commonly used supplements for pre-adipocytes. Dexamethasone has been used to culture ovine and mouse pre-adipocytes (90, 94), while T3 has been used for human and adult rat cells (91, 92). In the latter cases, the culture media were also supplemented with pantothenate and biotin.

The simplest defined culture conditions for adipocyte precursors are used to culture pre-adipocytes from newborn rats (*Protocol 5*). The medium consists of DMEM/F12 nutrient medium supplemented with insulin, transferrin, fibronectin, and FGF-2 (*Table 1*). Since cells freshly isolated from fat pads are directly plated in defined medium in the absence of serum, contamination of cultures with non-pre-adipocyte cells from the stromal vascular fraction does not occur. Under these conditions, 90% of the cells undergo differentiation in eight days.

Protocol 5

Primary culture of pre-adipocytes from newborn rat (93)

Equipment and reagents

- Filters: 253 μm and 80 μm
- Two-day-old rats
- Hank's balanced salt solution (Life Technologies)
- Collagenase (Sigma)
- BSA (Sigma)
- DMEM/F12 (Life Technologies)
- Fibronectin (UBI)
- Bovine insulin (Sigma)
- Human transferrin (Sigma)
- FGF-2 (UBI)

Method

1. Sacrifice rats by asphyxiation with CO_2, surgically remove inguinal fat pads, and place them in Hank's balanced salt solution (HBSS). Trim the fat pads of connective tissue and large blood vessels, and cut them into fragments for digestion with collagenase.
2. Incubate the fat pad fragments in 0.2% collagenase and 0.2% BSA in HBSS for 1 h at 37 °C. Disperse clumps of cells by pipetting. Filter the digest through 253 μm and 80 μm filters to remove large particulate matter.

Protocol 5 continued

3. Centrifuge the resulting cell suspension at 700 g for 10 min, and remove the floating adipocytes and the digestion buffer by aspiration. Resuspend the cell pellet in DMEM/F12 medium, and collect the pre-adipocytes by centrifugation. Wash the pre-adipocytes with DMEM/F12 twice more.
4. Plate the cells in 35 mm dishes in DMEM/F12 basal medium supplemented with the following components:
   - fibronectin, 2.5 μg/ml
   - bovine insulin, 10 μg/ml
   - human transferrin, 10 μg/ml
   - FGF-2, 2 ng/ml
5. Incubate the cells at 37 °C in a humidified atmosphere of 5% CO_2.

Adipocyte precursors from fat pads of adult rats (> three weeks) and from mice in this defined medium require the further addition of HDL and dexamethasone as was reported for adipocyte precursors from adult mice (94). These results suggest that the hormonal requirements for the proliferation and differentiation of adipocyte precursors depend on the age of the donor animal. This consideration should be kept in mind in developmental studies. Interestingly, no substantial differences were observed in the growth requirements of adipocyte precursors isolated from different fat depots such as inguinal and epididymal fat pads of the same animal. Thus, defined culture conditions allowing the proliferation and differentiation of adipocyte precursors are well established. A goal for future research is to define optimal culture conditions for the maintenance of fully mature adipocytes.

3.2.4 Muscle cells

Muscles of vertebrates are mainly of mesenchymal origin, and they can be classified as being either striated or non-striated. Skeletal and cardiac muscles are striated while smooth muscles are non-striated.

i. Skeletal and cardiac muscle cell cultures

Clonal growth and differentiation of skeletal muscle cells in culture was first reported by Konigsberg (95). The process of skeletal muscle cell differentiation *in vivo* is faithfully reproduced in culture. Therefore, skeletal muscle cells provide an excellent model for studying differentiation and morphogenesis. A procedure for establishing primary cultures of skeletal muscle cells is provided in *Protocol 6*.

Myoblasts enter the post-mitotic G_0 phase and myoblast fusion (fusion-burst) becomes evident within 48 hours after plating. Around the time of fusion-burst, transcription of muscle-specific genes is up-regulated (97, 98). The activation of muscle-specific genes is also observed in fusion-arrested myoblasts in low-calcium medium (97). Myofibrillogenesis takes place in multi-nucleated myotubes and spontaneous twitching can be observed within seven days of plating.

Protocol 6

Primary culture of skeletal muscle cells

Primary cultures can be initiated from skeletal muscle tissues of 11-day-old chick embryos, rat perinatal fetuses, or newborn rats as described below.

Reagents

- Trypsin (Sigma)
- Ca^{2+}- and Mg^{2+}-free PBS (Life Technologies)
- Collagenase (Sigma)
- Hank's balanced salt solution (Life Technologies)
- Soybean trypsin inhibitor (Sigma)
- Eagle's minimal essential medium (EMEM) (Life Technologies)
- Horse serum (HyClone)
- Embryo extract (Difco)
- Ascorbic acid (Sigma)

Method

1. Digest minced skeletal muscle tissue with 0.1% trypsin in Ca^{2+}- and Mg^{2+}-free PBS for 20 min at 37 °C. Satellite cells can also be isolated by digesting minced adult muscle with 0.2% (w/v) collagenase in Hank's salt solution followed by 0.1% (w/v) trypsin in PBS. Inactivate the trypsin with an equal volume of 0.1% soybean trypsin inhibitor in PBS. Cells enzymatically released from muscle tissue usually contain a significant number of fibroblasts in addition to myogenic cells.
2. Separate fibroblasts from myogenic cells using differential adhesion: incubate cells for 30 min on non-coated plastic culture dishes, and collect the non-adherent muscle cells. Repeat this differential cell adhesion step. Alternatively, D-arabinofuranosyl cytosine (Ara-C) can be used at 5–10 μM to eliminate dividing fibroblasts when mononuclear myoblasts begin to fuse in culture (96).
3. Plate myoblasts onto gelatin-coated dishes (see Chapter 5, *Protocol 2*) in EMEM supplemented with 10% horse serum and 2% embryo extract. It is advisable to add ascorbic acid to 100 μM to enhance collagen production.
4. Incubate the cells at 37 °C in a humidified atmosphere of 5% CO_2.

Skeletal muscle cell lines have been established. Among them mouse C2C12 (99) and rat L6E9 (100) cells are widely used. To induce muscle differentiation in these cells, culture medium is switched from growth medium of DMEM supplemented with 20% FCS to differentiation medium of 2–10% horse serum in MEM when the cells are approximately 80% confluent. Myofibrillogenesis is, however, poor in these cell lines relative to primary cell cultures. Myogenesis can also be induced in C3H10T1/2 mouse embryonic mesenchymal cells by exposing them to 3 μM 5-azacytidine (101). It is noteworthy that the myogenic regulatory gene MyoD was isolated from these cells (102). MM-1 serum-free medium of Florini and Roberts (103) supports the short-term proliferation of, and subsequent myotube formation by, the L6 myoblast cell line and primary cultures of newborn rat myoblasts.

Heart muscle cells can be isolated by trypsinization from ventricular muscles of chicken embryos (2–10 days of incubation), and fetal (12 days post-coitum or later) or newborn (0–3 days after birth) mice and rats. It is advisable to repeat the trypsin treatment two or three times to increase the yield of cells. The differential adhesion procedure can be used to eliminate fibroblasts. Non-adherent cells are pooled, washed, suspended in 5–10% FCS in MEM, and plated onto gelatin-coated plastic dishes. Myofibrillogenesis takes place in myocytes and spontaneous contractions can be observed. Heart muscle cell lines have not yet been established.

ii. Smooth muscle cell culture

Smooth muscle cells can be isolated from the aorta by explantation (104) or by enzymatic digestion (105) of the medial layer. Smooth muscle cells can also be isolated from the intima of the aorta; in this case endothelial cells must be removed from the surface by scraping. Primary cultures of adult human large vessel smooth muscle cells have been maintained in serum-free medium supplemented with human hepatoma-conditioned medium or under defined conditions in which the conditioned medium was largely replaced by PDGF (77) (*Table 1*). In the absence of serum, it was found that EGF, FGF-1, insulin, HDL, and PDGF were almost equally important for smooth muscle cell proliferation, and all five supplements were required for maximal proliferation. Procedures for establishing primary cultures of smooth muscle cells isolated from large blood vessels by enzymatic digestion (77) or by explantation (104) are provided in *Protocol 7*. Confluent smooth muscle cells typically exhibit a 'hill and valley' morphology. Cells grown to high densities express several differentiation-specific markers including vinculin, high molecular weight caldesmon, smooth muscle myosin, and α-actin.

Protocol 7

Primary culture of smooth muscle cells from large vessels (77)

Equipment and reagents

- Collagen-coated T25 flask (see Chapter 5, *Protocol 2*)
- Scalpel blade or a cotton swab
- Hepes buffer: 7.15 g Hepes, 7.07 g NaCl, 0.15 g KCl, 0.136 g KH_2PO_4, and 0.74 g dextrose per litre adjusted to pH 7.6
- Kanamycin (Sigma)
- Collagenase (Sigma)
- MCDB 107 basal medium (Kyokuto)
- Mouse EGF (UBI)
- Bovine insulin (Sigma)
- Dexamethasone (Sigma)
- HDL (Chemicon)
- FGF-1 (UBI)
- Heparin (Sigma)
- PDGF (UBI)
- DMEM (Life Technologies)

Protocol 7 continued

A. Isolation of cells by enzymatic digestion

1. Wash the vessel with sterile Hepes buffer containing 100 μg/ml kanamycin.
2. Open the vessel with a scalpel exposing the lumenal surface. Digest the vessel with 0.1% (w/v) collagenase in Hepes buffer at 37 °C for 30 min, and rinse the vessel with Hepes buffer containing kanamycin.
3. Scrape the lumenal surface with a scalpel blade or a cotton swab to remove endothelial cells, and wash the vessel with Hepes buffer containing kanamycin.
4. Incubate the vessel a second time with 0.1% collagenase solution for 30 min at 37 °C, and scrape the lumenal surface with a scalpel to harvest smooth muscle cells.
5. Collect the cells by centrifugation at 150 g for 5 min, resuspend them in culture medium, and plate them in a collagen-coated T25 flask. The culture medium consists of MCDB 107 basal medium with the following supplements:
   - mouse EGF, 10 ng/ml
   - bovine insulin, 5 μg/ml
   - dexamethasone, 1 μM
   - HDL, 50 μg/ml
   - FGF-1, 10 ng/ml
   - heparin, 10 μg/ml
   - PDGF, 2.5 U/ml
   - kanamycin, 100 μg/ml

   FCS may be added to this medium to concentrations of 2–10% (v/v).
6. Incubate the cells at 37 °C in a humidified atmosphere of 5% CO_2.

B. Isolation of cells by explantation (104)

1. Under a dissecting microscope, remove the adventitia and the outer media of vessel fragments in basal medium or PBS.
2. Wash the inner media and intima with basal medium or PBS containing 100 μg/ml kanamycin.
3. Cut the vessel into fragments approximating 1 mm^2, and transfer them to a culture vessel containing either DMEM supplemented with 10% FCS and 100 μg/ml kanamycin or the culture medium described above. Smooth muscle cells migrate within three to four weeks from explants in DMEM medium supplemented with 10% FCS, and they can be harvested by trypsin treatment.

Several smooth muscle cell lines have been established. A-10 and A7r5 are subclones of a cell line derived from the medial layer of fetal rat aorta (106). SM-3 was established from the medial layer of adult rabbit aorta (107). These cells proliferate in DMEM supplemented with 10% FCS, and they differentiate at confluence in MEM with 0.5% FCS. To maintain the cells in a proliferative undifferentiated state, they must be routinely subcultured in growth medium before reaching confluence.

3.3 Neuro-ectodermal cells

3.3.1 Stem cells for neurons and glia

Neural crest cells migrate from the dorsal neural tube of the developing embryo and give rise to neurons, glia, melanocytes, and other cell types in response to environmental signals (108). Lineage analyses of avian neural crest cells *in vitro* and *in vivo* have shown that these cells are multipotent (109–111). However, the temporal relationship between the determination of cell fate and cell differentiation has not been clearly defined for the various cell lineages.

A method of isolating and serially propagating rat neural crest cells has been described by Stemple and Anderson (112), which has allowed the developmental potential of individual cells to be examined *in vitro*. Neural crest cells that migrated from neural tubes explanted from rat embryos were harvested and plated at clonal density on plastic coated with fibronectin and poly-D-lysine in serum-free medium. Since mammalian neural crest cells express low affinity receptors for nerve growth factor (LNGFR) (114), $LNGFR^+$ cells were either first enriched by fluorescence-activated cell sorting or they were marked *in situ* with anti-LNGFR antibodies. In addition, $LNGFR^+$ cells expressed nestin, an intermediate filament protein characteristic of immature neuroepithelial cells (115). A complex serum-free medium containing chick embryo extract was therefore developed that would support the proliferation of undifferentiated $LNGFR^+$ $nestin^+$ cells (*Table 1*). In this medium individual clones doubled in size every 24 hours. After 9–14 days *in vitro*, cells within colonies differentiated as assessed by changes in morphology and expression of marker proteins. The developmental potentials of primary colonies and secondary colonies established after six days in culture were examined using antibodies to proteins expressed preferentially by neurons or glia (Schwann cells); non-neuronal non-glial cells were classified as 'O' (other) cells. The majority of primary neural crest cells were multipotent giving rise to colonies of neurons and Schwann cells, and the majority of primary colonies tested gave rise to at least one multipotent secondary colony while others were exclusively neuronal or glial. Furthermore, neural crest cells plated on fibronectin gave rise only to Schwann cells while those exposed to fibronectin and poly-D-lysine produced a majority of mixed colonies. Thus, an *in vitro* clonal assay has been used to demonstrate that individual migrating neural crest cells have stem cell properties, and they give rise to progeny of which some are also multipotent while others are committed to a particular differentiation pathway. The direction in which the cells differentiate is influenced by environmental factors such as substrate composition.

3.3.2 Oligodendrocyte-type 2 astrocyte progenitor cells

Oligodendrocyte-type 2 astrocyte (O-2A) progenitors are bipotential glial precursor cells that can differentiate into either oligodendrocytes or type 2 astrocytes in rat optic nerve cultures (116). These glial precursor cells can also be obtained through selective *in vitro* culture of mouse embryonic stem cells (see Section 3.5 and *Protocol 9*). Although fully differentiated oligodendrocytes and type 2 astrocytes do not divide, the precursor cells are of interest because they proliferate under defined culture conditions (*Table 1*), and the direction in which they differentiate is dictated by environmental factors. O-2A cells not exposed to mitogens *in vitro* do not proliferate, and differentiate prematurely into oligodendrocytes. O-2A cells treated with PDGF or grown in the presence of type 1 astrocytes divide a limited number of times as migratory bipolar cells and then undergo synchronous differentiation into oligodendrocytes (117, 118). Progenitors treated with appropriate inducers, such as FCS (116) and ciliary neurotrophic factor (CNTF) (119), differentiate into type 2 astrocytes. Later work showed that additional environmental cues influence the differentiation of O-2A progenitor cells. FGF-2 alone induces O-2A progenitors to differentiate into oligodendrocytes, but they continue to divide several times (120). By contrast, O-2A progenitor cells treated with both PDGF and FGF-2 proliferate as bipolar cells and do not differentiate for periods well beyond those during which they would have differentiated in response to either growth factor alone; the continuous presence of both polypeptides is necessary to maintain the cells in an undifferentiated state (120). Thus, for these cells the decision whether to differentiate or to continue undifferentiated self-renewal is controlled by two differentiation-inducing paracrine factors. This control mechanism, which was deciphered *in vitro*, is likely to be physiologically relevant since it has been found that the growth factor signals regulating the proliferation of O-2A progenitors and oligodendrocyte differentiation *in vivo* may be dependent on the electrical activity of adjacent neurons (121). It has been reported that CNTF, a natural product of type 1 astrocytes, increases the survival and maturation of O-2A-derived oligodendrocytes *in vitro* by reducing spontaneous and TNF-induced apoptoses (122).

3.3.3 Melanocytes

Melanocytes are a small population of melanin-producing skin cells of neuroectodermal origin that rarely proliferate *in vivo* and are difficult to maintain *in vitro*. Eisinger and Marko (123) used 12-*O*-tetradecanoylphorbol 13-acetate (TPA) with cholera toxin in FCS-supplemented medium to stimulate selectively the growth of melanocytes from human skin. When used in conjunction with 100 μg/ml geneticin (G418 sulfate) to eliminate fibroblasts (124), pure populations of melanocytes were obtained that could be serially passaged. These cells exhibited an increased plating efficiency on plastic dishes coated with basement membrane components (125) and their growth was stimulated 40- to 90-fold by extracts of melanoma, astrocytoma, and fibroblast cell lines but not by kidney or keratinocyte extracts nor by TGF-β, EGF, PDGF, melanocyte-stimulating hormone (MSH), and NGF (126). Melanocyte growth was also stimulated by extracts of bovine brain (127), bovine pituitary gland, and human placenta (128). These results led Halaban *et al.* (128) to find that in the presence of cAMP agonists, FGF-2, but not FGF-1, was mitogenic for melanocytes. FGF-2 could replace TPA in a serum-containing medium supplemented with TPA, isobutylmethyl xanthine (IBMX), and human placental extract (129), and it could replace bovine pituitary extract in a serum-free modified keratinocyte medium containing insulin, pituitary extract, ethanolamine, phosphoethanolamine, and hydrocortisone (131) (*Table 1*).

Although not included in either of these media, ECM components such as collagen or fibronectin or ECM itself would probably be beneficial in melanocyte cultures. Phorbol ester, insulin, and pituitary extract or FGF-2 but not agents that elevate cAMP levels were found to be essential in serum-free medium; melanocytes undergo 40–45 population doublings under these conditions.

3.4 Gonadal cells

3.4.1 Embryonic germ (EG) cells

Primordial germ cells are diploid precursor cells that give rise to eggs or sperm. In mouse development, primordial germ cells are derived from a small population of embryonic ectoderm cells sequestered prior to gastrulation. During embryonic development, these cells proliferate and migrate from the base of the allantois at the egg cylinder stage to the genital ridge. Primordial germ cells are potentially immortal in that they are a self-renewing population of cells. However, *in vivo* those cells destined to differentiate into eggs or sperm have a limited capacity to proliferate, and, until the early 1990s, culture conditions for primordial germ cells *in vitro* only permitted survival and growth for a few days (132, 133).

A greater understanding of the survival and growth requirements of primordial germ cells has led to the development of culture conditions that allow them to be serially passaged (134, 135). The first advance was the finding that primordial embryonic germ (EG) cells isolated from post-implantation mouse embryos would survive and proliferate longer on an irradiated feeder layer of STO embryonic mouse fibroblasts than on a plastic surface (136). The STO cell line is also widely used as a feeder layer for pluripotent embryo-derived stem (ES) cells obtained from pre-implantation mouse embryos (137). Secondly, STO cells were found to synthesize a polypeptide growth factor variously known as stem cell growth factor (SCF), mast cell growth factor (MCF), and steel factor, which is encoded by the steel (*sl*) gene and is known to bind to the receptor protein tyrosine kinase encoded by the c-kit proto-oncogene at the dominant white spotting (W) locus. Soluble SCF added to cultures of embryonic germ cells enhanced cell survival but was not mitogenic, and this effect was further increased by culturing the cells on feeder cells, such as STO, that expressed a transmembrane form of SCF (138–140). When embryonic germ cells were plated on feeder cells that did not express SCF, both survival and mitogenic effects were detected for soluble SCF and leukaemia inhibitory factor (LIF) (140), a second secreted product of STO cells (141). However, under these conditions the cells still exhibited a very limited lifespan *in vitro*. It was subsequently found that FGF-2 in conjunction with LIF, soluble SCF, and membrane-associated SCF provided by feeder cells allowed mouse embryonic germ cells to proliferate to the extent that they could be passaged at least 20 times (134, 135) (*Table 1*). Most of the serially passaged cells in three independent lines retained normal karyotypes with trisomy being the most frequent chromosomal abnormality (134).

With the ability to expand populations of undifferentiated embryonic germ cells *in vitro*, it became feasible to examine their developmental potential. Initial tests of cultured primordial germ cells have shown that they are pluripotent; they can form embryoid bodies that undergo further differentiation *in vitro*, they produce well differentiated teratocarcinomas in nude mice, and they participate in the formation of chimeric mice when injected into blastocysts (134). An important question that arose with the feasibility of propagating EG cells was whether or not these cells cultured from post-implantation embryos were functionally or developmentally equivalent either to pluripotent embryonal stem (ES) cells derived from the inner cell mass of pre-implantation blastocysts or to embryonal carcinoma (EC) cells of tumours arising from ectopically transplanted egg cylinders (142). Both ES and EC cells are able to colonize somatic tissues of chimeric mice. However, ES cells, but not EC cells, are able to give rise to functional gametes and thus can contribute to the germline of chimeras (137). ES cells have therefore been very valuable in establishing lines of transgenic and knockout mice. EG cells were found to resemble ES cells in that EG cell lines derived from day 9 post-coital 129/Sv mouse embryos (143) and day 8.5 post-coital C57BL/6 mouse embryos (144) were able to contribute to the germline of chimeric mice. Using the same culture conditions EG cell lines could not be derived from day 6.5 embryos or from day 15 or older embryos (144).

Human EG cell lines have been established from gonadal ridges and mesenteries of 5- to 9-week post-fertilization human embryos (145). The EG cells grew out of enzymatically disaggregated tissue samples plated on irradiated STO mouse fibroblast feeder layers in DMEM nutrient medium supplemented with 15% (v/v) FCS, 100 μM 2-mercaptoethanol, 2 mM glutamine, 1 mM sodium pyruvate, 1000 U/ml of recombinant human leukemia inhibitory factor (LIF), 1 ng/ml of FGF-2, and 10 μM forskolin. The cells were passaged under the same culture conditions every seven days for more than 20 passages. 95% of initiated cultures produced cells with morphological, biochemical, or immunochemical properties of pluripotent stem cells: the cells formed large compact multicellular colonies; the cells possessed high levels of alkaline phosphatase activity; a variable proportion of cells within colonies reacted with monoclonal antibodies to SSEA-1, SSEA-4, TRA-1-60, and TRA-1-81; and embryoid bodies formed by spontaneously differentiating colonies expressed immunoreactive markers from all three embryonic germ layers. Five cultures tested from passages 8–10 were karyotypically normal with three bearing XX and two bearing XY sex chromosome complements. The pluripotent nature of human EG cells indicates that they will be highly valuable in studies *in vitro* of cell type-specific differentiation. Based on the knowledge obtained from such studies, both EG cells and ES cells (see Section 3.5) could be used to generate differentiated cells or their progenitors for cellular transplantation.

3.4.2 Ovarian granulosa cells

During ovarian follicle growth *in vivo*, granulosa cells proliferate extensively and undergo changes in hormonal responsiveness and steroidogenesis. Isolated granulosa cells have been maintained in a functional state for several days in serum-free primary cultures where they produced oestrogen and progestins in response to follicle-stimulating hormone (FSH) (146). Cells were viable for upwards of 60 days under these conditions, but they did not proliferate well and could not be serially propagated (146). A serum-free medium containing insulin, FGF-2, lipoprotein, and BSA has been described that supports at least 20 population doublings of bovine granulosa cells in serial culture on collagen-coated plastic (147) (*Table 1*). These culture conditions are similar to those used by Savion *et al.* (148) except that collagen rather than ECM from bovine corneal endothelial cells was used as a culture substrate. In both cases serum extended the proliferative lifespan of granulosa cells in culture. However, as noted by Orly *et al.* (146), serum suppresses the differentiated function of FSH-induced steroidogenesis.

The potential to manipulate granulosa cells under relatively simple conditions *in vitro* makes possible studies of the relationship between proliferation and expression of differentiated properties and studies of the interactions between granulosa cells and other cell types of the developing follicle. J. P. Mather and colleagues extended the serum-free culture conditions established for granulosa cells by Orly *et al.* (146) to elucidate a role for activin-A, a homodimer of the βA subunit of inhibin, as a local mediator of FSH effects on granulosa cell proliferation and folliculogenesis (149, 150). Although the results of injecting activin into the ovarian bursa of 25-day-old rats suggested that activin inhibited granulosa cell proliferation and caused follicular degeneration (151), *in vitro* culture experiments demonstrated that activin was a direct mitogen for isolated granulosa cells, and the mitogenic effect of activin was neutralized by the activin-binding protein follistatin (149). FSH by itself had no effect on granulosa cell proliferation *in vitro*, but it synergized with activin in stimulating [^{3}H]thymidine incorporation by granulosa cells from 21-day-old rats but not from 14-day-old rats. These experiments indicated that activin, which was originally identified as a gonadal stimulator of pituitary FSH release, and not FSH itself was a direct mitogen for granulosa cells. In a further demonstration of the importance of activin in folliculogenesis, Li *et al.* (149) recreated follicle formation *in vitro* by treating co-cultures of granulosa cells and oocytes from 14-day-old rats with activin and FSH (*Protocol 8*). When placed in serum-free culture, granulosa cell–oocyte complexes attached to the substrate, and the granulosa cells formed a monolayer. Activin alone caused the granulosa cells to reaggregate around the oocytes in a single cell layer and take on the appearance of cumulus cells. Activin and FSH together stimulated further follicle-like development: a zona pellucida and a cumulus layer formed around each oocyte along with an associated multilayer ring of granulosa cells around an antrum-like space. In addition, the cultures accumulated high levels of progesterone and oestrogen, which was not the case for cultures treated with activin or FSH alone. Together, these results are an impressive demonstration that mechanisms underlying complex physiological processes can be teased apart and understood through the use of well defined *in vitro* culture systems.

Protocol 8

Reconstitution *in vitro* of rat ovarian follicles (149)

Equipment and reagents

- 27 gauge needles
- Poly-D-lysine-coated plastic cultureware (see Chapter 5, *Protocol 2*)
- 14-day-old female Sprague-Dawley rats
- Collagenase-dispase (Roche Molecular Biochemicals)
- DMEM/F12 (Life Technologies)
- BSA (Sigma)
- Bovine insulin (Sigma)
- Human transferrin (Sigma)
- Human fibronectin (Sigma)
- Recombinant human activin-A (Genentech)
- Human FSH (Sigma)

Method

1. Collect ovaries from euthanized rats. Remove surrounding tissues, and use 27 gauge needles to mince the ovaries.
2. Incubate the tissue pieces in 0.8% (w/v) collagenase-dispase in DMEM/F12 medium at 37 °C for 1 h, and disperse by gentle pipetting. Collect the cells by centrifugation, wash the cell pellet once with medium, and resuspend the cells in DMEM/F12 medium.
3. Layer the cell suspension on a discontinuous gradient of BSA (1%, 2%, and 3%, w/v) in DMEM/F12, and allow the cells to sediment at 1 g for 1 h. Harvest the follicles from the bottom of the gradient, and wash them once with medium.
4. Culture the follicles on poly-D-lysine-coated plastic in DMEM/F12 medium with the following supplements:
   - bovine insulin, 5 μg/ml
   - human transferrin, 5 μg/ml
   - human fibronectin, 5 μg/ml
   - recombinant human activin-A, 30 ng/ml
   - human FSH, 30 ng/ml
5. Incubate the cells at 37 °C in a humidified atmosphere of 5% CO_2.

3.5 Embryonal stem (ES) cells

The development of an *in vitro* system for the culture of developmentally uncommitted mouse ES cells allowed major advances in understanding mechanisms of development and differentiation (152). Culture of mouse ES cells that are capable of contributing to any tissue of the animal, including the germline, provides the capability to specifically target a gene for disruption or 'knockout' *in vitro* through homologous recombination, followed by derivation of germline chimeric mice resulting from introduction of these genetically altered ES cells into recipient embryos (153). In principle the progeny from such mice provide models to study the result of loss of expression of any targeted gene. The only limitation is that sufficient genomic sequence be available to generate a plasmid construct favouring homologous recombination. Thus far the approach has been successful only with mice, although the approach should be applicable to any animal for which enough totipotent stem cells exist in the early embryo to allow experimental manipulation. For the purposes of this section, 'ES cells' refers to early embryo-derived, genomically normal proliferating cultures that can be manipulated to differentiate into representatives of all three germ layers (ectoderm, endoderm, and mesoderm). Although the ultimate criterion for ES cells is the ability to contribute to the germline of chimeric animals, this criterion has not yet been fulfilled for ES cells of non-murine species.

The power of ES cell-based genetic approaches has led to the recognition of ES cells as essential models in studies of normal and abnormal biological processes, although development and utilization of stem cells requires a sophisticated appreciation of the use of a wide range of reagents and materials, including specialized culture media (154), feeder layers, hybridomas, molecular clones, expression vectors, polyclonal and monoclonal antibodies, and synthetic or recombinant peptides. Advances in plasmid development, for instance, have allowed the generation of a large collection of randomly-derived gene knockout ES clones, which may then be screened for the particular knockout of interest (155).

Attempts at the development of non-murine ES cells have been reported, with varying degrees of success, from a variety of species, including human and non-human primates, chicken, fish, sheep, pig, cow, rat, and hamster (156–158). In particular J. A. Thomson and colleagues have isolated ES cell lines from rhesus monkeys (159), common marmosets (160), and humans (161). An alternative approach to the development of animals with targeted gene disruption is the culture and genetic manipulation of primordial germ cells and ensuing reintroduction into the appropriate location in the developing embryo. Primordial germ cell cultures are covered in Section 3.4.1 of this chapter. Additionally, specific cell lineages have been derived by differentiation of ES cells to a variety of tissue types, including cardiomyocytes, endothelium, neurons, haematopoietic cells, liver, pancreas and other endoderm-derived endocrine lineages (162–164).

In addition to work with mammalian species, similar approaches are being taken toward the development of ES cell systems for other vertebrates, including the zebrafish, a recognized model for the study of developmental biology and genetics (165). An advantage of zebrafish over the mouse system is easy experimental access to large numbers of embryos in the early stages of development, allowing studies of the earliest developmental events and primary abnormalities resulting from disruption of normal gene expression. Often in the mouse ES system it is difficult to unambiguously determine primary effects of gene disruptions from more distal effects because early embryos expressing early mutations are difficult to obtain and study. The zebrafish differentiation-competent embryonal stem cell system allows differentiation toward several lineages by treatment with retinoic acid or substratum manipulation to, for instance, produce embryoid bodies. DNA transfection and reintroduction of cultured cells into recipient embryos have produced chimeric embryos in which cultured embryo cells contribute to the developing organism (166–168). In an examination of the expression of a variety of developmentally-regulated molecular markers, the embryonal stem cell marker pou-2 and the primordial germ cell marker vas were found to be maintained in the cultures (169).

Brüstle *et al.* (170) have recently demonstrated that oligodendrocytes and astrocytes (see Section 3.3.2) derived from mouse ES cells propagated *in vitro* were functional in an animal model of cell transplantation. The culture system used by Brüstle *et al.* to direct ES cell differentiation to glia is provided in *Protocol 9*. Embryoid bodies formed from J1 mouse ES cells (171) were cultured in a serum-free medium that promoted the survival of ES cell-derived neural precursor cells (172). Those cells were propagated sequentially in medium supplemented with FGF-2, followed by FGF-2 and EGF, and FGF-2 and PDGF, which selected for the outgrowth of glial progenitor cells (120). This population of bipolar cells reacted with antibodies to the A2B5 glial precursor antigen. When subjected to growth factor withdrawal, the cells differentiated into oligodendrocytes and astrocytes: within four days approximately 40% of the cells expressed oligodendrocyte-specific glycolipids while 35% of the cells expressed the astrocyte marker glial fibrillary acidic protein (GFAP). Murine glial progenitor cells cultured in the presence of FGF-2 and PDGF and transplanted into the spinal cords of myelin-deficient neonatal rats responded to local environmental cues by differentiating into myelin-forming cells that migrated within the spinal cord several millimetres from the site of implantation. Progenitor cells transplanted into the cerebral ventricle of day 17 rat embryo brains formed myelin sheaths in a number of brain regions. No tumours or non-neural tissues developed in the progenitor cell recipients. This research, which combined cell transplantation with growth factor-directed differentiation of totipotent embryonic stem cells *in vitro*, demonstrates that ES cells are a valuable source of somatic cell precursors for neural transplantation, and it further suggests that they will be valuable sources of other transplantable somatic cells for which *in vitro* differentiation strategies can be devised.

Protocol 9

Differentiation *in vitro* of mouse ES cells to glia (170)

Reagents

- DMEM/F12 basal medium (Life Technologies)
- Bovine insulin (Sigma)
- Human transferrin (Sigma)
- Selenium chloride (Sigma)
- Human fibronectin (Sigma)
- Polyornithine
- Progesterone (Sigma)
- Putrescine (Sigma)
- FGF-2 (UBI)
- Laminin (Sigma)
- Ca^{2+}- and Mg^{2+}-free Hank's buffered salt solution (Life Technologies)
- EGF (UBI)
- PDGF-AA (UBI)

Protocol 9 continued

Method

1. Propagate embryoid bodies formed from aggregated murine J1 ES cells in the absence of leukemia inhibitory factor (LIF) for five days in DMEM/F12 basal medium supplemented with 5 μg/ml bovine insulin, 50 μg/ml human transferrin, 30 nM selenium chloride, and 5 μg/ml human fibronectin.
2. Trypsinize the cells, plate them on dishes coated with 15 μg/ml polyornithine, and propagate them for five days in DMEM/F12 supplemented with 25 μg/ml insulin, 50 μg/ml transferrin, 20 nM progesterone, 100 μM putrescine, 30 nM selenium chloride, 10 ng/ml FGF-2, and 1 μg/ml laminin. Incubate the cells at 37 °C in a humidified atmosphere of 5% CO_2.
3. Harvest the cells by scraping in Ca^{2+}- and Mg^{2+}-free Hank's buffered salt solution, reduce them to a single cell suspension by pipetting, and split them at a 1:5 ratio in DMEM/F12 supplemented with 10 ng/ml FGF-2 and 20 ng/ml EGF. Add FGF-2 and EGF daily.
4. Prior to confluence replate the cells at a 1:5 ratio in DMEM/F12 containing 10 ng/ml FGF-2 and 10 ng/ml PDGF-AA. Add FGF-2 and PDGF-AA daily.
5. Induce differentiation into oligodendrocytes and astrocytes by growth factor withdrawal, and monitor differentiation by immunofluorescence using antibodies to A2B5 (Roche Molecular Biochemicals), O4 (Roche Molecular Biochemicals), and glial fibrillary acidic protein (ICN) with appropriate FITC-labelled second antibodies.

3.6 Leukocytes

3.6.1 B cell lineage: myelomas and hybridomas

Unlike T cells (see below) it is at present very difficult to maintain normal B cells in long-term culture. Thus long-term studies are limited to neoplastic cell lines (e.g. myelomas), the fusion products of such cells with normal B cells (i.e. hybridomas), or virally transformed cells which are usually obtained by treating B cells with Epstein–Barr virus (173). B cell growth factors are not normally required as culture additives by such cells, although they are produced as autocrine factors by some lines.

The hybridoma methodology developed by Köhler and Milstein (174) for producing monoclonal antibodies of predefined specificities relies to a great extent on *in vitro* cell culture; antibody production by plasma cells is immortalized by fusing splenocytes from immunized mice with a myeloma cell line and the resulting hybridomas are screened and propagated *in vitro*. Detailed protocols are not presented here as there is already a very extensive literature on the subject, including a chapter elsewhere in the Practical Approach Series (175), to which the reader should refer for practical details.

The original mouse cell lines used to generate hybridomas, and those still most commonly used today, were derived from the mouse MOPC 21 myeloma cell line P3 (176). Early attempts to simplify the production of monoclonal antibodies by culturing established hybridomas under serum-free conditions indicated that hybridomas would grow in basal medium supplemented with insulin and transferrin (177, 178). However, several parental myeloma cell lines were unable to survive and proliferate in similar defined media (179). As described below, a study of the differences in the growth requirements of myeloma cells and their hybridomas showed that myeloma lines clonally derived from P3 cells had in common an inability to synthesize cholesterol, which was not passed on to their hybridoma derivatives. This metabolic difference has been used to devise a more efficient method of selecting hybridomas (see below). Although several interleukins (IL-4 to IL-6) were initially identified on the basis of their mitogenic activity for B cells or plasmacytomas, exogenous sources of these growth factors are not required in myeloma or hybridoma cultures.

In contrast to established hybridoma cell lines, the myeloma cell lines NS-1-Ag4-1 (NS-1), P3-X63-Ag8 (X63), and X63-Ag8.653 (X63.653) did not survive at a density of 1×10^4 cells/ml in serum-free medium containing insulin, transferrin, ethanolamine, selenium, and 2-mercaptoethanol unless the medium was supplemented with human low density lipoprotein (LDL) at a concentration of 1–10 μg/ml (179, 180). Cholesterol, a major component of LDL, was found to replace LDL at 5–10 μg/ml in supporting the survival and growth of these myeloma cell lines if it were complexed with BSA (180, 181). In addition, in suboptimal concentrations of LDL, myeloma growth was enhanced by BSA–oleic acid (179). These results suggested that these myeloma cells, but not their hybridomas, were cholesterol auxotrophs, and that myeloma cells at low densities were stimulated by unsaturated fatty acids (*Table 1*). The biochemical lesion responsible for cholesterol auxotrophy in the NS-1, X63, and X63.653 myeloma lines was subsequently traced to 3-ketosteroid reductase, an enzyme involved in the demethylation of the cholesterol precursor lanosterol; distal enzymes in cholesterol biosynthesis were not involved as the downstream intermediate lathosterol was efficiently converted to cholesterol (182).

In addition to established hybridoma cell lines, newly formed NS-1 hybridomas did not require LDL or cholesterol for survival and growth. The recovery of nascent hybridomas from fusion products was similar in medium containing cholesterol in the form of serum or LDL and in medium with no cholesterol (179). These results implied that the hybridomas were obtaining the ability to synthesize cholesterol from their spleen cell parent. However, the yield of hybridomas was reduced in the absence of a source of unsaturated fatty acid such as BSA–oleic acid. The difference between NS-1 myeloma cells and NS-1 hybridomas in their need for cholesterol was used to devise a novel selection procedure for newly formed NS-1 hybridomas: fusion products were plated directly into cholesterol- and serum-free medium supplemented with BSA–oleic acid, in which only hybridomas would survive (183). This selection method gave five to ten times as many hybridoma colonies as selection in the corresponding HAT-supplemented medium, and it can be used with any parent myeloma cell line that is unable to synthesize cholesterol. This technique is an illustration of a practical application stemming directly from knowledge of the growth requirements of cells *in vitro*.

3.6.2 T cell lineage: IL-2-stimulated cytotoxic T cells

T cells are divided into helper and cytotoxic/suppressor subsets, which are distinguishable by function and by cell surface antigen expression (184). $CD4^+CD8^-$ helper T cells provide help for antibody-producing B cells, while $CD4^-CD8^+$ cytotoxic T cells kill cells infected with viruses or other pathogens. The functional activation of T cells requires concomitant stimulation by both lymphokines and MHC (major histocompatibility complex) molecule-bound antigen displayed by antigen-presenting cells such as macrophages and dendritic cells (Section 3.6.3). Helper T cells are activated by antigen bound to MHC class II molecules while cytotoxic T cells are activated by antigen bound to MHC class I molecules. Thus, T cell activation is both an antigen-specific and an MHC-restricted process (185). Both helper (186, 187) and cytotoxic (187, 188) T cell subsets are further functionally characterized by the patterns of lymphokines they secrete. Murine type 1 helper cells (T_H1), which mediate delayed type hypersensitivity reactions, produce interleukin-2 (IL-2), interferon-γ (INF-γ), and tumour necrosis factor-β (TNF-β) but little IL-4 or IL-5 (186); murine type 2 helper cells (T_H2) regulate antibody production and secrete IL-4, IL-5, and IL-6 but produce little IL-2 and INF-γ (186). Similar divergent patterns of INF-γ versus IL-4 and IL-5 production have been described in human helper T cells (187). Analyses of lymphokines produced by human (187) and murine (188) $CD8^+$ T cell clones have identified mutually exclusive patterns of secretion of IFN-γ and IL-10 (type 1) or IL-4 and IL-5 (type 2) that appear to correspond to cytotoxic and suppressor T cell phenotypes, respectively. Whether T cells express a type 1 or type 2 pattern of lymphokines depends on the lymphokines to which they were exposed during the activation process: IL-12 induces a type 1 T cell phenotype while IL-4 induces a type 2 phenotype (188, 189).

The ability to establish long-term cultures of mixed T cell populations and T cell clones that could then be subjected to functional and phenotypic analyses was largely dependent on the discovery and characterization of T cell growth factor (TCGF or IL-2) (190, 191). Long-term cultures of non-transformed human T cells were first established from bone marrow (192) or peripheral blood lymphocytes (PBL) (193) stimulated with the mitogen phytohaemagglutinin (PHA) and maintained in the presence of TCGF contained in culture medium conditioned by PHA-stimulated PBL. Long-term cultures of antigen-specific human cytotoxic T cells were generated by culturing responder PBL with X-irradiated allogeneic stimulator PBL in a one-way mixed lymphocyte culture and maintaining the resulting effector cells in TCGF-containing medium (194). In each of these cases T cell viability and proliferation were strictly dependent on TCGF. The subsequent cloning and expression of IL-2 cDNA gave a unique molecular identity to TCGF biological activity in lymphocyte-conditioned medium (195). Once recombinant IL-2 became available as a research reagent, it was found to act on B cells, natural killer (NK) cells, monocytes, macrophages, and oligodendrocytes (196).

Based on the demonstration that IL-2 could be used to generate and propagate antigen-specific cytotoxic T cell clones, IL-2 was used in attempts to generate tumour-specific cytotoxic T cells *in vitro* from mixed leukocyte populations obtained from cancer patients. Rosenberg and colleagues found that treatment with high concentrations of IL-2 *in vitro* induced tumoricidal activity in peripheral blood lymphocytes from patients and normal volunteers that was distinct from the NK cell activity of unstimulated PBL (197, 198). These lymphokine-activated killer (LAK) cells were heterogeneous but contained a large component of $CD8^+$ and $CD4^+$ T cells. LAK cells killed fresh autologous tumour cells but not autologous normal cells. However, LAK cells differed from antigen-specific cytotoxic T cells in that they were MHC non-restricted and killed allogeneic tumour cells *in vitro*. It has been proposed that MHC non-restricted cell killing by cytotoxic lymphocytes is dependent on cytokine production by lymphocytes or accessory cells induced by high levels of IL-2 (199). A protocol for generating LAK cells in serum-free medium (*Table 1*) from normal human peripheral blood cells is provided in *Protocol 10*. Okamoto *et al.* found that transferrin was required for IL-2-stimulated LAK cell production in the absence of human serum and that the ratio of $CD8^+$ T cells to $CD4^+$ T cells in the effector population increased when insulin was omitted from the medium (200). The cytotoxic activity of these LAK cells towards squamous carcinoma cells was three to five times that of LAK cells generated in the presence of human type AB serum.

Protocol 10

Generation of LAK cells from human peripheral blood (200)

Equipment and reagents

- Polypropylene centrifuge tubes (Falcon)
- Ficoll-Paque (Amersham Pharmacia Biotech)
- RPMI 1640 medium (Life Technologies)
- DMEM/F12 medium (Life Technologies)
- Kanamycin sulfate (Sigma)
- Human transferrin (Fe^{3+}-free) (Sigma)
- Recombinant human IL-2 (R and D Systems)
- PHA (Sigma)

A. Harvesting of peripheral blood lymphocytes (PBL) by sedimentation on a Ficoll-Paque gradient

1. Carefully underlay 20 ml of Ficoll-Paque in 50 ml polypropylene centrifuge tubes beneath 20 ml of human venous blood from normal volunteers.
2. Centrifuge the samples at 450 g for 20 min at 20 °C.
3. Remove the plasma without disturbing the interface.
4. Harvest PBL from the interface and transfer them to a new centrifuge tube.
5. Wash the PBL three times with 40 ml of RD nutrient medium (1:1 mixture by volume of RPMI 1640 and DMEM/F12), and collect the cells by centrifugation at 150 g for 10 min after each wash.

Protocol 10 continued

B. Serum-free LAK cell cultures

1 Resuspend PBL at 5×10^5 cells/ml in LAK medium (see below), and plate the cells in appropriate plastic culture vessels.
 (a) The basal medium consists of RD medium (see part A, step 5) containing sodium pyruvate, sodium bicarbonate, Hepes buffer, 90 μg/ml kanamycin sulfate.
 (b) Complete serum-free LAK medium consists of basal medium with the following supplements:
 - human transferrin (Fe^{3+}-free), 5 μg/ml
 - 2-aminoethanol, 10 μM
 - 2-mercaptoethanol, 10 μM
 - sodium selenite, 10 nM
 - recombinant human IL-2, 33 ng/ml

2 Incubate the cells at 37 °C in a humidified atmosphere of 5% CO_2 for three to seven days.

3 For longer culture periods add the T cell mitogen PHA to give a final concentration of 1 μg/ml in the medium.

Tumour-infiltrating lymphocytes (TIL) expanded *in vitro* in the presence of IL-2 were reported to have a higher degree of lytic activity towards autologous metastatic tumours than LAK cells (201). Some human TIL isolated from melanomas exhibited MHC class I-restricted cell killing after prolonged culture *in vitro* in IL-2, and they were predominately $CD3^+CD8^+$ T cells while TIL cultures with less specific cytotoxic activity were phenotypically heterogeneous (202). It is likely that the generation in culture of MHC-restricted TIL with cytotoxic activity towards autologous tumour cells reflects the selected expansion over time of antigen-specific cytotoxic T cells from a mixed population of leukocytes. TIL can be propagated in the continuous presence of IL-2 in medium developed for LAK cells and supplemented with LAK-conditioned medium (202, 203).

3.6.3 Dendritic cells

Mature or terminally differentiated dendritic cells (DCs) are potent antigen-presenting cells that control the activation of T and B lymphocytes (204). Immature dendritic cells located in peripheral tissues are weak antigen-presenting cells, but they are efficient at taking up and processing foreign antigens. DCs induced to mature by foreign antigens and local cytokines undergo changes in morphology and expression of cell surface proteins, and they migrate to lymphoid organs. Mature DCs express on their surfaces high levels of major histocompatibility (MHC) class I and class II antigens, which present processed foreign peptide antigens to T cell antigen receptors on cytotoxic and helper T cells, respectively, along with co-stimulatory molecules making them highly efficient activators of antigen-specific T cells. Activated T cells can then interact with B cells (to stimulate antibody production), macrophages, and target cells. Because of their central role in the initiation of cellular and humoral immune responses, dendritic cells have become a focal point for the development of immunological therapies for diseases.

Purified cytokines and *in vitro* culture methods proved crucial in elucidating the developmental origins of dendritic cells, in characterizing changes that occurred during DC maturation, and ultimately in isolating populations of mature dendritic cells for functional studies (204). There are currently thought to be three progenitor pools that give rise to dendritic cells: $CD34^+$ blood cell precursors, $CD34^-$ peripheral blood mononuclear cells, and $CD4^+$ $CD11c^-$ T cell precursors. The DCs derived from these different progenitors may have different functional properties (204). Granulocyte-macrophage colony stimulating factor (GM-CSF) in particular was important in demonstrating that some mouse dendritic cells derived from MHC class II negative myeloid precursor cells in bone marrow gave rise to granulocytes and macrophages (205). GM-CSF has been used to isolate dendritic cells from progenitors in mouse bone marrow (206) and peripheral blood (207), while GM-CSF combined with tumour necrosis factor-α (TNF-α) was used to generate DCs from $CD34^+$ precursor cells in human umbilical cord blood (208) and bone marrow (209). Stem cell factor (SCF/c-kit ligand) reportedly increased the number of DC precursors in human $CD34^+$ bone marrow cell cultures, and the effect of TNF-α in these cultures was to generate pure DC colonies in the presence of GM-CSF (210). Dendritic cells could be produced from mononuclear precursors in adult human peripheral blood by GM-CSF but only if macrophage differentiation were inhibited by interleukin (IL)-4 (211); in the absence of IL-4 most presumptive DC colonies differentiated into macrophages. Under this protocol (*Protocol 11*) $3–8 \times 10^6$ mature cytotoxic T cell-stimulatory DCs were obtained from 40 ml of whole blood, and they represented 6–15% of the mononuclear blood cells initially plated. However, these cells were not phenotypically stable unless they were exposed to macrophage-conditioned medium (212). Lymphoid-derived DCs have been obtained from murine thymic T cell precursors cultured in the presence of Flt3 ligand, SCF, TNF-α, IL-1β, IL-3, IL-7, and a ligating CD40 monoclonal antibody (213). In contrast to DCs derived from $CD34^+$ stem cells or monocytes the production of mature thymic DCs from T cell precursors was independent of exogenous GM-CSF. These DCs may induce tolerance instead of an active immune response (214).

Protocol 11

Generation of dendritic cells from human peripheral blood (211)

Reagents

- Lymphoprep (Nycomed Pharma AS)
- RPMI 1640 medium (Life Technologies)
- Gentamicin (Sigma)
- Human GM-CSF (R and D Systems)
- Human IL-4 (R and D Systems)

Protocol 11 continued

Method

1. Prepare mononuclear cells from 40–100 ml heparinized whole blood by sedimentation in Lymphoprep or Ficoll-Paque (see *Protocol 10*).
2. Incubate the mononuclear cells at 5–20 × 10^6 cells/35 mm dish in 3 ml of RPMI 1640 medium supplemented with 2 mM L-glutamine, 50 μM 2-mercaptoethanol, 20 μg/ml gentamicin, and 5–10% (v/v) FCS. After 2 h remove the non-adherent cells by aspiration; gently wash the adherent cells with medium, and aspirate any remaining non-adherent cells.
3. Culture the adherent cells at 37 °C in 5% CO_2 in 3 ml of medium supplemented with 800 U/ml human GM-CSF and 500 U/ml human IL-4. Large DC aggregates arise in five to seven days.
4. Incubate the dendritic cells for a further three days in macrophage-conditioned medium as described in ref. 212 to create terminally differentiated mature DCs.

3.7 Insect cells

3.7.1 *Drosophila* embryo cells

Unlike its mammalian culture counterpart, the *Drosophila* embryonic culture system has unique features that should be of use to investigators interested in early development and in the physiological consequences of mutations. The method described in *Protocol 12* is a modification of a method first developed by Seecof and Unanue (215). A similar method was independently described by Sang and colleagues (216, 217).

Protocol 12

Primary culture of *Drosophila* embryo cells

Equipment and reagents

- Dounce homogenizer
- Schneider's modified *Drosophila* medium (Life Technologies)
- 25 μm nylon mesh
- Bovine insulin (Sigma)

Method

1. Collect eggs on large agar plates for 2 h in population cages, and incubate them for 3 h at 25 °C to allow the embryos to reach the early gastrula stage.
2. Harvest the embryos, and dechorionate them in a 1:1 mixture of ethanol and 5% sodium hypochlorite. Rinse the eggs in sterile water.
3. Resuspend the sterile dechorionated embryos in 7 ml of Schneider's modified *Drosophila* medium supplemented with 5% (v/v) FCS, and disrupt them by gentle

Protocol 12 continued

homogenization in a Dounce homogenizer with a loose fitting pestle. The pestle should not be rotated during this procedure in order to minimize damage to the cells.

4. Pass the homogenate through a 25 μm nylon mesh to remove debris and large cell aggregates, and collect single cells from the filtrate by centrifugation at 1500 g for 5 min. The pellet can be resuspended and the filtration step repeated to further remove debris and aggregates from the cell preparation.
5. Resuspend the cell pellet in Schneider's modified *Drosophila* medium supplemented with 5% FCS and 200 ng/ml bovine insulin. Plate the cells in tissue culture dishes at a density of 10^6 cells/cm^2, and incubate at 25 °C. Insulin is an essential component of the culture medium, and it cannot be replaced by higher concentrations of FCS.

The *Drosophila* embryonic culture system is very different from primary cultures of vertebrate cells. It is notable for its reproducibility and ease of preparation as well as for the complexity and developmental potential of the cultured cells. The total time required to prepare *Drosophila* embryo cells for culture is approximately 30 minutes. One can easily obtain in the order of 1×10^7 embryonic cells from a single two hour egg collection using a healthy population cage containing 4000 young adult flies. Over a period of 24 hours the embryo cells multiply and overtly differentiate into a number of recognizable cell types. The types of cells that develop (216–218), their growth parameters (218, 219), and their differentiated characteristics have been described in the literature.

The complexity of cell types in embryo cell cultures, more than anything else, has prevented *Drosophila in vitro* cell biology from attaining the precision that *Drosophila* molecular genetics has achieved in recent years. The use of genetic analysis has not yet been applied extensively to embryo culture. However, new techniques are being introduced to simplify culture conditions and to take advantage of various mutants. Especially useful in this regard is the culture of individual embryos (220). Recently, methods have been established to reliably analyse stem cell composition and to culture specific cell types *in vitro* (221). In addition, fluorescence-activated cell sorting can be used to isolate genetically tagged cells to establish them preferentially in culture (222).

It should be noted that *Drosophila* embryo cultures are very sensitive to variations between batches of FCS. In our experience, the best way to choose serum is to screen for myoblast and neuronal differentiation. The degree of differentiation, as exemplified by the extent of multinucleated myotube formation, for example, is indicative of the ability of serum batches to support cell growth and differentiation in culture. This interesting observation leads us to conclude that the principles established for mammalian cells in culture (223, 224) will also hold for invertebrate cells. We believe that a fruitful contribution from cell biologists will provide a greater understanding of *Drosophila* biology that transcends DNA sequences.

Table 1 Culture media for specific cell types

Cell type (references)	Basal medium	Undefined supplements	Growth factors/ hormones	Other proteins	Lipids	Other supplements
Epithelial cells						
Keratinocytes (human) (15, 16, 22)	MCDB 153	70 μg/ml pituitary extract (optional)	5 μg/ml insulin, 5 ng/ml EGF, 10 ng/ml FGF-1, 10 ng/ml KGF (optional), 1.4 μM hydrocortisone	10 μg/ml transferrin (optional), collagen type I or fibronectin (coat) (optional)		10 μM ethanolamine, 10 μM phosphoethanolamine, 10 μg/ml heparin with FGF-1
Sub-maxillary gland epithelial cells (mouse) (T. Okamoto *et al.*)	MCDB 152		10 μg/ml insulin, 10 ng/ml EGF or 1 ng/ml FGF-1	5 μg/ml transferrin, collagen type I (coat)	1 mg/ml BSA–oleic acid	10 μM ethanolamine, 10 μM 2-mercaptoethanol, 10 nM sodium selenite, 10 μg/ml heparin with FGF-1
Prostate epithelial cells (rat) (29, 32)	WAJC 404	25 μg/ml pituitary extract or FGF-1	5 μg/ml insulin, 10 ng/ml EGF, 1 μg/ml prolactin, 1 μM dexamethasone, (optional), 10 ng/ml FGF-1	10 ng/ml cholera toxin	100 μg/ml BSA–oleic acid (optional)	10 μg/ml heparin with FGF-1
Mammary epithelial cells (human) (39)	MCDB 170	70 μg/ml pituitary extract (optional)	5 μg/ml insulin, 10 ng/ml EGF, 1 μg/ml prolactin, 1.4 μM hydrocortisone, 25 nM PGE_1	5 μg/ml transferrin		0.1 mM ethanolamine, 0.1 mM phosphoethanolamine
Mammary epithelial cells (mouse) (40)	DMEM/F12	100 μg/ml pituitary extract (optional)	10 μg/ml insulin, 10 ng/ml EGF, 10 ng/ml FGF-1 (optional)	10 μg/ml transferrin, 10 ng/ml cholera toxin (optional), 1–5 mg/ml BSA, collagen type I (coat)		10 μg/ml heparin with FGF-1
Hepatocytes (rat) (49)	Williams' E		10 nM insulin, 10 nM dexamethasone	fibronectin (coat)		5 kIU/ml aprotinin
Thymic epithelial cells (rat) (61)	WAJC 404A	2% calf serum (iron-supplemented)	10 μg/ml insulin, 10 nM dexamethasone, 10 ng/ml EGF	10 μg/ml transferrin, 20 ng/ml cholera toxin		
Mesenchymal cells						
Fibroblasts (human) (65)	MCDB 110		30 ng/ml EGF, 1 μg/ml insulin, 0.5 μM dexamethasone, 25 nM PGE_1, 2 μM $PGF_{2\alpha}$		6 μg/ml lecithin, 3 μg/ml cholesterol, 60 ng/ml vitamin E, 1 μg/ml sphingomyelin, 2 μg/ml vitamin E acetate	6.5 μM DTT, 0.65 μM glutathione, 10 μM phosphoenolpyruvate

Continued . . .

Table 1 Continued

Fibroblasts (human) (66)	MCDB 104		100 ng/ml EGF, 5 μg/ml insulin, 3 μg/ml PDGF, 55 ng/ml dexamethasone	5 μg/ml transferrin, fibronectin (coat)		
Fibroblasts (human) (18, 67)	RITC-80-7		10 ng/ml EGF, 1 μg/ml insulin, 10 ng/ml FGF-1 or FGF-2 (optional)	10 μg/ml transferrin, 5 mg/ml BSA, collagen type I (coat) or 10 μg/ml fibronectin		10 μg/ml heparin with FGF-1 or FGF-2
Endothelial cells (human) (75, 77, 81)	DMEM/F12 or MCDB 107	10% FCS	10 ng/ml FGF-1 or FGF-2, 10 ng/ml EGF, 10 ng/ml VEGF (optional)	5 μg/ml transferrin (optional), collagen type I or fibronectin (coat)	50 μg/ml HDL (optional)	10 μg/ml heparin with FGF-1, 90 μg/ml kanamycin sulfate
Pre-adipocytes (rat) (93, 94)	DMEM/F12		10 μg/ml insulin, 2 ng/ml FGF-2, 10 nM dexamethasone (adult)	10 μg/ml transferrin, 2.5 μg/ml fibronectin or collagen type I (coat)	5 μg/ml HDL (adult)	
Skeletal myoblasts (rat) (103)	Ham's F12		6 μg/ml insulin, 100 nM dexamethasone	10 μM fetuin		
Smooth muscle cells (human) (77)	MCDB 107	50 μg/ml HepG2 CM (optional), 10% FCS (optional)	10 ng/ml EGF, 5 μg/ml insulin, 10 ng/ml FGF-1, 1 μM dexamethasone, 2.5 U/ml PDGF	collagen type I (coat)	50 μg/ml HDL	10 μg/ml heparin with FGF-1
Neuro-ectodermal cells						
Neural crest cells (rat) (112)	L-15 CO2 (ref. 113)	10% chick embryo extract	5 μg/ml insulin, 20 nM progesterone, 39 pg/ml dexamethasone, 100 ng/ml MSH, 10 ng/ml PGE_1, 67.5 ng/ml T3, 100 ng/ml EGF, 4 ng/ml FGF-2, 20 ng/ml 2.5S NGF	100 μg/ml transferrin, fibronectin (coat)	1 mg/ml BSA, 10 ng/ml oleic acid	16 μg/ml putrescine, 30 nM selenous acid, 1 μg/ml biotin, 25 ng/ml cobalt chloride, 35 ng/ml retinoic acid, 5 μg/ml vitamin E, 63 μg/ml hydroxybutyrate, 3.6 mg/ml glycerol, poly-D-lysine
0-2A progenitors (rat) (120)	DMEM		0.6 ng/ml progesterone 0.4 ng/ml thyroxine, 0.3 ng/ml T3, 10 ng/ml PDGF, 10 ng/ml FGF-2	0.5 μg/ml bovine transferrin, 100 μg/ml human transferrin, 100 μg/ml BSA		16 μg/ml putrescine, 0.4 ng/ml sodium selenite, 5.6 mg/ml glucose, poly-D-lysine
Melanocytes (human) (131)	MCDB 153	25 μg/ml pituitary extract (optional)	5 μg/ml insulin, 0.5 μM hydrocortisone, 10 ng/ml FGF-2			0.1 mM ethanolamine, 0.1 mM phosphoethanolamine, 10 ng/ml PMA

Continued . . .

Table 1 Continued

Cell type (references)	Basal medium	Undefined supplements	Growth factors/ hormones	Other proteins	Lipids	Other supplements
Melanocytes (human) (129)	Ham's F10	10% newborn calf serum, 20 μg/ml placental extract (ref. 130)	10 ng/ml FGF-2 or 85 nM TPA			0.1 mM IBMX
Gonadal cells						
Embryonic germ (EG) cells (mouse) (134)	DMEM	15% FCS	10 ng/ml FGF-2, 20 ng/ml LIF, 60 ng/ml SCF			STO feeder cells (irradiated)
Granulosa cells (bovine) (147)	DMEM/F12		10 ng/ml FGF-2, 5 μg/ml insulin	2.5 mg/ml BSA, collagen type I (coat)	25 μg/ml lipoprotein	500 ng/ml aprotinin
Granulosa cells (rat) (149)	DMEM/F12		5 μg/ml insulin, 30 ng/ml activin A, 30 ng/ml FSH	5 μg/ml fibronectin, 5 μg/ml transferrin		Poly-D-lysine
Embryonic stem cells						
J1 ES cell line (mouse) (170)	DMEM/F12	20% FCS	1000U/ml LIF			0.1 mM 2-mercaptoethanol, embryonic fibroblast feeder cells (irradiated)
J1 embryoid bodies (mouse) (170)	DMEM/F12		25 μg/ml insulin, 10 ng/ml FGF-2, 20 nM progesterone	1 μg/ml laminin, 50 μg/ml transferrin		30 nM selenium chloride, 100 μM putrescine, 15 μg/ml polyornithine (coat)
Leukocytes						
B cell lineage (mouse) myelomas and hybridomas (180, 183)	RPMI 1640/ DMEM or IMDM		5 μg/ml insulin	5 μg/ml transferrin	5 μg/ml LDL or BSA–cholesterol (myeloma), 0.5–1 mg/ml BSA–oleic acid (optional)	10 μM ethanolamine, 10 μM 2-mercaptoethanol, 10 nM sodium selenite
T cell lineage (human) LAK cells (200)	RPMI 1640/ DMEM		33 ng/ml human IL-2	5 μg/ml transferrin, 1 μg/ml PHA		10 μM ethanolamine, 10 μM 2-mercaptoethanol, 10 nM sodium selenite
Dendritic cells (human peripheral blood) (211, 212)	RPMI 1640	5–10% FCS, macrophage CM	800 U/ml GM-CSF, 500 U/ml IL-4			50 μM 2-mercaptoethanol, 20 μg/ml gentamicin
Insect cells						
Embryo cells (*D. melanogaster*) (221)	Schneider's	5% FCS	200 ng/ml insulin			

4 Conclusion

From the preceding discussions of a small number of cell types it is clear that much progress has been made over the past forty years in culturing specialized cells and in defining their growth and differentiation requirements. Much of this success followed directly from the realization that commonly used undefined culture supplements such as serum and tissue extracts were masking or inhibiting physiological needs and responses exhibited by cells *in vivo*, and it was led by demonstrations that basal media could be optimized for individual cell types and that serum could be replaced by physiologically relevant sets of hormones, growth factors, transport proteins, lipids, and attachment factors. As mammalian cell culture has become an essential research tool in the biological sciences, these early developments in culturing specialized cells have had, and will continue to have, an impact on many fundamental discoveries in cellular physiology and molecular biology. In addition, the ability to culture normal and aberrant cells from a variety of tissues will prove essential to the success of the emerging field of functional genomics.

As optimal culture conditions for a variety of individual cell types have become defined, the accumulated knowledge has made defining the growth requirements of cells *in vitro* a more rational process. It is now possible to build upon a core of commonly required media supplements with components known to stimulate the growth or enhance the function of related cells. For example, insulin, transferrin, ethanolamine, selenium, and a source of unsaturated fatty acids are useful starting supplements for most cells, while EGF, FGF-1, and dexamethasone are often beneficial for epithelial cells. Although not all cells can currently be serially passaged *in vitro*, the common themes in growth requirements developed empirically for classes of cells can be applied to prolong the lifespan or maintain the differentiated functions in culture of almost any cell type of interest.

The researcher must be mindful of the fact that specific culture media and culture conditions select for cell types that are able to proliferate over those that cannot. In addition, populations of cells that proliferate *in vitro* for a number of generations are prone to genetic and phenotypic changes which may be directly related to their need to adjust to a new environment. Normal rodent cells propagated in serum-containing medium have often been observed to undergo a period of 'crisis' in which the majority of cells die and a minority of aneuploid cells survive and continue to proliferate as immortal but not fully transformed cells. By contrast, under some serum-free conditions (see for example Section 3.1.2), some types of rodent cells proliferate for many generations without crisis and maintain a diploid karyotype. These results suggest that, as culture conditions are optimized, cells are placed under less severe selection pressures, which result in fewer heritable changes being selected over time within a population of cells. Similarly, cells in culture being transferred to an optimized defined medium should not require a gradual weaning process that selects for cells able to survive under a new set of circumstances. The hallmark of an adequate culture medium formulation is that it immediately supports high levels of viability and cell growth, and clonal growth assays set the most exacting standards for the adequacy of a medium.

It is noteworthy that advances in culturing specific cell types were often preceded by the discovery of novel growth factors or growth mediators, which then became widely available to the research community. The EGF and FGF family of growth factors and a number of interleukins are examples of growth factors which were found to act on a variety of target cells once they had been initially characterized as mitogens for specific cell types. As novel growth factors have mainly been discovered through efforts to define mitogens for specific cells, it is likely that new growth factors will be discovered as a wider variety of specialized cells are propagated *in vitro*. Further progress in culturing specialized cells will accrue in tandem with an increased understanding of the physiology of those cells *in vivo* and a greater appreciation of the interactions between cells within tissues. The reconstitution of tissues *in vitro* from isolated cells would represent a stringent test of the level of that understanding. Hypotheses on the paracrine, autocrine, or endocrine regulation of particular cells can be further studied in intact animals through genetic ablation by homologous recombination in embryonic stem cells and by genetic crosses between strains of knockout mice harbouring different gene deletions.

Acknowledgements

The authors thank the National Institutes of Health and the National Center for Research Resources, the Council for Tobacco Research-USA, the US Army Medical Research and Materiel Command, the Japan Society for the Promotion of Science, and the Japanese Ministry of Education, Science, and Culture for financial support. Since the writing of the first edition of this chapter, Tomoyuki Kawamoto and Izumi Hayashi have succumbed to diseases in the prime of life; they are missed by their friends, colleagues, and co-authors.

References

1. Eagle, H. (1955). *Science*, **122**, 501.
2. Eagle, H. (1955). *J. Exp. Med.*, **102**, 595.
3. Evans, V. J., Bryant, J. C., Fioramoti, M. C., McQuilkin, W. T., Sanford, K. K., and Earle, W. R. (1956). *Cancer Res.*, **16**, 77.
4. Buonassisi, V., Sato, G. H., and Cohen, A. I. (1962). *Proc. Natl. Acad. Sci. USA*, **48**, 1184.
5. Yasamura, Y., Tashjian, A. H. Jr., and Sato, G. H. (1966). *Science*, **154**, 1186.
6. Sato, G. H., Zaroff, L., and Mills, S. E. (1960). *Proc. Natl. Acad. Sci. USA*, **46**, 963.
7. Barnes, D. W., Sirbasku, D. A., and Sato, G. H. (ed.) (1984). *Cell culture methods for cell biology*, Vols 1–4. Alan R. Liss, New York.
8. Ham, R. G. and McKeehan, W. L. (1979). In *Methods in enzymology* (ed. W. Jacoby and I. Pastan), Vol. 58, pp. 44–93. Academic Press, San Diego, CA.
9. Barnes, D. W. and Sato, G. H. (1980). *Cell*, **22**, 649.
10. Barnes, D. W. (1987). *Biotechniques*, **5**, 534.

11. Jacoby, W. B. and Pastan, I. H. (ed.) (1979). *Methods in enzymology*, Vol. 58. Academic Press, San Diego, CA.
12. Freshney, R. I. (1994). *Culture of animal cells. A manual of basic technique* (3rd edn). Alan R. Liss, New York.
13. Baserga, R. (ed.) (1989). *Cell growth and division: a practical approach*. IRL Press, Oxford.
14. Rheinwald, J. G. and Green, H. (1975). *Cell*, **6**, 331.
15. Tsao, M. C., Walthall, B. J., and Ham, R. G. (1982). *J. Cell. Physiol.*, **110**, 219.
16. Boyce, S. T. and Ham, R. G. (1983). *J. Invest. Dermatol.*, **81** (Suppl.), 33.
17. O'Keefe, E. J., Chui, M. L., and Payne, R. E. (1988). *J. Invest. Dermatol.*, **90**, 767.
18. Shipley, G. D., Keeble, W. W., Hendrickson, J. E., Coffey, R. J. Jr., and Pittlekow, M. R. (1989). *J. Cell. Physiol.*, **138**, 511.
19. Hennings, H., Michael, D., Cheng, C., Steinert, P., Holbrook, K., and Yuspa, S. (1980). *Cell*, **19**, 245.
20. Gilchrist, B. A., Calhoun, J. K., and Maciag, T. (1982). *J. Cell. Physiol.*, **112**, 197.
21. Coffey, R. J. Jr., Derynck, R., Wilcox, J. N., Bringman, T. S., Goustin, A. S., Moses, H. L., *et al.* (1987). *Nature*, **328**, 817.
22. Finch, P. W., Rubin, J. S., Miki, T., Ron, D., and Aaronson, S. A. (1989). *Science*, **245**, 752.
23. Wigley, C. B. and Franks, L. M. (1976). *J. Cell Sci.*, **20**, 149.
24. Loo, D. T., Fuquay, J. I., Rawson, C. L., and Barnes, D. W. (1987). *Science*, **236**, 200.
25. Rawson, C. L., Loo, D. T., Duimstra, J. R., Hedstrom, O. R., Schmidt, E. E., and Barnes, D. W. (1991). *J. Cell Biol.*, **113**, 671.
26. Okamoto, T., Myoken, Y., Yabumoto, M., Osaki, T., Fujiti, Y., Whitney, R. G., *et al.* (1996). *Biochem. Biophys. Res. Commun.*, **221**, 795.
27. Huggins, C. and Russell, P. S. (1946). *Endocrinology*, **39**, 1.
28. Parker, M. G., Scrace, G. T., and Mainwaring, W. I. P. (1978). *Biochem. J.*, **170**, 115.
29. McKeehan, W. L., Adams, P. S., and Rosser, M. P. (1984). *Cancer Res.*, **44**, 1998.
30. Isaccs, J. T., Werssman, R. M., Coffey, D. S., and Scott W. W. (1980). In *Models for prostate cancer* (ed. G. N. Murphy), pp. 311–23. Alan R. Liss, New York.
31. Peehl, D. M. and Ham, R. G. (1980). *In Vitro*, **16**, 526.
32. McKeehan, W. L., Adams, P. S., and Fast, D. (1987). *In Vitro Cell. Dev. Biol.*, **23**, 147.
33. Peehl, D. M., Wong, S. T., and Stamey, T. A. (1988). *In Vitro Cell. Dev. Biol.*, **24**, 530.
34. Yan, G., Fukabori, Y., Nikolaropoulos, S., Wang, F., and McKeehan, W. L. (1992). *Mol. Endocrinol.*, **6**, 2123.
35. Feng, S., Wang, F., Matsubara, A., Kan, M., and McKeehan, W. L. (1997). *Cancer Res.*, **57**, 5369.
36. Allegra, J. C. and Lippman, M. E. (1978). *Cancer Res.*, **38**, 3823.
37. Barnes, D. W. and Sato, G. H. (1979). *Nature*, **281**, 388.
38. Smith, H. S., Lan, S., Ceriani, R., Hackett, A. J., and Stampfer, M. R. (1981). *Cancer Res.*, **41**, 4637.
39. Hammond, S. L., Ham, R. G., and Stampfer, M. R. (1984). *Proc. Natl. Acad. Sci. USA*, **81**, 5435.
40. Tomooka, Y., Imagawa, W., Nandi, S., and Bern, H. A. (1983). *J. Cell. Physiol.*, **117**, 290.
41. Seglen, P. O. (1976). *Methods Cell Biol.*, **13**, 29.
42. Hook, M., Rubin, K., Oldberg, A., Obrink, B., and Vaheri, A. (1977). *Biochem. Biophys. Res. Commun.*, **79**, 726.
43. Bissell, D. M. and Guzelian, P. S. (1981). *Ann. NY Acad. Sci.*, **349**, 85.
44. McGowan, J. A., Strain, A. J., and Bucher, N. L. R. (1981). *J. Cell. Physiol.*, **180**, 353.
45. Luetteke, N. C., Michalopoulos, G. K., Teixido, J., Gilmore, R., Massague, J., and Lee, D. C. (1988). *Biochemistry*, **27**, 6487.
46. Kan, M., Huang, J., Mansson, P., Yasumitsu, H., Carr, B., and McKeehan, W. L. (1989). *Proc. Natl. Acad. Sci. USA*, **86**, 7432.

47. Nakamura, T., Nishizawa, T., Hagiya, M., Seki, T., Shimonishi, M., Sugimura, A., *et al.* (1989). *Nature*, **342**, 440.
48. Carr, B. I., Hayashi, I., Branum, E. L., and Moses, H. L. (1986). *Cancer Res.*, **46**, 2330.
49. Nakamura, T., Asami, O., Tanaka, K., and Ichihara, A. (1984). *Exp. Cell Res.*, **155**, 81.
50. Isom, H. C., Secott, T., Georgoff, I., Woodworth, C., and Mummaw, J. (1985). *Proc. Natl. Acad. Sci. USA*, **82**, 3252.
51. Miyazaki, M., Handa, Y., Oda, M., Yabe, T., Miyano, K., and Sato, J. (1985). *Exp. Cell Res.*, **159**, 176.
52. Guguen-Guillouzo, C., Clement, B., Baffet, G., Beaumont, C., Morel-Chany, E., Glaise, D., *et al.* (1983). *Exp. Cell Res.*, **143**, 47.
53. Friedman, S. L. and Roll, F. J. (1987). *Anal. Biochem.*, **161**, 207.
54. Sigal, S. H., Brill, S., Fiorino, A. S., and Reid, L. M. (1992). *Am. J. Physiol.*, **263**, G139.
55. Pyke, K. W. and Gelfand, E. W. (1974). *Nature*, **251**, 421.
56. Papiernik, M., Nabarra, B., and Back, J. F. (1975). *Clin. Exp. Immunol.*, **90**, 439.
57. Rimm, I. J., Bhan, A. K., Schneeberger, E. E., Schlossman, S. F., and Reinherz, E. L. (1984). *Clin. Immunol. Immunopathol.*, **31**, 56.
58. Sun, T. T., Bonitz, P., and Burns, W. H. (1984). *Cell. Immunol.*, **83**, 1.
59. Potworowski, E. F., Turcotte, F., Beauchemin, C., Hugo, P., and Zelechowska, M. G. (1986). *In Vitro*, **22**, 557.
60. Glimcher, L. H., Kruisbeek, A. M., Paul, W. E., and Green, I. (1983). *Scand. J. Immunol.*, **17**, 1.
61. Piltch, A., Naylor, P., and Hayashi, J. (1988). *In Vitro Cell. Dev. Biol.*, **24**, 289.
62. Kuriharcuch, W. and Green, H. (1978). *Proc. Natl. Acad. Sci. USA*, **75**, 6107.
63. Hayflick, L. and Moorhead, P. S. (1961). *Exp. Cell Res.*, **25**, 589.
64. McKeehan, W. L., McKeehan, K. A., Hammond, S. L., and Ham, R. G. (1977). *In Vitro*, **13**, 399.
65. Bettger, W. J., Boyce, S. T., Walthall, B. J., and Ham, R. G. (1981). *Proc. Natl. Acad. Sci. USA*, **78**, 5588.
66. Phillips, P. D. and Cristofalo, V. J. (1981). *Exp. Cell Res.*, **134**, 297.
67. Kan, M. and Yamane, I. (1982). *J. Cell. Physiol.*, **111**, 155.
68. Greenwood, D., Srinivasan, A., McGoogan, S., and Pipas, J. M. (1989). In *Cell growth and division: a practical approach* (ed. R. Baserga), pp. 37–60. IRL Press, Oxford.
69. Jaffe, E. A., Nachman, R. L., Becker, C. G., and Minick, C. R. (1973). *J. Clin. Invest.*, **52**, 2745.
70. Schwartz, S. M. (1978). *In Vitro*, **14**, 966.
71. Tauber, J.-P., Cheng, J., Massoglia, S., and Gospodarowicz, D. (1981). *In Vitro*, **17**, 519.
72. Folkman, J., Haudenschild, C. C., and Zetter, B. R. (1979). *Proc. Natl. Acad. Sci. USA*, **76**, 5217.
73. Gimbrone, M. A. Jr., Cotran, R. S., and Folkman, J. (1974). *J. Cell Biol.*, **60**, 673.
74. Gospodarowicz, D., Brown, K. D., Birdwell, C. R., and Zetter, B. R. (1978). *J. Cell Biol.*, **77**, 774.
75. Maciag, T., Hoover, G. A., Stemerman, M. B., and Weinstein, R. (1981). *J. Cell Biol.*, **91**, 420.
76. Mansson, P.-E., Malark, M., Sawada, H., Kan, M., and McKeehan, W. L. (1990). *In Vitro Cell. Dev. Biol.*, **26**, 209.
77. Hoshi, H., Kan, M., Chen, J.-K., and McKeehan, W. L. (1988). *In Vitro Cell. Dev. Biol.*, **24**, 309.
78. Myoken, Y., Kan, M., Sato, G. H., McKeehan, W. L., and Sato, J. D. (1990). *Exp. Cell Res.*, **191**, 299.
79. Leung, D. W., Cachianes, G., Kuang, W. J., Goeddel, D. V., and Ferrara, N. (1989). *Science*, **246**, 1306.

80. Keck, P. J., Hauser, S. D., Krivi, G., Sanzo, K., Warren, T., Feder, J., *et al.* (1989). *Science*, **246**, 1309.
81. Myoken, Y., Kayada, Y., Okamoto, T., Kan, M., Sato, G. H., and Sato, J. D. (1991). *Proc. Natl. Acad. Sci. USA*, **88**, 5819.
82. Yu, Y. and Sato, J. D. (1999). *J. Cell. Physiol.*, **178**, 235.
83. Gerber, H. P., McMurtrey, A., Kowalski, J., Yan, M., Keyt, B. A., Dixit, V., *et al.* (1998). *J. Biol. Chem.*, **273**, 30336.
84. Ng, C. W., Poznanski, W. J., Borowiecki, M., and Reimer, G. (1971). *Nature*, **231**, 445.
85. Adebonojo, F. O. (1973). *Biol. Neonate*, **23**, 366.
86. Bjorntorp, P., Karlsson, M., Petroft, P., Pettersson, P., Sjostrom, L., and Smith, U. (1978). *J. Lipid Res.*, **19**, 316.
87. Serrero, G., McClure, D. B., and Sato, G. H. (1979). In *Hormones and cell culture*. Cold Spring Harbor Conference on Cell Proliferation, Vol. 6 (ed. G. H. Sato and R. Ross), pp. 523–30. Cold Spring Harbor Laboratory Press, NY.
88. Serrero, G. and Khoo, J. C. (1982). *Anal. Biochem.*, **120**, 351.
89. Gaillard, D., Negrel, R., Serrero, G., Cermolacce, E., and Ailhaud, G. (1984). *In Vitro*, **20**, 79.
90. Broad, T. E. and Ham, R. (1983). *Eur. J. Biochem.*, **135**, 33.
91. Deslex, S., Negrel, R., Etienne, J., and Aihaud, G. (1986). *Int. J. Obesity*, **11**, 19.
92. Deslex, S., Negrel, R., and Ailhaud, G. (1987). *Exp. Cell Res.*, **168**, 15.
93. Serrero, G. and Mills, D. (1987). *In Vitro Cell. Dev. Biol.*, **23**, 63.
94. Litthauer, D. and Serrero, G. (1992). *Comp. Biochem. Physiol.*, **101A**, 59.
95. Konigsberg, I. R. (1963). *Science*, **140**, 1273.
96. Fischbach, G. D. (1972). *Dev. Biol.*, **28**, 407.
97. Paterson, B. and Strohman, R. C. (1972). *Dev. Biol.*, **29**, 113.
98. Delvin, R. B. and Emerson, C. P. Jr. (1978). *Cell*, **13**, 599.
99. Yaffe, D. and Saxel, O. (1977). *Nature*, **270**, 725.
100. Yaffe, D. (1968). *Proc. Natl. Acad. Sci. USA*, **61**, 477.
101. Taylor, S. M. and Jones, P. A. (1979). *Cell*, **17**, 771.
102. Davis, R. L., Weintraub, H., and Lasser, A. B. (1987). *Cell*, **51**, 987.
103. Florini, J. R. and Roberts, S. B. (1979). *In Vitro*, **15**, 983.
104. Ross, R. (1971). *J. Cell Biol.*, **50**, 172.
105. Chamley-Campbell, J., Campbell, G. R., and Ross, R. (1979). *Physiol. Rev.*, **59**, 1.
106. Kimes, B. W. and Brandt, B. L. (1976). *Exp. Cell Res.*, **98**, 349.
107. Sasaki, Y., Seto, M., and Komatsu, K.-I. (1990). *FEBS Lett.*, **276**, 161.
108. LeDourin, N. M. (1980). *Nature*, **286**, 663.
109. Sieber-Blum, M. and Cohen, A. (1980). *Dev. Biol.*, **80**, 96.
110. Bronner-Fraser, M. E. and Fraser, S. E. (1988). *Nature*, **335**, 161.
111. Frank, E. and Sanes, J. R. (1991). *Development*, **111**, 895.
112. Stemple, D. L. and Anderson, D. J. (1992). *Cell*, **71**, 973.
113. Hawrot, E. and Patterson, P. H. (1979). In *Methods in enzymology* (ed. W. Jacoby and I. Pastan), Vol. 58, pp. 574–84. Academic Press, San Diego, CA.
114. Bernd, P. (1986). *Dev. Biol.*, **115**, 415.
115. Lendahl, U., Zimmerman, L. B., and McKay, R. D. G. (1990). *Cell*, **60**, 585.
116. Raff, M. C., Miller, R. H., and Noble, M. (1983). *Nature*, **303**, 390.
117. Noble, M., Murray, K., Stroobant, P., Waterfield, M., and Riddle, P. (1988). *Nature*, **333**, 560.
118. Raff, M., Lillien, L. E., Richardson, W. D., Burne, J. F., and Noble, M. D. (1988). *Nature*, **333**, 562.
119. Hughes, S., Lillien, L. E., Raff, M. S., Rohrer, H., and Sendtner, M. (1988). *Nature*, **335**, 70.

120. Bögler, O., Wren, D., Barnett, S. C., Land, H., and Noble, M. (1990). *Proc. Natl. Acad. Sci. USA*, **87**, 6368.
121. Barres, B. A. and Raff, M. C. (1993). *Nature*, **361**, 258.
122. Louis, J.-C., Magal, E., Takayama, S., and Varon, S. (1993). *Science*, **259**, 689.
123. Eisinger, M. and Marko, O. (1982). *Proc. Natl. Acad. Sci. USA*, **79**, 2018.
124. Halaban, R. and Alfano, F. D. (1984). *In Vitro*, **20**, 447.
125. Gilchrest, B. A., Albert, L. S., Karassik, R. L., and Yaar, M. (1985). *In Vitro*, **21**, 114.
126. Eisinger, M., Marko, O., Ogata, S.-I., and Old, L. J. (1985). *Science*, **229**, 984.
127. Wilkens, L., Gilchrist, B. A., Szabo, G., Weinstein, R., and Maciag, T. (1985). *J. Cell. Physiol.*, **122**, 350.
128. Halaban, R., Ghosh, S., and Baird, A. (1987). *In Vitro Cell. Dev. Biol.*, **23**, 47.
129. Halaban, R., Langdon, R., Birchall, N., Cuono, C., Baird, A., Scott, G., *et al.* (1988). *J. Cell Biol.*, **107**, 1611.
130. Halaban, R., Ghosh, S., Duray, P., Kirkwood, J. M., and Lerner, A. B. (1986). *J. Invest. Dermatol.*, **87**, 95.
131. Pittelkow, M. R. and Shipley, G. D. (1989). *J. Cell. Physiol.*, **140**, 565.
132. De Felici, M. and McLaren, A. (1983). *Exp. Cell Res.*, **144**, 417.
133. Wabik-Sliz, B. and McLaren, A. (1984). *Exp. Cell Res.*, **154**, 530.
134. Matsui, Y., Zsebo, K., and Hogan, B. L. M. (1992). *Cell*, **70**, 841.
135. Resnick, J. L., Bixler, L. S., Cheng, L., and Donovan, P. J. (1992). *Nature*, **359**, 550.
136. Donovan, P. J., Stott, D., Cairns, L. A., Heasman, J., and Wylie, C. C. (1986). *Cell*, **44**, 831.
137. Robertson, E. J. (ed.) (1987). *Teratocarcinomas and embryonic stem cells: a practical approach*. IRL Press, Oxford.
138. Godin, I., Deed, R., Cooke, J., Zsebo, K., Dexter, M., and Wylie, C. C. (1991). *Nature*, **352**, 807.
139. Dolci, S., Williams, D. E., Ernst, M. K., Resnick, J. L., Brannan, C. I., Lock, L. F., *et al.* (1991). *Nature*, **352**, 809.
140. Matsui, Y., Toksoz, D., Nishikawa, S., Nishikawa, S.-I., Williams, D., Zsebo, K., *et al.* (1991). *Nature*, **353**, 750.
141. Smith, A. G., Heath, J. K., Donaldson, D. D., Wong, G. G., Moreau, J., Stahl, M., *et al.* (1988). *Nature*, **336**, 688.
142. McLaren, A. (1992). *Nature*, **359**, 482.
143. Stewart, C. L., Gadi, I., and Bhatt, H. (1994). *Dev. Biol.*, **161**, 626.
144. Labosky, P. A., Barlow, D. P., and Hogan, B. L. M. (1994). *Development*, **120**, 3197.
145. Shamblott, M. J., Axelman, J., Wang, S., Bugg, E. M., Littlefield, J. W., Donovan, P. J., *et al.* (1999). *Proc. Natl. Acad. Sci. USA*, **95**, 13726.
146. Orley, J., Sato, G. H., and Erikson, G. F. (1980). *Cell*, **20**, 817.
147. Hoshi, H., Takagi, Y., Kobayashi, K., Onodera, M., and Oikawa, T. (1991). *In Vitro Cell. Dev. Biol.*, **27A**, 578.
148. Savion, N., Lui, G.-M., Laherty, R., and Gospodarowicz, D. (1981). *Endocrinology*, **109**, 409.
149. Li, R., Phillips, D. M., and Mather, J. P. (1995). *Endocrinology*, **136**, 849.
150. Li, R., Philips, D. M., Moore, A., and Mather, J. P. (1997). *Endocrinology*, **138**, 2648.
151. Woodruff, T. K., Lyon, R. J., Hansen, S. E., Rice, G. C., and Mather, J. P. (1990). *Endocrinology*, **127**, 3196.
152. Evans, M. J. and Kaufman, M. H. (1981). *Nature*, **292**, 154.
153. Bradley, A. and Luo, G. (1998). *Nature Genet.*, **20**, 322.
154. Sato, J. D. and Kan, M. (1998). In *Current protocols in cell biology* (ed. J. S. Bonifacino, M. Dasso, J. B. Hartford, J. Lippincott-Schwartz, and K. M. Yamada), pp. 1.2.1–1.2.14. John Wiley & Sons, Inc., New York.

155. Zambrowicz, B. P., Freidrich, G. A., Buxton, E. C., Lilleberg, S. L., Person, C., and Sands, A. T. (1998). *Nature*, **392**, 608.
156. Talbot, N. C., Powell, A. M., and Rexroad, C. C. (1995). *Mol. Rep. Dev.*, **42**, 35.
157. Etches, R. J. (1998). *Br. Poult. Sci.*, **39**, 5.
158. Thomson, J. A. and Marshall, V. S. (1998). *Curr. Top. Dev. Biol.*, **38**, 133.
159. Thomson, J. A., Kalishman, J., Golos, T. G., Durning, M., Harris, C. P., Becker, R. A., *et al.* (1995). *Proc. Natl. Acad. Sci. USA*, **92**, 7844.
160. Thomson, J. A., Kalishman, J., Golos, T. G., Durning, M., Harris, C. P., and Hearn, J. P. (1996). *Biol. Reprod.*, **55**, 254.
161. Thomson, J. A., Itskovitz-Eldor, J., Shapiro, S. S., Waknitz, M. A., Swiergiel, J. J., Marshall, V. S., *et al.* (1998). *Science*, **282**, 1145.
162. Bain, G., Kitchens, D., Yao, M., Huettner, J. E., and Gottlieb, D. I. (1995). *Dev. Biol.*, **168**, 342.
163. Gottlieb, D. I. and Huettner, J. E. (1999). *Cells, Tissues and Organs*, **165**, 165.
164. Robertson, S., Kennedy, M., and Keller, G. (1999). *Ann. NY Acad. Sci.*, **872**, 9.
165. Helmrich, A. and Barnes, D. (1999). In *Methods in cell biology*, pp. 29–37. Academic Press, New York.
166. Sun, L., Bradford, C. S., and Barnes, D. W. (1995). *Mol. Mar. Biol. Biotech.*, **4**, 43.
167. Sun, L., Bradford, S., Ghosh, C., Collodi, P., and Barnes, D. (1995). *Mol. Mar. Biol. Biotech.*, **4**, 193.
168. Sun, L., Bradford, S., Ghosh, C., Collodi, P., Barnes, D. (1996). *J. Mar. Biotech.*, **3**, 211.
169. Singh, N., Fischer, K., Hedstrom, O., and Barnes, D. W. (2001). *Mar. Biotech.*, **3**, 27.
170. Brüstle, O., Jones, K. N., Learish, R. D., Karram, K., Choudhary, K., Wiestler, O. D., *et al.* (1999). *Science*, **285**, 754.
171. Li, E., Bestor, T. H., and Jaenisch, R. (1992). *Cell*, **69**, 915.
172. Okabe, S., Forsberg-Nilsson, K., Spiro, A. C., Segal, M., and McKay, R. D. G. (1996). *Mech. Dev.*, **59**, 89.
173. Steinitz, M., Klein, G., Koshimies, S., and Makel, O. (1977). *Nature*, **269**, 420.
174. Köhler, G. and Milstein, C. (1975). *Nature*, **256**, 495.
175. Harbour, C. and Fletcher, A. (1991). In *Mammalian cell biotechnology: a practical approach* (ed. M. Butler), pp. 109–38. IRL Press, Oxford.
176. Horibata, K. and Harris, A. W. (1970). *Exp. Cell Res.*, **60**, 61.
177. Chang, T. H., Steplewski, Z., and Koprowski, H. (1980). *J. Immunol. Methods*, **39**, 369.
178. Murakami, H., Masui, H., Sato, G. H., Sueoka, N., Chow, T. P., and Kano-Sueoka, T. (1982). *Proc. Natl. Acad. Sci. USA*, **79**, 1158.
179. Kawamoto, T., Sato, J. D., Le, A., McClure, D. B., and Sato, G. H. (1983). *Anal. Biochem.*, **130**, 445.
180. Sato, J. D., Kawamoto, T., and Okamoto, T. (1987). *J. Exp. Med.*, **165**, 1761.
181. Sato, J. D., Kawamoto, T., McClure, D. B., and Sato, G. H. (1984). *Mol. Biol. Med.*, **2**, 121.
182. Sato, J. D., Cao, H.-T., Kayada, Y., Cabot, M. C., Sato, G. H., Okamoto, T., *et al.* (1988). *In Vitro Cell. Dev. Biol.*, **24**, 1223.
183. Myoken, Y., Okamoto, T., Osaki, T., Yabumoto, M., Sato, G. H., Takada, K., *et al.* (1989). *In Vitro Cell. Dev. Biol.*, **25**, 477.
184. Lanzavecchia, A. and Sallusto, F. (2000). *Science*, **290**, 92.
185. Zinkernagel, R. and Doherty, P. C. (1974). *Nature*, **248**, 701.
186. Mossman, T. R. and Coffman, R. L. (1989). *Ann. Rev. Immunol.*, **7**, 145.
187. Salgame, P., Abrams, J. S., Clayberger, C., Goldstein, H., Convit, J., Modlin, R. L., *et al.* (1991). *Science*, **254**, 279.
188. Croft, M., Carter, L., Swain, S. L., and Dutton, R. W. (1994). *J. Exp. Med.*, **180**, 1715.
189. Swain, S. L., Weinberg, A. D., English, M., and Huston, G. (1990). *J. Immunol.*, **145**, 3796.

190. Ruscetti, F. W. and Smith, K. A. (1981). *Adv. Immunol.*, **31**, 137.
191. Smith, K. A. (1988). *Science*, **240**, 1169.
192. Morgan, D. A., Ruscetti, F., and Gallo, R. C. (1976). *Science*, **193**, 1007.
193. Ruscetti, F. W., Morgan, D. A., and Gallo, R. C. (1977). *J. Immunol.*, **119**, 131.
194. Gillis, S., Baker, P. E., Ruscetti, F. W., and Smith, K. A. (1978). *J. Exp. Med.*, **148**, 1093.
195. Taniguchi, T., Matsui, H., Fujita, T., Takaoka, C., Kashima, N., Yoshimoto, R., *et al.* (1983). *Nature*, **302**, 305.
196. Gillis, S. (1989). In *Fundamental immunology* (2nd edn) (ed. W. E. Paul), pp. 621–38. Raven Press, NY.
197. Grimm, E. A., Mazumder, A., Zhang, H. Z., and Rosenberg, S. A. (1982). *J. Exp. Med.*, **155**, 1823.
198. Rayner, A. A., Grimm, E. A., Lotze, M. T., Chu, E. W., and Rosenberg, S. A. (1985). *Cancer*, **55**, 1327.
199. Grimm, E. A. and Owen-Schaub, L. (1991) *J. Cell. Biochem.*, **45**, 335.
200. Okamoto, T., Tani, R., Yabumoto, M., Sakamoto, A., Takada, K., Sato, G. H., *et al.* (1996). *J. Immunol. Methods*, **195**, 7.
201. Rosenberg, S. A., Speiss, P., and Lafreniere, R. (1986). *Science*, **233**, 1318.
202. Topalian, S. L., Solomon, D., and Rosenberg, S. A. (1989). *J. Immunol.*, **142**, 3714.
203. Topalian, S. L., Muul, L. M., Solomon, D., and Rosenberg, S. A. (1987). *J. Immunol. Methods*, **102**, 127.
204. Banchereau, J. and Steinman, R. M. (1998). *Nature*, **392**, 245.
205. Inaba, K., Inaba, M., Deguchi, M., Hagi, K., Yasumizu, R., Ikehara, S., *et al.* (1993). *Proc. Natl. Acad. Sci. USA*, **90**, 3038.
206. Inaba, K., Inaba, M., Romani, N., Aya, H., Deguchi, M., Ikehara, S., *et al.* (1992). *J. Exp. Med.*, **176**, 1693.
207. Inaba, K., Steinman, R. M., Witmer-Pack, M., Aya, K., Inaba, M., Sudo, T., *et al.* (1992). *J. Exp. Med.*, **175**, 1157.
208. Caux, C., Dezutter-Dambuyant, C., Schmitt, D., and Banchereau, J. (1992). *Nature*, **360**, 258.
209. Reid, C. D. L., Stackpole, A., Meager, A., and Tikerpae, J. (1992). *J. Immunol.*, **149**, 2681.
210. Young, J. W., Szabolcs, P., and Moore, M. A. S. (1995). *J. Exp. Med.*, **182**, 1111.
211. Romani, N., Gruner, S., Brang, D., Kampgen, E., Lenz, A., Trockenbacher, B., *et al.* (1994). *J. Exp. Med.*, **180**, 83.
212. Romani, N., Reider, D., Heuer, M., Ebner, S., Kampgen, E., Eibl, B., *et al.* (1996). *J. Immunol. Methods*, **196**, 137.
213. Saunders, D., Lucas, K., Ismaili, J., Wu, L., Maraskovsky, E., Dunn, A., *et al.* (1996). *J. Exp. Med.*, **184**, 2185.
214. Suss, G. and Shortman, K. (1996). *J. Exp. Med.*, **183**, 1789.
215. Seecof, R. L. and Unanue, R. L. (1968). *Exp. Cell Res.*, **50**, 654.
216. Shields, G. and Sang, J. H. (1970). *J. Embryol. Exp. Morphol.*, **23**, 53.
217. Shields, G., Dubendorfer, A., and Sang, J. H. (1975). *J. Embryol. Exp. Morphol.*, **33**, 159.
218. Seecof, R. L., Alleaume, N., Teplitz, R. L., and Gerson, I. (1971). *Exp. Cell Res.*, **69**, 161.
219. Salvaterra, P. M., Bournias-Vardiabasis, N., Nair, T., Hou, G., and Lieu, C. (1987). *J. Neurosci.*, **7**, 10.
220. Cross, D. P. and Sang, J. H. (1978). *J. Embryol. Exp. Morphol.*, **435**, 161.
221. Hayashi, I. and Perez-Megallanes, M. (1994). *In Vitro Cell. Dev. Biol.*, **30A**, 202.
222. Krasnow, M. A., Cumberledge, S., Manning, G., Herzenberg, L. A., and Nolan, G. P. (1991). *Science*, **251**, 81.
223. Sato, G. H. (1975). In *Biochemical action of hormones* (ed. G. Litwack), Vol. III, pp. 391–6. Academic Press, New York.
224. Hayashi, I. and Sato, G. H. (1976). *Nature*, **259**, 132.

Chapter 8
Cloning

John Clarke
Haemophilia Centre, St. Thomas' Hospital, Lambeth Palace Road, London SE1 7EH, UK.

Alison J. Porter
Lonza Biologics plc, 228 Bath Road, Slough, Berkshire SL1 4DY, UK.

Robin Thorpe
Division of Immunobiology, National Institute for Biological Standards and Control, Blanche Lane, South Mimms, Hertfordshire EN6 3QG, UK.

John M. Davis
Research & Development Department, Bio Products Laboratory, Dagger Lane, Elstree, Hertfordshire WD6 3BX, UK.

1 Introduction

A clone is a population of cells which are descended from a single parental cell. Clones may be derived from continuous cell lines or from primary cultures, but in either case the purpose of cloning is the same: to minimize the degree of genetic and phenotypic variation within a cell population. This is done by isolating a single cell under suitable conditions and then allowing it to multiply to produce a large enough number of cells for the required purpose.

1.1 Development of techniques

Cell cloning techniques were devised very early during the development of modern cell culture. Sanford *et al.* (1) isolated single cells within capillary tubes and Wildy and Stoker (2) isolated single cells in droplets of medium under liquid paraffin. In both cases, conditioning factors could accumulate in proximity to the isolated cells and promote clonal growth. The serial dilution procedure developed by Puck and Marcus (3) was less technically demanding. These authors also used feeder layers of X-irradiated, non-replicating cells to supply growth factors. Cooper (4) devised a procedure involving the distribution of small volumes of a dilute suspension of cells among the wells of 96-well tissue culture trays and MacPherson (5) devised a technique in which individual cells were taken up in extra-fine Pasteur pipettes and inoculated into the wells of 96-well trays. The simpler and more rapid 'spotting' technique utilizing these trays was introduced by Clarke and Spier (6). Other techniques include colony formation in semi-solid media (7), colony formation on glass coverslip fragments (8), and separation of cells by fluorescence-activated cell sorting (9).

1.2 Uses of cloning

Cloning cells from continuous lines has a number of applications:

(a) Many continuous lines are genetically unstable and their properties may alter during passage. Cloning can be used to isolate cultures with properties more closely resembling those of the original population, or conversely may be used to isolate variants. Examples of the latter may include karyological and biochemical variants, and cells which exhibit different levels of product secretion or different susceptibilities to viruses. Treatment with mutagens can be used to attempt to increase the proportion of variant cells (see, for example, Chapter 6, *Protocol 12*).

(b) Variation within a continuous cell line may be studied by examining the properties of panels of clones established at different passage levels.

(c) Cells transfected with DNA (see Chapter 6) do not necessarily form populations with homogeneous genetic constitutions and cloning can enable cells to be selected and cultures developed with the required characteristics.

(d) In applied biotechnology, it may be desirable from a regulatory stand-point that a cell line used to derive a product can be defined as originating from a single cell.

In general, the cloning of primary cells is less successful than the cloning of established cell lines, as they tend to have a low colony-forming efficiency (CFE, see Section 3.1) and 'normal' cells can only undergo a limited number of population doublings (see Chapter 5) which may prevent the generation of a sufficient number of cells for future use. Nevertheless, it is sometimes possible to isolate specific cell types from a mixed primary population, and to develop clones large enough for subsequent studies and free of unwanted cell types (often fibroblasts) which might otherwise overgrow the culture.

1.3 Limitations of cloning

It must be stressed that cloning does *not* guarantee subsequent homogeneity of the derived cell population. Many cell lines are phenotypically heterogeneous in culture and often this is an inherent characteristic of these lines which cannot be eliminated even by multiple rounds of cloning. Some cloned cell lines are capable of differentiation *in vitro* and the phenotype or balance of phenotypes within the population may change depending on environmental conditions (e.g. presence or absence of growth factors, inducers, etc.). Furthermore, the very genetic instability that may have necessitated the cloning will in many instances not be eliminated by cloning, with the result that although initially genetically homogeneous, the clones again become heterogeneous as the cells multiply, and may need to be re-cloned at regular intervals to minimize the degree of heterogeneity. Also, in any cell population, there will be a tendency for mutations to accumulate with time. Thus the homogeneity of a cloned cell population will depend on the intrinsic properties of the cell line, the elapsed time in culture since cloning, and also possibly on the culture environment.

As outlined above, many different approaches have been taken to the problem of isolating single cells under conditions in which they will continue to proliferate, and details of many of these methods are given in the protocols in this chapter. Each method has its advantages and disadvantages, but all share one common problem; there is no way to be *certain* that the derived population is the progeny of one cell. As there is always a slight but finite chance that a second cell is present along with the assumed single cell, cloning always deals with the *probability* of a colony being a true clone. A reliable estimation of this probability is important in certain situations; for example, when a cell line is to be used to generate a product for public use.

Limiting dilution is a commonly-used method which relies on the assumption that cells are distributed according to the Poisson distribution to calculate a probability of monoclonality. More than one round of cloning by this method is normally carried out to increase the probability of obtaining a monoclonal colony (11). However, a number of factors may lead to incorrect estimates of the probability of monoclonality. First, the assumption that cells are distributed according to the Poisson distribution was based on models using artefacts such as ion exchange beads in place of cells (12). It has to be questioned how well these beads represent cells. For example, unlike the beads, some cell species tend to adhere to each other and this could affect the distribution, resulting in an increased proportion of colonies arising from two or more cells. Secondly, even if the cells are distributed according to the Poisson distribution, the colonies which develop from them may not follow this mathematical model (13).

Some examples of possible problems are as follows:-

(a) One cell could have a toxic effect on a neighbouring cell(s), leading to a higher proportion of the growing colonies actually being monoclonal than would be predicted by the Poisson distribution.

(b) Colonies arising from more than one cell may have an increased chance of survival over a colony arising from a single cell, therefore increasing the proportion of colonies arising from two or more cells. There is some evidence that this may occur with certain cells (14).

(c) Where microscopic observation is being used to eliminate those colonies arising from more than one cell, colonies may mistakenly be identified as arising from a single cell if the cells are in close proximity, or if some cells are temporarily dormant (depending on when the observation is performed). This would result in more colonies than estimated *not* being monoclonal.

(d) If the colonies are being screened for a particular phenotype, then the probability of a colony which displays the desired characteristic actually having arisen from a single cell decreases with the rarity of the desired phenotype (15).

(a) and (b) may be particularly problematic if occurring with cells displaying a poor CFE, and which consequently have been distributed at a high average number of cells per well.

Thus whilst the limiting dilution technique of cloning is attractive because of the mathematical analysis to which it may be subjected, in some systems the numbers obtained may bear little relationship to what is actually occurring in the cloning plate because the assumptions on which they are based are invalid. It is therefore suggested that in many cases the careful use of a suitable technique in which one observes a single cell being seeded (see, for example, *Protocol 4*) will give greater assurance that true clones are isolated. Such techniques are, of course, very labour-intensive, can only be used with limited numbers of cells, and may not be applicable in all situations.

2 Special requirements of cells growing at very low densities

By its very nature, cloning requires that cells should grow and multiply at very low population densities, at least during the first few divisions after plating. This is, however, a very non-physiological situation; *in vivo*, animal cells would normally exist at population densities of up to 10^9 cells/ml in the presence of autocrine and paracrine factors and hormones, as well as (in most cases) considerable cell–cell and cell–matrix interaction. Many of these environmental factors are important for cell proliferation, and thus the culture conditions during cloning must attempt to mimic the essentials of the *in vivo* environment, but in a situation where individual proliferating cells can be isolated from one another. This places much greater demands on the culture medium and its additives than is the case for normal culture.

All culture media and sera to be used for cloning should be tested for their ability to support growth at low population density of the cell line to be cloned. The use of a test for CFE (see *Protocol 1*) is a good, quantitative way of doing this, and facilitates optimization of the medium/serum/additive mixture. In general, enriched media containing cell growth factors and other growth-enhancing substances (see ref. 16 and Chapter 7) may improve clonal growth. FCS at high concentrations (e.g. 20% or more) will also often be found to enhance clonal growth, but it is important to test individual batches with individual cell lines as there is enormous batch to batch variation and a batch which is very effective at supporting the clonal growth of one cell line may be inhibitory to another.

Low CFE and slow initial cell proliferation may also be improved by adding conditioned medium to the cultures. This is medium harvested from normal cultures (generally of the cells being cloned), after growth to 50% confluency or 50% normal maximum concentration (see also Chapter 6, Section 2.4.3). Improvements may also be obtained by the use of layers of feeder cells treated to prevent replication (see, for example, Chapter 6, *Protocol 7*).

It must be stressed that actively growing cultures are essential in order to obtain as high a CFE as possible, and consequently cloning must only be performed using cells from such cultures (i.e. for established cell lines, cells in log phase - see Chapter 5, Section 7.2).

3 Cell cloning procedures

3.1 Choice of technique

The procedure to be used should be selected with particular reference to the CFE of the cell population. CFE is defined as:

number of colonies obtained/number of individual cells plated

and is usually expressed as a percentage. The CFE may vary from less than 1% for some primary cells, to practically 100% for some established cell lines. CFE may be assessed using *Protocol 1*.

Protocol 1

Determination of colony-forming efficiency

As described, this procedure is only suitable for cells which will attach (at least to some extent) to the surface of the tissue culture vessel. Cells growing free in suspension will require the use of a semi-solid medium (see *Protocols 7* and *8*) in order to localize the growing colonies.

Equipment and reagents

- Tissue culture flasks, Petri dishes, or multiwell dishes (4- or 6-well)
- Inverted microscope
- Equipment for performing a viable cell count (see Chapter 5, *Protocol 4*)
- Complete medium suitable for the cells under test (including serum, other supplements, and conditioned medium, as appropriate)

Method

1 Resuspend adherent cells by standard subculture method; use suspension cells directly from culture.

2 Count viable cells. Prepare suspensions of 100, 1000, and 10 000 cells/ml in all combinations of medium and serum under investigation.

3 Inoculate suitable volumes of the cell suspensions into multiwell trays, Petri dishes, or tissue culture flasks.

4 Incubate at appropriate temperature (dependent on the species of origin of the cells - see Chapter 1, Section 3.1.2). If using Petri dishes or multiwell trays, use a humidified atmosphere containing the appropriate CO_2 concentration for the medium being used (see Chapter 3, *Table 1*). Flasks should be pre-gassed with a CO_2/air mixture.

5 Using an inverted microscope, inspect the cells after four days and then after every

Protocol 1 continued

two days; with a marker pen, mark on the base of the container the positions of colonies that appear.

6. When distinct colonies have formed, count the total number and calculate the CFE:
   (colonies formed/total cells seeded) × 100%.
7. If the CFE appears very low or cell growth appears slow, consider repeating the test using feeder cells and/or modified medium constituents.
8. If only a limited number of clones are required and well separated colonies form, these may be isolated for subculture at this stage by the use of cloning rings (see *Protocol 5*).

For cells with a high CFE, the 'spotting' procedure (*Protocol 3*) is very effective and results in the isolation of single cells in wells of 96-well tissue culture trays. The procedure is easy to perform and allows direct verification that only single cells are present. The 'micro-manipulation' procedure (*Protocol 4*) also allows this, and facilitates the direct selection of individual cells with particular observable characteristics. However, it is technically more difficult and time-consuming.

However, if CFE is low (less than 5%), the above techniques would require a large number of 96-well trays to generate a significant number of clones, and regular inspection of all the wells would be very protracted. In this situation, the various other techniques described would be more convenient, and these can in fact be used with cells of high or low CFE simply by adjusting the number of cells placed in each plate or well.

Some of the techniques described below are applicable to any cell type (Section 3.2), whilst some are only suitable for use with attached cells (Section 3.3), and others are generally only suitable for cells capable of growing in suspension (Section 3.4).

3.2 Methods applicable to both attached and suspension cells

3.2.1 Limiting dilution

This is probably the most widely used method of cloning, and depends on plating cells in multiwell plates at a sufficient dilution such that there is a high probability that any colony which grows subsequently is derived from a single cell.

i. Theoretical considerations

If a number of single cells are distributed randomly and independently into a large number of wells, then the fraction of the total number of wells which will contain a particular number of cells is described by the Poisson distribution:

$$F_r = \frac{(c/w)^r}{r!} e^{-(c/w)}$$

where r is the number of cells in a well, F_r the fraction of the total number of wells containing r cells, c the total number of cells distributed, and w the total number of wells into which they were distributed. The ratio c/w can be replaced by the term u, which thus represents the average number of cells per well. The above equation then becomes:

$$F_r = \frac{u^r}{r!} e^{-u}$$

It is u, the average number of cells per well, that is the parameter usually used to describe the way a limiting dilution cloning is performed (e.g. 'The cells were cloned at 0.1 cells per well'). By knowing u, it is possible to calculate the fraction of wells containing a given number of cells.

Thus the fraction of wells containing no cells is:

$$F_0 = \frac{u^0}{0!} e^{-u} = e^{-u}$$

Similarly, the fraction containing one cell is:

$$F_1 = \frac{u^1}{1!} e^{-u} = ue^{-u}$$

The fraction containing two cells is:

$$F_2 = \frac{u^2}{2!} e^{-u} = \frac{u^2}{2} e^{-u}$$

The fraction containing three cells is:

$$F_3 = \frac{u^3}{3!} e^{-u} = \frac{u^3}{6} e^{-u}$$

and so on.

By using these equations it is possible to estimate the fraction of wells containing a given number of cells in any cloning performed at a known average number of cells per well, as illustrated in *Table 1*.

Table 1 Fraction of wells containing a given number of cells (F_r): variation with average number of cells per well (u)

No. of cells per well (r)	**Average number of cells per well (u)**		
	1.0	**0.3**	**0.1**
0	0.368	0.741	0.905
1	0.368	0.222	0.090
2	0.184	0.033	0.005
3	0.061	0.003	0.000
4	0.015	0.000	0.000

If it is assumed that growth will occur in any well receiving at least one cell, then the probability that any colony (chosen at random) is actually derived from only a single cell is equal to the ratio of the number of colonies derived from one cell to the number of colonies in total. This is numerically equivalent to:

$$\frac{F_1}{1 - F_0} = \frac{ue^{-u}}{1 - e^{-u}}$$

Table 2 Probability that any colony picked at random is derived from a single cell (i.e. is a true clone)

	Average number of cells per well (u)		
	1.0	**0.3**	**0.1**
Probability	0.582	0.857	0.951

and representative figures can be calculated from the data in *Table 1*. This is illustrated in *Table 2*. It should be noted that even at an average of 0.1 cells per well there is still about a 5% chance that any particular colony will not be a clone. Thus where clonality is important, it is recommended that cells cloned by limiting dilution are actually cloned more than once at a low average number of cells per well. The statistics of multiple rounds of cloning are dealt with in ref. 11. It should be stressed that all the foregoing is dependent on the initial assumptions being true, namely that only single cells are being distributed (i.e. that there are no clusters of cells), and that colonies are selected at random. Clearly, if wells are monitored during cell growth and those showing more than one focus of growth are excluded, then the chances of picking a clone are increased. However, the problems and limitations already discussed in Section 1.3 should always be borne in mind.

For a fuller account of limiting dilution techniques, including derivation of the relevant statistics from first principles, see ref. 17.

ii. Practical aspects

A general procedure for cloning cells by limiting dilution is given in *Protocol 2*.

Protocol 2

Cloning by limiting dilution

Equipment and reagents

- 96- and 24-well tissue culture trays
- 25 cm^2 and 75 cm^2 tissue culture flasks
- Equipment for performing a viable cell count (see Chapter 5, *Protocol 4*)
- Inverted microscope
- Complete medium (including serum and any other additives required)
- Actively growing cells

Method

1. Resuspend adherent cells by standard subculture method; use suspension cells directly from culture.
2. Perform viable cell count.
3. For an expected CFE of less than 5%, prepare 10 ml of each of three cell suspensions with concentrations of 5000, 1000, and 500 cells/ml. For a CFE of 5–10%, prepare suspensions with concentrations of 500, 100, and 50 cells/ml. For a CFE of over 10%, prepare suspensions with concentrations of 50, 10, and 5 cells/ml.

Protocol 2 continued

4. Add 100 μl of one cell suspension to each well of a 96-well tray. Repeat for each dilution, using separate trays. If feeder layers are required, these should be prepared in the trays in advance.
5. Add 100 μl of complete medium to each well. Alternatively, if considered necessary, add 100 μl of conditioned medium, or 100 μl of 1:1 conditioned medium/fresh medium, to each well.
6. Incubate, at the appropriate temperature for the cells, in a humidified atmosphere containing the correct CO_2 concentration for the medium in use.
7. Using the inverted microscope, inspect the wells after four days and then at two day intervals.
8. Mark the wells in which a single colony appears. Ensure that there is only one centre of growth as the colony develops.
9. As the colony grows, it may be advisable to feed the wells at intervals with medium. Typically, if growth is vigorous, 50% of the medium may be changed every four to seven days. In media containing phenol red, the medium should be changed sufficiently frequently that it never turns bright yellow, as excess acidity will kill the cells. If growth is not vigorous, medium changes should be infrequent, as these may retard the growth of fastidious cells by reducing the accumulation of conditioning factors in the medium. It may be necessary to add conditioned medium at 20% or 50% of the total medium volume. The precise regime will depend on the observed behaviour of the colonies, the known requirements of the cells, and the characteristics of the medium. Feeding may disrupt cell colonies which are not strongly adherent, giving rise to new foci of proliferation. With such cells, feeding should be delayed, if possible, until the operator is confident that only one colony is growing in the well.
10. Select those plates in which only a limited number of wells show cell growth. When colonies of a suitable size have formed, resuspend the cells by trypsinization if adherent, or careful pipetting if non-adherent.
11. Subculture each colony into an individual well in a 24-well tray, which may contain feeder cells and/or conditioned medium if considered necessary. The total volume of medium per well should not exceed 1 ml.
12. Incubate and feed the cells as before.
13. When sufficient cells are present (i.e. when more than 50% of the area of the well is covered), subculture into 25 cm^2 flasks, and subsequently into 75 cm^2 flasks.
14. Stocks of cloned cells should be frozen in liquid nitrogen as soon as a sufficient number of cells is available (see Chapter 5, Section 9).
15. In many cases, it will be necessary to repeat the foregoing procedure once or twice, for the reasons discussed previously.

3.2.2 'Spotting' technique

This technique is simple and convenient, lending itself well to aseptic technique and giving low risk of adventitious contamination. An enhanced antibiotic regime is not usually required.

Protocol 3

Cloning of cells by 'spotting' in 96-well plates

Equipment and reagents

- Sterile Pasteur pipettes
- Inverted microscope
- 96- and 24-well tissue culture trays
- 25 cm^2 and 75 cm^2 tissue culture flasks
- Complete medium (including conditioned medium if required)

Method

1. Resuspend adherent cells by standard subculture method; use suspension cells directly from culture.
2. Perform viable cell counts. Adjust the cell concentration to between 500 and 1000 cells/ml using complete medium.
3. Insert the tip of a Pasteur pipette into the cell suspension. Allow a small volume to rise into the tip by capillary action.
4. Tap the tip of the loaded pipette in the centre of the base of each well of a 96-well tissue culture tray. The action should deposit a droplet of about 1 μl of suspension. The droplet should not touch the sides of the well.[a]
5. Examine the wells microscopically for the presence of a single cell.[b] Use ×50 magnification initially, then ×100 to confirm. Mark the wells that contain a single cell, using a waterproof marker. Occasionally, surface characteristics of the plastic bases of individual wells prevent droplets from spreading fully, causing dark areas to appear around the perimeter. This will also occur where a droplet touches the side. Cells within such areas may sometimes be seen with a higher magnification and/or by adjusting the illumination. If the whole droplet still cannot be rigorously inspected, that well must not be used.
6. When each tray has been inspected, add 200 μl of medium to each marked well. Alternatively, if deemed necessary, add 100 μl of complete medium plus 100 μl of conditioned medium, or 200 μl conditioned medium containing resuspended feeder cells.
7. Follow *Protocol 2*, steps 6–14.

[a] Due to the small size of the droplets, drying can occur during the microscopic examination of the wells in step 5 if not performed rapidly enough. Thus when the technique is first used it is recommended that droplets are only placed in half of the wells prior to microscopic examination. Once medium has been added to the appropriate wells (step 6) return to step 4 and place droplets in the remaining unused wells.

[b] It is beneficial to have a second observer verify the presence of only a single cell at step 5. A video monitor connected to the microscope may facilitate this.

3.2.3 Cloning by micro-manipulation

This technique permits the observation of the seeding of a single cell, and therefore gives a reliable method for obtaining colonies which are true clones. It also enables the selection of cells with particular morphologies or other observable characteristics. Some dexterity and practice will be required to achieve consistent results. A sufficient uninterrupted period of time should be set aside to carry out the procedure.

Protocol 4

Cloning by micro-manipulation

Equipment and reagents

- Inverted microscope with CCD standing in a Class II MSC
- Micro-manipulator with micro-syringe and accessories (Research Instruments Ltd.)
- Preformed micropipettes with 30° tip angle (Research Instruments Ltd.)
- Sterile, bacteriology-grade Petri dishes (60 mm and 35 mm diameter)
- Video monitor connected to the CCD
- 96-well cell culture cluster – half-area plates (Corning Costar, code 3696)
- 96- and 24-well tissue culture trays
- 25 cm^2 and 75 cm^2 tissue culture flasks
- Medium (including serum and/or conditioned medium as required)

Method

A micro-manipulator can be fitted to the side of the stage of an inverted microscope. A preformed micropipette, angled at the tip, is attached to the micro-manipulator which moves it horizontally and vertically in order to place it where desired. To permit the precise control of liquid aspirated into and expelled from the micropipette, an oil-filled micro-syringe and flexible tubing is attached to it.

1. Set up the microscope with its monitor, micro-manipulator, and micro-syringe etc. according to the manufacturer's instructions.
2. Add 100 μl of medium to six wells well separated from one another in a sterile 96-well cell culture cluster – half-area plate. Keep this plate in a CO_2 incubator until required. (The wells should be well separated from one another in order to avoid confusion and inadvertent cross-contamination.)
3. Resuspend adherent cells by standard subculture method: use suspension cells directly from culture.
4. Prepare a cell suspension containing a low enough concentration of cells such that only one cell is normally visible in the microscope's field of view.
5. Add 5 ml of this suspension to a 60 mm Petri dish. Add medium alone to two further 35 mm Petri dishes.
6. Aseptically attach a new sterile micropipette to the micro-manipulator and micro-syringe.

Protocol 4 continued

7 Place the first 35 mm Petri dish (containing only medium) on the microscope stage.

8 Lower the micropipette into the medium and, using the micro-syringe, draw a small volume into it.

9 Raise the micropipette and replace the 35 mm dish with the 60 mm dish containing the cell suspension.

10 Using the microscope, examine the cell suspension and locate a suitable cell which is well separated from others.[a]

11 Lower the micropipette into the cell suspension and manoeuvre the tip towards the chosen cell using the controls of the micro-manipulator. Using the micro-syringe, gently draw the cell into the tip of the micropipette.

12 Raise the micropipette out of the 60 mm Petri dish, and replace the dish with the second 35 mm dish containing only medium.

13 Lower the micropipette into the medium, and gently expel the cell. Examine the contents of the dish in the vicinity of the micropipette to confirm that only a single cell was aspirated, then gently draw it back into the micropipette.[a]

14 Raise the micropipette and replace the Petri dish with the half-area plate.

15 Lower the micropipette into one of the wells, and expel the single cell into it.

16 Raise the micropipette and return the half-area plate to the incubator.

17 Repeat steps 6–16 for the other five wells which contain medium in the half-area plate, using a new micropipette and 35 mm Petri dishes each time.

18 Leave the half-area plate for approximately half an hour (to allow the cells to settle to the bottom of the wells) before carefully removing 50 μl of medium from the top of each well.

19 Using an inverted microscope, confirm the presence of a single cell in each well.[a]

20 Repeat steps 2–19 for any further plates required.

21 Continue by following *Protocol 2*, steps 9–14.

[a] It is beneficial to have a second person verify the presence of only a single cell at steps 10, 13, and 19. The video monitor assists in this process.

3.2.4 Fluorescence-activated cell sorting

This technique can be used to separate cells on the basis of their light scattering properties and the particular surface molecules which they express (9). In conjunction with other techniques it can also separate cells according to their rate of secretion of molecules of interest (18, 19). In each case, the molecules are detected by the use of specific ligands (e.g. antibodies) labelled with a fluorochrome. A stream of microdroplets containing the cells is passed through a laser beam. Light scattering at low angle and at 90° is detected, along with the fluorescence of the fluorochrome(s) excited by the laser. Cells with light scattering and fluorescence parameters falling within the predetermined limits are electrostatically deflected for collection.

The technique can be adapted to deflect single cells into the wells of multi-well plates. However, this method of cloning requires extremely expensive and sophisticated equipment and a highly skilled operator, is prone to adventitious contamination, and, like all the other (very much cheaper) cloning methods, still carries a finite chance that a colony thus isolated will not in fact be derived from a single cell, due to coincidence (the presence of more than one cell in the droplet(s) deflected).

3.3 Methods for attached cells

3.3.1 Cloning rings

In this technique, cells are grown at low cell density in conventional plastic or glass tissue culture vessels, but once discrete colonies have formed cloning rings are used to isolate individual colonies and permit their trypsinization and removal for subculture. Cloning rings are small, hollow cylinders, generally made from glass, stainless steel, or PTFE, and can be of any size which is convenient. They may be cut from suitable tubing and the ends smoothed with a file or stone. Alternatively, they may be purchased from Bellco (cloning cylinders, code 2090) who supply them in a variety of sizes in either stainless steel or borosilicate glass.

Stainless steel rings of 8 mm inside diameter, 2 mm wall thickness, and *c.* 12 mm height have been found to be particularly satisfactory as their weight seats them firmly in the silicone grease, permitting a good seal to be maintained during the cell detachment procedure (see below). However, the height of the cloning rings to be used must be chosen with reference to the dimensions of the vessel within which they will be used, as they must not be so tall that they prevent closure of the vessel (see *Protocol* 5, step 10).

Protocol 5

Cloning of attached cells using cloning rings

Equipment and reagents

- Sterile forceps
- 24-well tissue culture trays
- 4- or 6-well tissue culture trays (alternatively, tissue culture grade Petri dishes or 'peel apart' flasks, e.g. 'Ezin' flasks from Life Technologies, may be used)
- Inverted microscope
- Cloning rings (Bellco)
- Silicone grease
- Complete medium including serum (and conditioned medium if required)

Method

1 Sterilize the cloning rings and silicone grease by dry heat in separate glass Petri dishes or foil-covered glass beakers.

Protocol 5 continued

2. For established cell lines or primary cultures: seed cells into 4- or 6-well trays (or selected alternative container) at 10 000, 1000, 100, and 10 cells/well. These quantities may be varied according to the expected CFE of the cells. They should be increased in proportion to the size of the container if Petri dishes, peel-apart flasks, or standard tissue culture flasks are used. The intention is to obtain discrete, well separated colonies.
3. For hybrid or recombinant cells: culture under appropriate selective conditions (see Chapter 6). This will generally be done at the end of a hybridization or transfection procedure. The number of cells seeded per dish will be governed by the proportion of hybrids or recombinants expected, and their CFE in the selective medium.
4. For all cells: incubate, at the appropriate temperature for the cells, in a humidified atmosphere containing the correct CO_2 concentration for the medium in use.
5. Inspect after four days and then at two day intervals. Mark the position of colonies that appear to have arisen from one growth centre and are well separated from other cells.
6. Remove the medium from the cultures.
7. Using sterile forceps, pick up a cloning ring and dip its base in silicone grease. Ensure that the grease is evenly distributed.
8. Place the ring around a marked colony and press down, moving slightly to obtain a good seal.
9. Repeat steps 7 and 8 for the other selected colonies.
10. Fill the rings with trypsin solution. Leave for 20 sec, then remove most of the solution, leaving just a thin film.[a] Use a new pipette for each ring. Close the culture vessel and incubate at 37 °C, inspecting periodically to monitor the detachment process.
11. When the cells have detached, fill each ring with complete medium. Carefully pipette the medium up and down to suspend the cells, and transfer the suspension to a well of a 24-well tray. These trays may contain feeder cells or conditioned medium if necessary. The total volume of medium per well should not exceed 1 ml. Repeat the process with the contents of each ring, using a new pipette each time. Place the tray in the incubator and incubate as in step 4.
12. When sufficient cells are present (more than 50% of the area of the well is covered), subculture into 25 cm^2 flasks, and subsequently into 75 cm^2 flasks.
13. Stocks of cloned cells should be frozen in liquid nitrogen as soon as a sufficient number of cells are available (see Chapter 5, Section 9).

[a] Cells that are difficult to detach from the substrate may require more trypsin to be left in the cloning ring, up to a maximum of *c.* 100 μl (in an 8 mm i.d. ring).

3.3.2 Petriperm dishes

Another cloning technique applicable only to attached cells involves growing cells at low density on Petriperm dishes (Heraeus). The bottom of these dishes is made of a thin FEP (fluorinated ethylene propylene) foil, which can be produced with either a hydrophobic or hydrophilic surface. The latter ('Hydrophil') would normally be used for cell cloning. If the cells are plated at a sufficiently low density such that the resulting colonies are well separated, individual colonies can be removed whilst still attached to the substrate simply by cutting the FEP foil with a scalpel.

Protocol 6

Cloning on Petriperm dishes

Equipment and reagents

- Petriperm 'Hydrophil' dishes (Heraeus)
- Two pairs of sterile forceps
- Sterile scalpel (with thin, pointed blade)
- Sterile Petri dishes (90 mm or 100 mm diameter)
- Sterile 50 ml (centrifuge) tubes
- Sterile 5 ml or 10 ml pipettes
- Basal medium
- FCS
- Conditioned medium (if required)
- PBS
- Trypsin (Sigma)
- Microtitre plates

A. Plating the cells

1. Using sterile forceps, aseptically remove the Petriperm dishes from their package and place each one in a 90 mm or 100 mm sterile Petri dish. This is necessary to ensure that the bottom of the Petriperm dish remains sterile during incubation.
2. Trypsinize the inoculum as appropriate for the cells in use.
3. Perform a viable count on the cells.
4. Using basal medium, dilute the cells to 1000 cell/ml.[a]
5. To a 50 ml centrifuge tube, add:
   (a) 10 ml of FCS.
   (b) 10 ml of conditioned medium (replace with basal medium if not required).
   (c) 29 ml of basal medium.
   (d) 1 ml of diluted cell suspension.
6. Close the tube and mix gently by inversion.
7. Pipette 5 ml of this cell suspension into each of the Petriperm dishes. This must be done carefully, avoiding, for example, whirlpools that would tend to concentrate cells in the centre of the dish.
8. Incubate the dishes and examine regularly for the growth of colonies.

Protocol 6 continued

B. Picking the colonies

1. Add an appropriate amount of medium to the receiving vessels, and gas if necessary.[b]
2. Place a pair of sterile forceps, and a sterile scalpel bearing a thin, pointed blade, under the lid of a large Petri dish. These items should be returned to the dish when not in use in order to preserve their sterility.
3. Remove the medium from the Petriperm dish. Rinse once with PBS, discard the rinse, and replace the lid.
4. Hold the plate vertically, with the bottom facing you, so that the colonies are visible.
5. Using the sterile scalpel, cut around each colony leaving a small intact bridge of FEP foil.
6. Using the sterile forceps, tear away the pieces of foil bearing the colonies and place each of them in the well of a microtitre plate. Work rapidly and keep the microtitre plate covered to prevent drying of the colonies.
7. Trypsinize the cells in the microtitre wells, monitoring the trypsinization under the microscope.
8. After trypsinization, inactivate the enzyme by the addition of a small volume of serum, serum-containing medium, or other inhibitor.
9. Transfer the cells from each well into a separate receiving vessel or culture well (prepared in step 1), and incubate as appropriate.

[a] This cell concentration is sufficient for cells with a high CFE (50–100%), yielding 50–100 colonies per plate. Proportionately higher concentrations should be used with cells displaying a lower CFE.

[b] The size of tissue culture vessel used to receive each colony after trypsinization will depend on the degree to which the cells can be safely diluted. Some colonies can be transferred straight into 25 cm^2 flasks, but this is unusual and it would be more common to use 35 mm Petri dishes or wells of a multiwell plate. If in doubt, it is safer to err on the side of smaller volume, and examine the cells more frequently.

3.4 Methods for suspension cells

The following methods work on the principle of confining cell progeny to a region close to where they originated (thus forming a colony) by growing them in a medium of high viscosity. Both methods are frequently used for the cloning of hybridomas as well as other cell lines and both, with suitable modifications, can also be applied to normal cells which will grow in suspension (i.e. haemopoietic progenitor cells).

3.4.1 Cloning in soft agar

The method described here uses a base layer of 0.5% agar, overlaid with cells suspended in 0.28% agar. If desired, these concentrations can be changed somewhat without affecting the performance of the system, and the base layer can even be omitted for some cells.

Protocol 7

Cloning in soft agar

Equipment and reagents

- 90 mm Petri dishes
- 24-well tissue culture trays
- Sterile pipettes
- Inverted microscope
- Sterile Pasteur pipettes
- Agar (tested for non-toxicity on the cells in use)
- Medium containing 15% FCS

Method

1. Prepare a 0.5% (w/v) stock agar solution by adding 2 g of agar powder to 30–40 ml of tissue culture grade water, autoclaving, then cooling to 55 °C. Add medium (containing 15% FCS) at 55 °C, to make a total volume of 400 ml. This should be sufficient to clone ten cell lines.
2. To each of four 90 mm Petri dishes add 12–14 ml of molten 0.5% agar. Allow agar to gel and dry for 20 min at room temperature.
3. Add 0.8 ml of medium to each of four wells in a 24-well culture tray.
4. To the first of these wells, add 2–4 $\times 10^3$ cells in 0.2 ml of medium.
5. Pipette this suspension up and down a few times to mix the cells, then transfer 0.2 ml to the next well.
6. Repeat step 5 to make two more serial 1:5 dilutions. Discard 0.2 ml from the final well.
7. Allow stock 0.5% agar solution to cool to 47–45 °C, then add 1 ml to each well, mix, and transfer the contents of each well to separate agar-containing Petri dishes. Incubate as appropriate for the cell line and medium being used.
8. Using an inverted microscope, check daily for cell growth, and discard plates containing too many colonies, or no cells. When colonies contain 20–100 cells, pick them individually from the plate using a Pasteur pipette. Transfer them to 1 ml of medium (containing 15% FCS) in separate wells of a 24-well tray for further growth.[a]

[a] The best size of colony to transfer, and the volume into which it should be transferred, will vary from cell line to cell line. The figures given here generally work well with mouse hybridomas.

3.4.2 Cloning in methylcellulose

In principle this is similar to cloning in agar, but has the advantage that neither the cells nor the medium need be subjected to the elevated temperatures required to keep agar liquid. Although not specifically described here, methylcellulose can also be used as an overlayer on an agar base.

Protocol 8

Cloning in methylcellulose

Equipment and reagents

- Wide-mouthed conical glass flask
- 100 ml Duran bottles (Schott)
- 35 mm bacteriological-grade Petri dishes
- 100 mm Petri dishes
- Sterile Pasteur or capillary pipettes
- Powdered basal medium
- Methylcellulose powder (e.g. Methocel MC, 3000–5500 mPa.s, Fluka)
- Conditioned medium

A. Preparation of 2% (w/v) methylcellulose stock solution

1. Prepare a 2× solution of the basal medium to be used (i.e. prepare from powder as for single strength medium, but in half the volume).
2. Sterilize (either by hot air or autoclaving) a 500 ml wide-mouthed conical flask capped with aluminium foil and containing a large magnetic stirrer bar. If autoclaved, the flask should only contain a minimal amount of water after sterilization.
3. Weigh the flask, complete with stirrer-bar and cap. Note this weight, then add to the flask 100 ml of tissue culture grade water.
4. Boil gently over a Bunsen burner for 5 min. Remove the foil cap and sprinkle 4 g of methylcellulose powder onto the surface of the water. It is important that none of it touches the walls of the flask. Recap the flask.
5. Heat the contents until they just start to boil. Remove the Bunsen burner and swirl the flask gently to help mixing until the suspension (which is white and opaque at this stage) has ceased to boil.
6. Repeat step 5 until 5 min have elapsed since the suspension first started to boil.
7. Remove the flask from the heat and plunge it into an ice/water slurry. Swirl continuously until all the contents of the flask have become viscous. This should be accompanied by a partial clearing of the flask's contents, which should now be translucent.
8. Add 100 ml of the 2× basal medium prepared earlier. Swirl the flask until the entire contents are mobile then place on a magnetic stirrer and stir at 4 °C for 1 h.
9. Weigh the flask and add sterile water until the total weight of the flask, stirrer-bar, cap, and contents is equal to 204.5 g plus the starting weight noted in step 3.
10. Stir overnight at 4 °C, then dispense (by careful pouring) into two 100 ml Duran bottles. Store at −20 °C.

B. Plating the cells

1. Thaw the 2% stock methylcellulose solution and allow to equilibrate in a 37 °C water-bath.
2. Using low speed centrifugation, pellet the cells to be cloned. Resuspend in a mixture

Protocol 8 continued

containing (by volume) 53.3% FCS, 26.6% conditioned medium, and 20% basal medium, to a cell concentration of about 110 cells/ml.[a]

3. Place 7.5 ml of this suspension in a 50 ml centrifuge tube and, using a sterile syringe without a needle, add 12.5 ml of the warmed methylcellulose solution. Mix initially by inversion, then by gentle vortexing.
4. Using a syringe fitted with a 19 gauge needle, place 1 ml aliquots of this cell suspension in 35 mm bacteriological-grade Petri dishes. Tip the dishes to distribute the suspension over the whole surface. The volumes given in step 3 should be sufficient for 18 or 19 dishes.
5. Place two of these dishes, along with a third, open 35 mm Petri dish containing distilled water, in a 100 mm Petri dish and incubate under appropriate conditions in a humidified incubator.

C. Picking the colonies

1. Examine the plates at regular intervals (e.g. every two to three days) for the presence of colonies visible to the naked eye.
2. Examine colonies under the microscope. Check the colony is of a suitable size for picking (usually 0.5–1.0 mm diameter) and is comprised of healthy cells with a good morphology. Check that there are no overlapping colonies, or other colonies within about 1.5 mm.
3. Mark the position of colonies suitable for picking using a felt-tipped pen on the underside of the Petri dish.
4. Remove individual colonies from the plate by aspirating into a Pasteur pipette or capillary pipette. (It may be helpful to view the process with the aid of a free-standing magnifying glass or dissecting microscope.)
5. Place the cells in a suitable tissue culture vessel (35 mm dish, or well of a 24- or 48-well plate) with at most 1 ml of tissue culture medium. It may be helpful to continue to use conditioned medium and high FCS concentrations at this stage.
6. Expand the cells as required. Check for the presence of the required properties and freeze a number of ampoules of any suitable cells as soon as possible.

[a] This cell concentration should be sufficient for a cell line with a high CFE (50–100%), yielding 20–40 colonies per plate. Proportionately higher concentrations should be used with cells displaying a lower CFE.

This whole procedure (as applied to the cloning of hybridomas) has been described in greater detail elsewhere (20).

References

1. Sanford, K. K., Earle, W. R., and Likely, G. D. (1948). *J. Natl. Cancer Inst.*, **9**, 229.
2. Wildy, P. and Stoker, M. (1958). *Nature*, **181**, 1407.
3. Puck, T. T. and Marcus, P. I. (1955). *Proc. Natl. Acad. Sci. USA*, **41**, 432.

4. Cooper, J. E. K. (1973). In *Tissue culture – methods and applications* (ed. P. F. Kruse and M. K. Patterson), p. 266. Academic Press, London.
5. MacPherson, I. A. (1973). In *Tissue culture – methods and applications* (ed. P. F. Kruse and M. K. Patterson), p. 241. Academic Press, London.
6. Clarke, J. B. and Spier, R. E. (1980). *Arch. Virol.*, **63**, 1.
7. MacPherson, I. and Montagnier, L. (1964). *Virology*, **23**, 291.
8. Paul, J. (1975). *Cell and tissue culture* (5th edn), p. 255. Churchill-Livingstone, Edinburgh.
9. Shapiro, H. M. (1988). *Practical flow cytometry* (2nd edn), p. 110. Alan R. Liss, New York.
10. Rittenberg, M. B., Buenafe, A., and Brown M. (1986). In *Methods in enzymology* (ed. J. J. Langone and H. Van Vunakis), Vol. 121, p. 327. Academic Press, New York.
11. Coller, H. A. and Coller, B. S. (1986). In *Methods in enzymology* (ed. J. J. Langone and H. Van Vunakis), Vol. 121, p. 412. Academic Press, London.
12. Coller, H. A. and Coller, B. S. (1983). *Hybridoma*, **2**, 91.
13. Underwood, P. A. and Bean, P. A. (1988). *J. Immunol. Methods*, **107**, 119.
14. McCullough, K. C., Butcher, R. N., and Parkinson, D. (1983). *J. Biol. Stand.*, **11**, 183.
15. Staszewski, R. (1984). *Yale J. Biol. Med.*, **57**, 865.
16. Ham, R. G. and McKeehan, W. L. (1979). In *Methods in enzymology* (ed. W. B. Jakoby and I. H. Pastan), Vol. 58, p. 44. Academic Press, London.
17. Lefkovits, I. and Waldmann, H. (1999). *Limiting dilution analysis of cells of the immune system* (2nd edn). Oxford University Press, Oxford.
18. Weaver, J. C., McGrath, P., and Adams, S. (1997). *Nature Med.*, **3**, 583.
19. Holmes, P. and Al-Rubeai, M. (1999). *J. Immunol. Methods*, **230**, 141.
20. Davis, J. M. (1986). In *Methods in enzymology* (ed. J. J. Langone and H. Van Vunakis), Vol. 121, p. 307. Academic Press, London.

Chapter 9
The quality control of cell lines and the prevention, detection, and cure of contamination

B. J. Bolton and P. Packer
European Collection of Cell Cultures, Centre for Applied Microbiology and Research, Porton Down, Salisbury, Wiltshire SP4 0JG, UK.

A. Doyle
The Wellcome Trust, 210 Euston Road, London NW1 2BE, UK.

1 Introduction

The use of animal cell cultures is widespread and encompasses a large range of scientific disciplines. Central to this application of tissue culture technique is the routine maintenance of cells in a way in which the researcher can be assured that the material is contaminant-free. Quality control procedures cannot be viewed as an 'optional' extra and are not solely for use when the material is first received or derived in the laboratory. Guidelines must be set for continuous monitoring with the emphasis placed on day-to-day vigilance; in addition, full awareness by technical staff of the potential for problems to occur is a necessity. The overall state of health of a culture is one of the most significant of the criteria applied as part of routine observation, as not all contaminants are overt. It should always be borne in mind that an infection can become widespread amongst cultures before gross indications are seen and subsequent remedial steps taken.

Once a culture becomes contaminated, what then? There was at one time a widely held belief that the only way to remove mycoplasma contamination, for example, was by autoclaving the cultures. This drastic measure is only acceptable if replacements are available and a correct stock management system of master and working cell banks (1) means that fresh stocks can be put into use once laboratories have been decontaminated. However, if the problem occurs with unique material, then attempts can be made to eradicate the contamination.

In the case of virus contamination, where the only available technique would be an unlikely combination of cell cloning with the possibility of parallel treatment with specific antisera, there is little prospect of deriving a contaminant-free subclone. With bacteria, fungi, and mycoplasma there is a much better chance of success and information is provided concerning suitable eradication methods.

2 Obtaining the basic material

2.1 Importance of cell culture collections

Public cell culture collections came about in response to a widespread need for well characterized, microbe-free seed stocks derived from cultures supplied by cell line originators. The explosion of research on viruses in the mid to late 1950s, enabled by the extensive use of cell culture, led to exploitation of the technique, often without awareness of the critical need for cell line quality control. By 1960 the problems of cellular and microbial contamination of cell lines had become so acute that scientists in the USA banded together to establish a bank of tested cells (2). They realized the fundamental requirement for the application of sensitive and extensive quality control testing procedures. These efforts coupled with improving preservation technology then extended internationally.

Over the past 30 years considerable sums of money have been invested in cell banking programmes, over and above the research costs of the developed cell lines. These funds have been willingly provided by granting agencies in the expectation of insuring their investment in research that uses cells as model systems.

While the rationale for development and use of well organized collections is understood by many laboratory scientists, poorly characterized cell stocks for use in research studies are still exchanged all too frequently. Thus it is important to restate the potential pitfalls associated with the use of cell stocks obtained and processed casually, in order to increase and reinforce awareness of the problem within the scientific community.

Numerous instances of the exchange of cell lines contaminated with cells of other species have been documented and published by others (see Chapter 5, Section 3.4). Similarly, the problem of intra-species cross-contamination among cultured human cell lines has been recognized for a considerable time and emphasized more recently (3, 4). The effect of invalidation of results and consequent loss of scarce research funds as a result of these problems is incalculable.

Although bacterial and fungal contamination represent an added concern, in most instances they are overt and easily detected and are therefore of less serious consequence than the more insidious contamination by mycoplasma. That the presence of these micro-organisms in cultured cell lines often completely negates research findings has been emphasized over the years (5). Still, the difficulties of detection and prevalence of contaminated cultures in the research community suggest that repeated restatements are warranted.

Other sections within this chapter will go into more detail about the various techniques available for the characterization of cell lines and the detection of contamination in cell cultures.

2.2 Resource centres

With the increased biotechnological use of animal cells, there has been a parallel increase in the number of culture collections and other resource centres. A comprehensive list of the resource centres worldwide is given in ref. 5. However, the larger culture collections who offer cell line supply and related services are listed in Appendix 2.

2.3 Cell culture databases

All of the resource centres listed in Appendix 2 and some of the more specialized collections in ref. 5 produce hard copy catalogues of material held. However, no matter how often new editions are produced they are very soon out of date. Access to on-line data systems of the above culture collections provides the potential customer with the most up-to-date listing of material available. The American Type Culture Collection (ATCC) database is available on-line at www.atcc.org as is The European Collection of Cell Cultures (ECACC) catalogue at www.ecacc.org.uk (see also Appendix 3).

Both ECACC and ATCC have recently produced their main catalogues on CD-ROM which provide for easier distribution and enable updates to be easily provided.

2.3.1 CABRI (Common Access Biotechnological Resources and Information)

The EU funded CABRI has brought together information from the major microbial and animal cell resource centres in Europe to provide the customer with a single Internet gateway to a large range of biological material. In addition to the usual catalogue information, there is also a wealth of detail related to quality control procedures and technical information.

In identifying cell lines through CABRI only a single search is necessary without going through all the individual databases of the collections concerned. Currently ECACC, DSMZ, and IST are the cell line catalogues incorporated. The database is available on www.cabri.org.

3 Quarantine and initial handling of cell lines

The most common source of microbial contamination in the cell culture laboratory is from cell lines received from other laboratories. Commercially prepared media and media supplements now undergo stringent quality control procedures by their producers which minimize (but do not completely eliminate) the chance of them contributing any significant microbial hazard. However, care should be taken to set up quality control procedures when preparing media in-house from powders or liquid concentrates.

With the above in mind it is important to have in place the correct facilities and procedures for handling incoming cell lines, as outlined below.

3.1 Accessioning scheme

It is important that every laboratory has defined procedures for the handling of new or incoming cell lines and that all staff are familiar with them. A flow chart of the scheme which operates at ECACC is shown in *Figure 1* with the essential points being:

(a) Cultures should be handled in Class II microbiological safety cabinets offering operator protection as the source of the cell lines may not be known (6, 7).

(b) Cultures should be handled in a quarantine laboratory separate from the main tissue culture area (see below).

(c) A token freeze should be made as soon as possible.

(d) Initial characterization (e.g. species verification) and microbial quality control should be performed.

(e) After satisfying the above conditions the cultures may be transferred to the main tissue culture area for production of master and working banks.

It is recommended that these procedures are followed irrespective of the source of the cell line, i.e. whether it be a recognized culture collection or research laboratory. In the long-term it can save a lot of time and money.

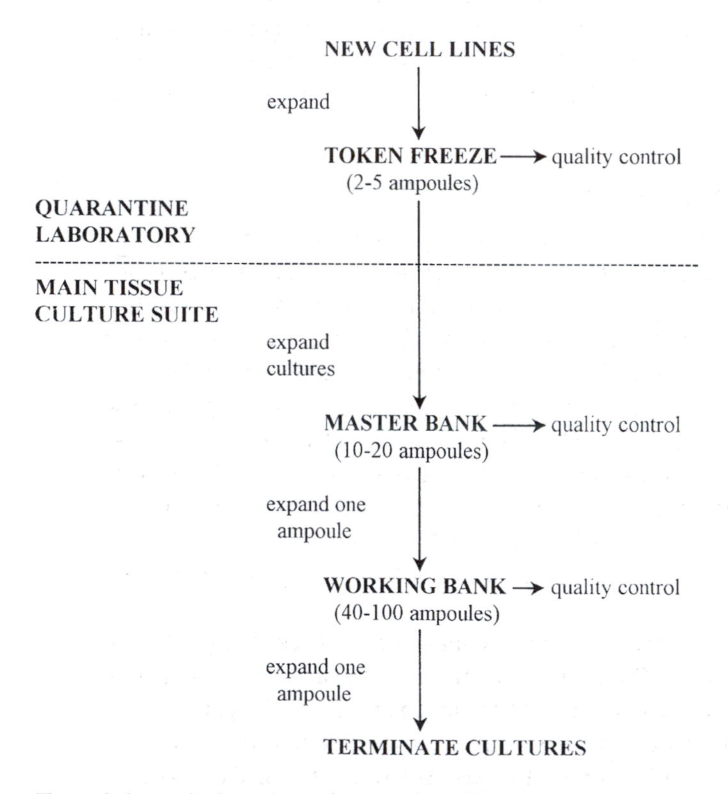

Figure 1 Accessioning scheme for incoming cell lines

3.2 Laboratory design

Design criteria for cell culture laboratories are given in Chapter 1. However, it is worth stressing that when setting up a cell culture laboratory, either within existing facilities or from new, an area (preferably a whole laboratory), should be set aside for cell reception purposes. This should include the following features:

(a) The laboratory should be as far away from clean areas as possible.

(b) If possible the whole laboratory should operate under negative pressure with respect to the rest of the tissue culture suite.

(c) It should as far as possible be self-contained in that it has its own incubator, water-bath, microscope, etc.

(d) Staff should have separate lab coats specifically for working in the quarantine laboratory – ideally a different colour from those used elsewhere.

3.3 Handling of cell lines

Many laboratories frequently use antibiotics routinely in their cell cultures. This is very bad practice for a number of reasons, but mainly because it can suppress bacterial contamination to a level unable to be detected by eye and furthermore can encourage the spread of antibiotic-resistant strains. Also, while not actually killing mycoplasma, antibiotics, particularly the aminoglycosides, can reduce the numbers below the level of detection. Therefore it is essential that all quality control tests are carried out on cell cultures subcultured for a minimum of two passages in antibiotic-free media. Full details of microbial tests are given in Section 4.

3.4 Production of cell banks

When a culture is received in the laboratory it is important that a strict accessioning scheme is followed, similar to that described in Section 3.1. Starter cultures or ampoules received should be propagated according to the conditions provided by the originator and a token freeze produced (usually up to five ampoules). It is cultures derived from this token freeze which should be subjected to detailed microbial quality control and characterization. If these tests provide satisfactory results then a master cell bank (or seed stock) can be made. All steps prior to this last one should be carried out in the quarantine laboratory. It is on the master cell bank that the major authentication efforts, such as isoenzyme analysis, karyology, and DNA fingerprinting (see Section 5), should be applied, as it is ampoules from this bank which are used to provide the working or distribution stock.

As shown in *Figure 1*, one ampoule from the master bank is used to produce a working or distribution bank. The number of ampoules in each bank is dependent on how often it will be used. As a lot of time and expense will be put into the production and characterization of these banks, particularly master banks, the numbers required should not be underestimated. For some industrial processes it is not uncommon to have master banks containing as many as 200 ampoules and working banks of up to 1000 ampoules. When the working bank is depleted a fresh one is made from an ampoule from the master bank and similarly a new master bank from the token freeze stock. If no more token freeze stock remains then one of the current master bank ampoules should be used.

The level of characterization and quality control which should be carried out at each stage will be dependent on what the cells are to be used for and what the necessary regulatory bodies require (see Section 6). However, it is recommended that for any purpose each stage should be tested for microbial contamination (bacteria, yeasts and other fungi, and mycoplasma) and the cell line's species verified as an absolute minimum.

4 Microbial quality control

Microbial quality control is an essential part of all routine cell culture and should not be neglected. It is concerned with the testing of culture medium and cell lines for a variety of micro-organisms, including bacteria, yeasts and other fungi, mycoplasma, and viruses.

4.1 Sources of microbial contamination

These can be divided into four broad categories:

- poor laboratory conditions
- inadequately trained personnel
- non-quality controlled cell lines
- contaminated media components

The first point above is covered quite extensively in Section 3.2 and elsewhere in this book.

Concerning training, it is important that all staff are aware of the various possible routes of contamination and are familiar with the techniques of good cell culture practice and aseptic technique. The most common and often underestimated source of contamination is from cell lines received from external sources. It cannot be overemphasized that when new cell lines are acquired they should be handled in quarantine conditions (see Section 3.1) until they can be shown to be free from microbial contamination. As already mentioned above, many laboratories routinely use antibiotics in cell culture medium which can allow low level contaminants to go undetected and thus to spread to other cultures. Therefore for these reasons it is important that cell lines are obtained from recognized culture collections which guarantee contaminant-free cultures.

If a microbial contamination problem arises in the laboratory then it is important to discover the source to prevent it continuing to cause problems. The most common sources of the different groups of micro-organisms are given in *Table 1*.

Table 1 Common sources of microbial contamination

Organism	Source
Bacteria	• clothing, skin, hair, aerosols (e.g. due to sneezing or pipetting), insecure caps on media and culture flasks • air currents • humidified incubators • purified water • insects • plants • contaminated cell lines
Fungi (excluding yeasts)	• fruit • damp wood or other cellulose products, e.g. cardboard • humidified incubators • plants
Yeasts	• bread • humidified incubators • operators
Mycoplasma	• contaminated cell lines • serum • medium • operators

4.2 Testing for bacteria, yeasts, and other fungi

If cell lines are cultured in antibiotic-free media, contamination by bacteria, yeasts, or other fungi can usually (although not invariably) be detected by an increase in turbidity of the medium or by a change in pH. Reagents added during the preparation of cell culture medium, e.g. fetal bovine serum, contribute to the risk of contamination, so it is good practice to set up quality control checks on culture medium prior to its use. The two methods generally used for the detection of bacteria and fungi involve microbiological culture (*Protocol 1*) or direct observation using Gram's stain (8).

4.2.1 Detection by microbiological culture

Two types of microbiological culture medium should be used. Those suggested by the US Code of Federal Regulations (9) and the European Pharmacopoeia (10) are:

(a) Fluid thioglycollate medium for the detection of aerobic and anaerobic forms.

(b) Soya bean-casein digest (tryptose soya broth) for the detection of aerobes, facultative anaerobes, and fungi.

Protocol 1

Detection of bacteria and fungi by cultivation

Reagents

- Tryptose soya broth (Oxoid)
- Fluid thioglycollate medium (Oxoid)

Method

1. Inoculate 1 ml of test material into each of four universals (15 ml), two containing tryptose soya broth and the other two fluid thioglycollate.
2. Incubate one pair (i.e. one tryptose and one thioglycollate) at 35 °C and the other at 25 °C.
3. If bacteria are present the broths generally become turbid within one to two days, although incubation may require as long as two weeks.
4. Positive controls may be used, e.g. *Bacillus subtilis* (aerobe) and *Clostridium sporogenes* (anaerobe), although this does imply the use of specialist facilities some distance from routine tissue culture laboratories. Type strains are available from the UK National Collection of Type Cultures (NCTC) and the ATCC.

4.3 Testing for mycoplasma

Mycoplasma infection of cell cultures was first observed by Robinson *et al.* in 1956 and the incidence since then of mycoplasma-infected cell cultures has been found to vary from laboratory to laboratory. It is very important to use mycoplasma-free cell lines as mycoplasma can:

- affect the rate of cell growth
- induce morphological changes
- cause chromosome aberrations
- influence amino acid and nucleic acid metabolism
- induce cell transformation
- affect transfer across the cell membrane

Regulatory bodies insist that cell cultures used for the production of pharmaceutical products, reagents for diagnostic kits, or therapeutic agents are free from mycoplasma infection (9, 10).

4.3.1 Hoechst 33258 DNA staining method

The fluorochrome dye Hoechst 33258 binds specifically to DNA, causing fluorescence when viewed under ultraviolet (UV) light. A fluorescence microscope equipped for epifluorescence with a 340–380 nm excitation filter and 430 nm suppression filter is required. Due to the fundamental requirement for mycoplasma-free cell lines in tissue culture a detailed outline of this technique is provided in *Protocol 2*.

Protocol 2

Hoechst 33258 DNA staining method using Vero cells as an indicator cell line

Equipment and reagents

- Fluorescence microscope (set up as above)
- Hoechst (bis-benzimide) 33258 (Sigma)
- Vero cells (ECACC, code 84113001, or ATCC, code CCL 81)

Method

All cell cultures should be passaged at least twice in antibiotic-free medium prior to testing. Failure to do this may lead to false negative results.

1. Prepare two 22 mm sterile coverslips in 35 mm diameter tissue culture dishes for each sample to be tested and an additional two dishes for each test run as a negative control. To each dish add sufficient Vero cells to give a semi-confluent culture within 24 h. It is important to use a Vero stock which is guaranteed negative for mycoplasma.
2. Harvest adherent cells by the usual subculture method and resuspend in the original culture medium at approximately 5×10^6 cells/ml. Suspension cells are tested directly at a similar concentration.
3. Add 2–3 ml of test cells to each of two 22 mm sterile coverslips in 35 mm diameter tissue culture dishes which were seeded with Vero cells in step 1. Incubate one dish for 24 h and the other for 72 h at 37 °C in a humidified atmosphere of 5% CO_2 and 95% air.[a] Also incubate a control dish for 24 h and 72 h.
4. Before fixing, examine the cells on an inverted microscope (×40 magnification) for evidence of microbial contamination.
5. Add to the medium 2 ml of freshly prepared Carnoy's fixative (methanol/glacial acetic acid, 3:1, v/v) dropwise from the edge of the culture dish taking care not to sweep unattached cells to one side of the dish, and leave for 3 min at room temperature.
6. Pour the fixative into a waste bottle and immediately add another 2 ml of fixative; leave for a further 3 min at room temperature.
7. Dry the coverslip in air on the inverted tissue culture dish lid for 30 min.
8. Make a stock solution of Hoechst 33258 stain (10 mg/100 ml) and store it protected from light. Dilute this to the working concentration just before use. In a fume cupboard and wearing gloves, add 2 ml freshly prepared Hoechst 33258 stain (100 μg/litre in distilled water) and leave for 5 min, shielding the coverslip from the light.
9. Still in the fume cupboard, decant the stain, which is toxic and, therefore, must be disposed of according to local safety procedures.
10. Add one drop of mountant to a glass slide and mount the coverslip, cell side down.

Protocol 2 continued

11 Examine using UV fluorescence under oil immersion at ×1000 magnification. Uncontaminated cells show only brightly fluorescing cell nuclei, whereas mycoplasma-infected cultures contain small cocci or filaments which may or may not be adsorbed on to the cells, see *Figure 2*.

[a] Or other CO_2 concentration appropriate for the medium in use.

(a)

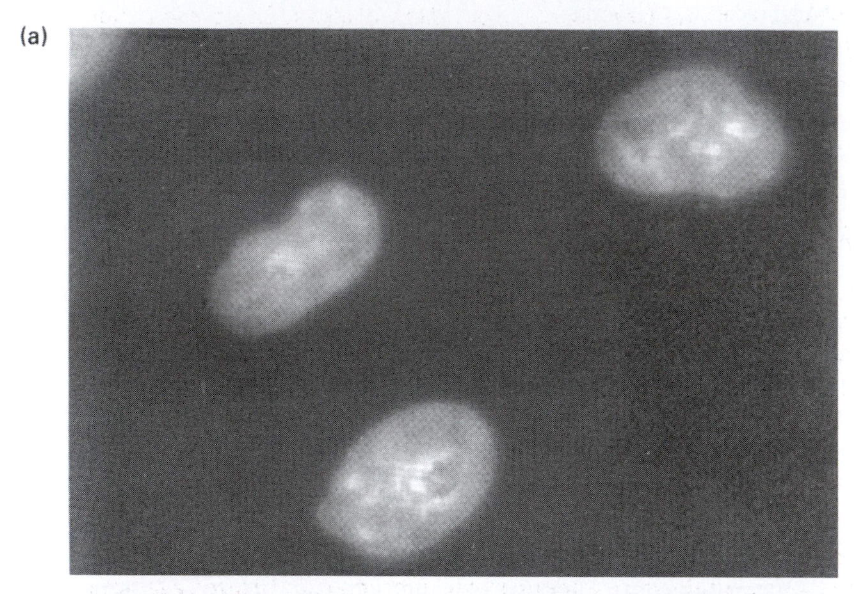

(b)

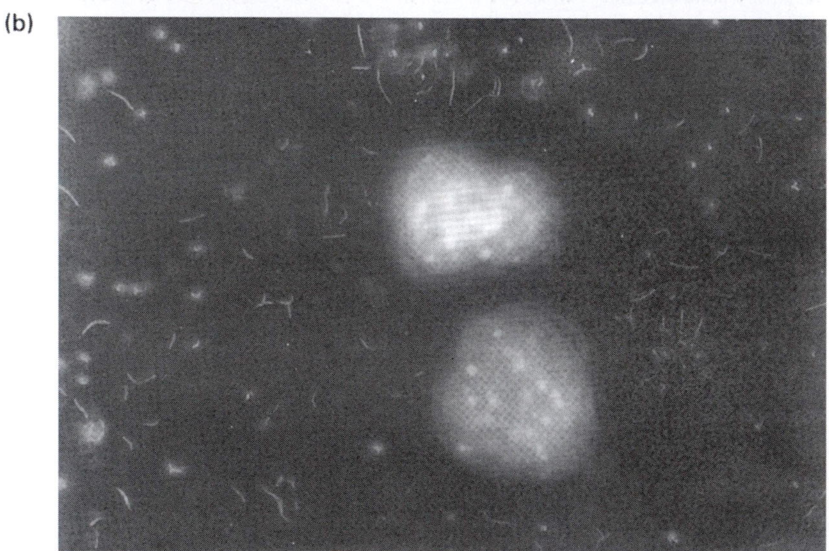

Figure 2 Cell preparations of the Vero cell line stained with Hoechst 33258 DNA stain and viewed at ×1000 magnification using a fluorescence microscope. (a) Mycoplasma negative. (b) Mycoplasma positive.

For reasons of standardization, the assay is performed using a monolayer culture as an indicator onto which the test sample is inoculated. Vero cells are used as an indicator because they have a low cytoplasm/nucleus ratio which permits better differentiation between contaminated and uncontaminated cultures. This method also has the advantage of being able to screen serum, cell culture supernatants, and other reagents which do not contain cells.

Some preparations may show extracellular fluorescence caused by disintegrating nuclei. Fluorescent debris is usually not of a uniform size and is too large to be mycoplasma. Contaminating bacteria or fungi will also stain if present, but will appear much larger than mycoplasma. Compare with positive and negative slides which can be prepared from positive control strains [e.g. *Mycoplasma hyorhinis* and *M. orale*, available from NCTC (Nos. NC10130 and NC10112 respectively)]. They should be inoculated at *c.* 100 c.f.u./dish.

Positive controls have to be handled in specialist facilities quite separate from the tissue culture laboratory. However for a small tissue culture unit which cannot for whatever reasons set aside specialist facilities, positive slides can be supplied commercially, e.g. by ECACC.

The main advantages of this method are the speed (less than one day) at which results are obtained, and that the non-cultivable *M. hyorhinis* strains can be detected. However, this method is not as sensitive as the culture method (see Section 4.3.2). It is generally considered that approximately 10^4 mycoplasma per ml are required to produce a clear positive result by the Hoechst stain technique.

4.3.2 Microbiological culture

Most mycoplasma cell culture contaminants will grow on standardized agar and broth media, with the exception of certain strains of *M. hyorhinis*. The methods are discussed in detail in ref. 11. As is the case with microbial positive controls (see *Protocol 1*), the culture methods described below should be performed in a laboratory quite separate from the main tissue culture area.

Both agar and broth medium should be checked prior to use for their ability to grow species of mycoplasma known to contaminate cell lines, e.g. *Acholeplasma laidlawii*, *M. arginini*, *M. fermentans*, *M. hominis*, *M. hyorhinis*, and *M. orale*. Type strains are available from the National Collection of Type Cultures or the ATCC. For further discussion on interpretation of results see ref. 11. The main advantage of the method is that, theoretically, one viable organism per inoculum can be detected, compared with 10^4 per ml for the Hoechst staining method.

4.3.3 Other methods of mycoplasma testing

Regulatory requirements often limit the methods of detection to the Hoechst 33258 DNA staining and microbiological culture methods. However, for those laboratories which have no need to follow such requirements, the methods available in kit form may prove useful. Those most widely used are outlined below.

i. Mycoplasma TC kit

This is produced by Gen-Probe, California (available from Lab Impex in the UK), requires no culture procedure, and gives same-day results. The protocol requires incubation of a ^{3}H-labelled, single-stranded DNA probe with either cell culture supernatant or cell extracts. Hydroxyapatite is utilized to separate hybridized from unhybridized probe prior to scintillation counting. The DNA probe used is homologous to mycoplasma ribosomal RNA (rRNA), which hybridizes with different species of mycoplasma, but not with mammalian cellular or mitochondrial rRNAs. The disadvantage with this method is that it is expensive, relative to the Hoechst DNA staining method. However, it does provide quick and reliable results.

ii. Myco-Tect kit

This is produced by Gibco BRL (Life Technologies) and requires co-cultivation of the cells to be tested with 6-methylpurine deoxyriboside. If mycoplasma are present, 6-methylpurine deoxyriboside is broken down into toxic components which cause the cells to die. This method is comparable in length but is more complex than the Hoechst DNA staining method.

iii. Mycoplasma detection kit

This is produced by Roche and involves an enzyme immunoassay. The test is relatively quick, giving results overnight. It has one major disadvantage in that it only detects four species, *M. arginini*, *M. hyorhinis*, *A. laidlawii*, and *M. orale*. However, it does have the advantage of identifying the species of the contaminant.

iv. Polymerase chain reaction (PCR)

Several different PCR primers are commercially available for the detection of mycoplasma. Two of these kits, manufactured by Stratagene (Mycoplasma Plus PCR Primer Set, Cat. No. 302008) and Takara (PCR Mycoplasma Detection Set, Cat. No. 6601) are currently being evaluated by ECACC. These PCR technologies depend upon amplification of a specific, yet highly conserved, region of the mycoplasma genome such that all species of mycoplasma should be detected without cross-reaction with other prokaryotes, particularly those bacterial groups (such as Gram positive eubacteria) most closely related to mycoplasma.

ECACC has an on-going programme of running PCR in parallel with its routine methods of Hoechst DNA stain and microbiological culture. The specificity of the PCR method using both primer sets appears to be excellent and all positive results have been confirmed by the traditional methods. The theoretical analytical sensitivity of PCR should be high.

Stratagene confirms reactivity of its primer sets with all six of the common contaminating species mentioned in Section 4.3.2 (manufacturer's data). Takara confirms reactivity with four of these species (manufacturer's data) but not *A. laidlawii*. ECACC's provisional data confirms this result.

The PCR-based assay system is very convenient and the whole test can be completed in less than one day. An added advantage is that the sampling method is non-destructive, as supernatants from growing cultures can be tested directly as can frozen ampoules after thawing without the need for further subculture. A further advantage of PCR is that it enables speciation of the mycoplasma contaminant which helps to identify the original source.

4.4 Virus testing

There are three major considerations in discussing the topic of virus contamination of cell lines. The first of these is the safety of the laboratory staff handling cell lines and the elimination of risk in experimental and other procedures. The second is the contamination status of any products derived from cells and applied therapeutically to patients. This problem also impinges on regulatory matters which are discussed in more detail in Section 6. Finally, there is the validity of any experimental data produced using contaminated cell lines.

The source of viral contamination can be from the original tissue used to prepare the cell line or, once established, from growth medium, from other infected cultures, or as a result of *in vivo* passage; there is also a slight risk of contamination arising from laboratory personnel. For a full review of operator-induced contamination in cell culture systems see ref. 12. Each aspect will be discussed and an indication given of the likelihood of risk with particular consideration to human pathogens, e.g. hepatitis B virus (HBV), human immunodeficiency virus (HIV), and human T cell lymphotropic virus (HTLV). The range of testing procedures available will also be discussed.

4.4.1 Tissue-derived viral contamination

The factors influencing the choice of starting material for cell line derivation should take into account the possibility of existing virus infection. This is dependent upon the species of origin, the tissue taken, and the clinical history of the animal/patient. This must include an evaluation of viruses endemic to a population. A list of the most commonly occurring viruses in humans is provided in *Table 2*.

A risk assessment must be made in the light of the incidence of viral infection in the population from which clinical specimens are derived, and in laboratory staff. For example, 80–90% of the population have experienced Epstein–Barr virus (EBV) infection, and may carry the virus in cells derived from peripheral blood, lymph node, or spleen. Therefore, there is a high risk of EBV contamination associated with such material. This is counterbalanced by the fact that most laboratory staff will also be EBV seropositive. However, adequate precautions must be taken to ensure that seronegative staff are not exposed to risk of infection.

Laboratory staff involved in receiving samples of human material are at a high risk of exposure to HBV contaminated material. However, staff may be protected by vaccination and correct handling of clinical materials. Similarly the incidence of HIV/HTLV infection in non-high risk groups (in the UK, this includes

Table 2 Some viruses which can occur in humans

Virus	Tissue involved	Persistent *in vivo?*
Herpes simplex virus-1	general	+
Herpes simplex virus-2	general	+
Human cytomegalovirus	general	+
Epstein–Barr virus	general	+
Hepatitis B	general	+
Hepatitis C	general	+
Human herpes virus-6	general	+
Human immunodeficiency virus-1	general	+
Human immunodeficiency virus-2	general	+
Human T-cell lymphotropic virus-I	general	+
Human T-cell lymphotropic virus-II	general	+
Adenovirus	general	±
Reovirus	general	–
Rubella	general	–
Measles	general	+
Mumps	general	–
Human parvovirus	general	+
Varicella–Zoster	general	+
Respiratory syncytial	respiratory	–
Influenza A	respiratory	–
Influenza B	respiratory	–
Parainfluenza	respiratory	–
Rhinovirus	respiratory	–
Coronavirus	respiratory	–
Poliovirus	enteric	–
Coxsackie A	enteric	–
Coxsackie B	enteric	–
Echovirus	enteric	–
Rotavirus	enteric	–
Norwalk virus	enteric	–
Calicivirus	enteric	–
Astrovirus	enteric	–
Papillomavirus	Skin/epithelium	+
Poxvirus	Skin/epithelium	–

those other than homosexuals, bisexuals, and intravenous drug users) remains low, and therefore the risk of receiving virally contaminated clinical material is low. Therefore, the geographical location of sources of human clinical material and virus infections endemic to that area must be considered.

The viruses listed in *Table 3* have been shown to be present in rodent populations and of greatest concern are the first five, namely hantavirus, lymphocytic

Table 3 Potentially pathogenic viruses considered to be possible contaminants of rodent cell lines

Virus	**Species affected**	**Detected by MAP/RAP test?**[a]
Hantavirus	mouse, rat	√
Lymphocytic choriomeningitis virus	mouse	√
Rat rotavirus	rat	×
Reovirus type 3	mouse, rat	√
Sendai virus	mouse, rat	√
Ectromelia virus	mouse	√
K virus	mouse	√
Kilham rat virus	rat	√
Lactate dehydrogenase virus	mouse	√
Minute virus of mice	mouse, rat	√
Mouse adenovirus	mouse	√
Mouse cytomegalovirus	mouse	√
Mouse hepatitis virus	mouse	√
Mouse polio virus	mouse	√
Mouse rotavirus e.g. epizootic diarrhoea of infant mice	mouse	√
Pneumonia virus of mice	mouse, rat	√
Polyoma virus	mouse	√
Rat coronavirus	rat	×
Retroviruses	mouse, rat	×
Sialodacryoadenitis virus	rat	√
Thymic virus	mouse	√
Toolan virus	rat	√

[a] MAP, mouse antibody production; RAP, rat antibody production.

choriomeningitis virus (LCM), rat rotaviruses, reovirus type 3, and Sendai virus, all of which can infect humans. In fact, laboratory-derived infection of humans has been reported in some cases (for hantaviruses and LCM) (13).

4.4.2 Serum-derived viral contamination

Bovine viral diarrhoea virus (BVDV), infectious bovine rhinotracheitis virus, and parainfluenza are the major concerns in fetal bovine serum. It is known that 50–90% of cattle in the USA are infected with BVDV. The current methodology available to detect contamination is based on fluorescence/antibody techniques which lack sensitivity. Now, PCR technology is being validated and better standards of screening are becoming available (see below).

4.4.3 Methods of detection

A range of methods exists for detection of viruses, from the classical haemadsorption methods to more modern PCR technology. A brief resumé of techniques is given below. No single method will detect all viruses; for example, not all viruses cause haemadsorption, and not all viruses produce cytopathic effects. Therefore, exhaustive screening for viral contaminants is usually uneconomic.

i. Co-cultivation

In this method an extract of a test cell line is incubated with semi-confluent monolayers of a range of cell lines susceptible to a wide variety of viruses. The co-cultivations are maintained by passaging for two to four weeks and the host cell lines are regularly checked for the presence of cytopathic effects and haemadsorbing agents. The following host cell lines are susceptible to a wide range of viruses and are often used depending on the species of the test cell line:

- BHK21 (hamster)
- WI 38 (human)
- HeLa (human)
- Vero (monkey)
- MDCK (canine)
- JM: sensitive to HIV1
- H9: sensitive to HIV1
- fresh T cells: sensitive to HTLV-I

ii. Electron microscopy

Transmission electron microscopy has been used not only to detect viral contaminants but also to identify (to family level) and give an indication of the quantity of the contaminants. This method allows a wide range of viruses and virus-like particles to be identified by their characteristic morphology. This method has been particularly applied to mouse–mouse hybridomas and recombinant CHO cell lines, due to their use in the production of monoclonal antibodies and recombinant biological therapeutic proteins. For a detailed discussion and protocols, see ref. 14.

iii. In vivo *methods*

In vivo methods for the detection of viral contaminants may involve inoculation of materials by a variety of routes into a range of laboratory animals of different ages. The animals are subsequently examined for evidence of adverse effects. For more details see ref. 14.

The murine, rat, and hamster antibody production tests are designed to detect the range of rodent viruses shown in *Table 3*. Briefly, animals are inoculated with test material and subsequently examined for evidence of production of antibodies to the above list of viruses, or in the case of lactate dehydrogenase virus, for raised levels of lactate dehydrogenase activity.

iv. Cell culture assays for murine retroviruses

Murine leukaemia viruses (MuLV) may be ecotropic (infect only rodent cells), xenotropic (infect only cells other than rodent), or amphotropic (infect both cell types) depending on viral surface molecules. Xenotropic MuLV may be detected by the formation of foci in S^+L^- mink cells (15) and ecotropic MuLV may be detected by the formation of syncytia and vacuoles in XC cells (16).

v. Reverse transcriptase assay for retrovirus detection

The method involves precipitation of potential virus particles and protein from cell-free samples with polyethylene glycol. The extract is assayed for reverse transcriptase (RT) activity by incorporation of [^{3}H]TTP on to a poly(rA).p(dT) primer/template. Presence of host DNA polymerase, which may give a high background incorporation of the RNA template, is detected by incorporation of [^{3}H]TTP on to a poly(dA).p(dT) primer template. The assays are usually carried out in duplicate, using both RNA and DNA templates and both Mn^{2+}- and Mg^{2+}-containing buffers (*Protocol 3*).

Protocol 3

Detection of retroviruses in cell supernatants by reverse transcriptase assay

Reagents

- [^{3}H]TTP (1.1 TBq/mmol, 30 Ci/mmol) (Amersham)
- Positive control reverse transcriptase preparations (Roche) (see step 7)
- Virus solubilizing buffer: 0.8 M NaCl, 0.5% (v/v) Triton X-100, 0.3 mg/ml phenylmethylsulfonyl fluoride, 20% (v/v) glycerol, 50 mM tris(hydroxymethyl)-aminomethane (Tris)–HCl pH 8.0, 4 mM dithiothreitol (DTT)
- Solution A (DNA template): 60 mM Tris–HCl pH 7.5, 1.3 mM DTT, 0.1 mM ATP, 0.6 A_{260} U/ml poly(dA).p(dT), 12 mM $MgCl_2$ (store at −20 °C)
- Solution B (RNA template): as solution A except that poly(dA).p(dT) template primer is replaced by poly(rA).p(dT) at the same concentration (store at −20 °C)
- Solution C (DNA template): 60 mM Tris–HCl pH 8.5, 1.3 mM DTT, 0.1 mM ATP, 0.6 A_{260} U/ml poly(dA).p(dT), 0.3 mM $MnCl_2$ (store at −20 °C)
- Solution D (RNA template): as solution C except that poly(dA).p(dT) is replaced by poly(rA).p(dT) (store at −20 °C)
- Virus precipitation solution: 30% (w/v) PEG (8000) in phosphate-buffered saline (PBS) pH 7

Method

1. Prepare cell-free supernatant (4 ml) by centrifugation at 400 *g* at 4 °C for 10 min. Take the supernatant from cells in log phase of growth as dying cells release host DNA polymerase which gives high backgrounds.
2. Precipitate virus and protein from supernatant by addition of 2.5 ml virus precipitation solution followed by incubation at 4 °C for 18 h.
3. Collect precipitate by centrifugation at 800 g at 4 °C for 30 min, discard supernatant, and resuspend pellet in 0.3 ml of virus solubilizing buffer.
4. Store samples in liquid nitrogen until required.

Protocol 3 continued

5. Pipette 20 μl aliquots of [^{3}H]TTP into 1.5 ml microcentrifuge tubes (four for each test sample and two for each positive control). Place cotton wool plugs in the neck of the tubes and dry in a 37 °C incubator.
6. Add 150 μl of solution A to the first tube, 150 μl of solution B to the second, 150 μl of solution C to the third, 150 μl of solution D to the fourth.
7. For the positive controls, e.g. avian myeloblastosis virus (AMV) RT, add 150 μl of solution A to one tube and 150 μl of solution B to the second; for Moloney murine leukaemia virus (MoMLV) RT add 150 μl of solution C to one tube and 150 μl of solution D to the second.
8. Seal all tubes and place on a vortex mixer to mix gently. Allow any aerosol to settle for 5 min before opening the tubes. Keep on ice.
9. Divide the contents of each tube into two aliquots (75 μl each).
10. Add 20 μl of test sample to appropriate tubes.
11. For positive controls add 30 U of AMV RT or 30 U of MoMLV RT in 20 μl virus solubilizing buffer.
12. Seal tubes and vortex. Incubate for 90 min in a water-bath at 37 °C.
13. Terminate the reaction by addition of 0.5 ml cold 10% trichloroacetic acid (TCA) to each tube to precipitate any incorporated radioactivity.
14. Set up a Millipore filter box (or equivalent) with 2.4 cm GF/C filters (Whatman) and pre-wash filters with 20 ml of cold 20% TCA.
15. Pour contents of each tube onto a separate filter and rinse out each tube twice with 1 ml of cold 20% TCA, and pour rinse onto relevant filter.
16. Wash each filter four times with 20 ml cold 5% TCA.
17. Wash each filter once with 20 ml of cold 70% ethanol.
18. Remove filters and dry at room temperature.
19. Add each filter to a scintillation vial containing 5 ml of scintillation fluid.
20. Count in a beta-counter for 1 min.
21. Compare the activity bound to control (host polymerase) with assay filters (RNA templates). If their values are similar or sample activity is below control activity then a negative result is given.
22. Positive controls should give an activity greater than 100 000 c.p.m. and an RNA–DNA template incorporation ratio of at least 5.

vi. PCR method for detection of BVDV and other viruses

BVDV crosses the placenta in pregnant cows, and therefore is a common contaminant of FCS. More recently, PCR-based techniques have been developed which provide more rapid results than the previously used co-cultivation and immunoassay methods. PCR following an initial RT step can be carried out in a single day and gives clear positive/negative results. This method is currently being used at ECACC to determine the BVDV status of its cell lines. For a detailed description and protocols, see ref. 17.

PCR can be used to detect any virus whose nucleic acid sequence is known. However, uncharacterized virus strains with divergent target sequences may be missed by PCR due to inadequate hybridization of primers. Also, the PCR technique does not give information on the potential infectivity of the viral contaminant detected.

4.5 Bovine spongiform encephalopathy

Bovine spongiform encephalopathy (BSE) is a new concern for the tissue culture scientist. It first became apparent in 1986 and is believed to have been caused by the use of contaminated feedstuffs for cattle. Consideration should be given to the risk of handling FCS or putatively infected cell cultures. Prior to starting any work a full risk assessment should be carried out in consultation with the recent ACDP publication on transmissible spongiform encephalopathy agents (18). The risk factors include level of infectivity, route of exposure, and the species barrier. It is also important to consider the inactivation of such agents - see Chapter 2, Section 9.2.

Serum from mature sheep with clinical scrapie contains no detectable infectious agent and hence by analogy it is considered highly unlikely that the BSE agent would occur in FCS from infected cattle. This, combined with the fact that the agent is only present in high concentrations in neural tissue (although it can be found in other tissues such as lymphocytes), plus the problem of the species barrier, means that BSE is not considered to present a high risk in FCS (19). However, most companies supplying FCS can supply from source countries without reported cases of BSE, or at least a reduced incidence. However there is insufficient supply from such countries (e.g. New Zealand, Australia, USA) to satisfy the world demand, with consequences already discussed in Chapter 3, Section 3.4.2.

4.6 Elimination of contamination

If a cell culture is found to be contaminated with bacteria, fungi, mycoplasma, or viruses then the best method of elimination is to discard the culture and obtain fresh cultures from your own stocks or from a reputable source. It is extremely important to find out the source of the contamination, e.g. whether it is from medium, culture reagents, a faulty safety cabinet, poor aseptic technique, etc., to prevent it happening again (see Section 4.1). If the source is found to be within the laboratory then the best way of approaching this would be to fumigate the whole laboratory. If such facilities are not available then it is possible to purchase individual fumigators for Class II MSCs (Hi-Tech). The other option is to swab all surfaces and equipment with a suitable disinfectant.

In the case of irreplaceable cell stocks it will be necessary to attempt to eliminate the contamination using antibiotics (*Protocol* 4). This approach is possible for bacteria, mycoplasma, yeast, and other fungi but there are no reliable methods for the eradication of viruses from infected cell cultures. With a cell culture-derived product, once a contaminant is identified it is possible to demonstrate that the method of downstream processing removes the contaminant virus from the final product. This may require detailed validation by 'spiking' experiments to ensure that the methodology used is capable of eliminating the contaminant. In any such case, it will be important to discuss the problem with the relevant regulatory authority at the earliest opportunity, as even the use of the best available validated technology of virus removal/inactivation may not convince them to license your product; much will depend on the virus involved and the proposed use of the product.

Mycoplasma easily develop resistance to antibiotics particularly Ciprofloxacin and Mycoplasma Removal Agent (MRA), the only effective antibiotics currently available. Very recently a new product has been introduced to the market by Minerva Biolabs called Mynox (Cat. No. Myn-200/500/1000). The manufacturer claims that this works by a biophysical mechanism different to antibiotics, therefore preventing development of resistant strains.

Protocol 4

Elimination of contamination by antibiotic treatment

1. Culture cells in the presence of the chosen antibiotic (see *Table 4*) for 10–14 days or at least three passages. If cells are heavily contaminated with bacteria or fungi, give the cells a complete medium change prior to antibiotic treatment. Perform each passage at the lowest cell density at which growth occurs.
2. If the contaminant is still detectable then repeat step 1 using a different antibiotic.
3. If the contamination appears to be eradicated, then retest the cells after culture in antibiotic-free medium, after 5–7 days for bacteria and fungi, or 25–30 days for mycoplasma. Only if they test negative after this period should the contaminant be considered eliminated.
4. If contamination is detected again, repeat the above steps with another antibiotic.

5 Authentication

Authentication of cell lines to confirm their identity and species of origin is an essential requirement in the management of cell stocks. The occurrence of cross-contamination between cell cultures is an often neglected subject and is exemplified by the presence of HeLa marker chromosomes in a number of commonly used cell lines (3). This problem is most likely to occur in cultures taken to high passage numbers or where slowly growing cultures are maintained for long periods of time, and in laboratories using many cell lines. Cross-contamination may well go unnoticed in laboratories using many cultures of identical morphology, e.g. fibroblast cultures. In order to avoid the problem at the outset, cell lines must be obtained from well documented and quality controlled sources,

Table 4 Antibiotics commonly used in elimination of microbial contamination

Antibiotic	Working concentration	Active against
Amphotericin B	2.5 mg/litre	yeasts, and other fungi
Ampicillin	2.5 mg/litre	bacteria, (Gp, Gn)
Cephalothin	100 mg/litre	bacteria, (Gp, Gn)
Ciprofloxacin	10–40 mg/litre	mycoplasma
Gentamicin	50 mg/litre	bacteria (Gp, Gn), mycoplasma
Kanamycin	100 mg/litre	bacteria (Gp, Gn), yeasts
Mycoplasma removal agent[b]	0.5 mg/litre	mycoplasma
Neomycin	50 mg/litre	bacteria, (Gp, Gn)
Nystatin	50 mg/litre	yeasts, and other fungi
Penicillin-G[c]	100 000 U/litre	bacteria, (Gp)
Polymyxin B	50 mg/litre	bacteria, (Gn)
Streptomycin sulphate[c]	100 mg/litre	bacteria, (Gn, Gp)
Tetracyclin	10 mg/litre	bacteria, (Gn, Gp)

Gn, Gram-negative; Gp, Gram-positive.

[a] Although it is stated in the literature that gentamicin is active against mycoplasma, the authors' experience is that most mycoplasma species are resistant.

[b] Supplied exclusively by ICN Biomedicals.

[c] Usually supplied as a mixture of penicillin-G and streptomycin sulphate.

such as the original depositor or an established culture collection, rather than passed from one laboratory to another (see Section 2).

It is important to emphasize that cell authentication is an essential part of quality assurance for both research and commercial use of cell cultures, and should be a primary concern for everyone in this field. Three methods are used by ECACC for cell line authentication: isoenzyme analysis, cytogenetic analysis, and DNA fingerprinting. Outlines of each are given in the following sections.

5.1 Isoenzyme analysis

Isoenzymes are polymorphic enzyme variants which catalyse the same reaction but have different electrophoretic mobilities. A standardized method, the Authentikit system (Innovative Chemistry), is available in kit form. In its basic form this provides for detection of seven different enzyme reactions. However, in many cases a good indication of the species of origin can be obtained with results for just four enzymes, glucose-6-phosphate dehydrogenase, nucleoside phosphorylase, malate dehydrogenase, and lactate dehydrogenase. An example of the system is shown in *Figure 3*.

As the number of enzymes studied increases, a composite picture is built up and confidence in the species of origin identified is increased. However, it is difficult to achieve unique identification of a cell line using this method. Mixtures of cells from two species can also be detected provided the level of the contaminant cell line is at least 20%. Thus isoenzyme analysis enables rapid speciation of cell cultures and is of great practical use in routine testing.

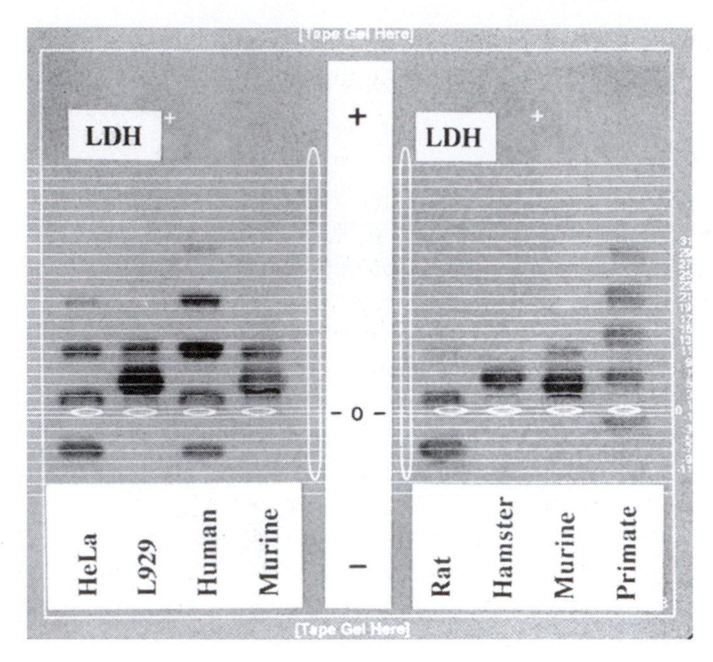

Figure 3 Isoenzyme analysis of six cell lines compared with two standards, HeLa (lane 1) and L929 (lane 2). The test enzyme is lactate dehydrogenase and the system used is Authentikit (Innovative Chemistry).

5.2 Cytogenetic analysis

Cytogenetic analysis is used to establish the common chromosome complement or karyotype of a species or cell line (20, 21). Using this technique it is possible to detect changes in cell cultures and the occurrence of cross-contamination between cell lines. This methodology has therefore been important in the past for the quality control of cell lines used to produce biologicals, although it is being demanded less now, particularly for continuous cell lines. However, it is a complicated and laborious technique and requires the operator to have a high degree of training and experience. There are many different methods described for staining and banding chromosomes; these include Giemsa (G) banding, G-11 banding, and quinacrine (Q) banding. Of these G banding is the most commonly used due to its wide ranging applications. In this method, treatment of chromosomes with trypsin and Giemsa stain reveals banding patterns (G-bands) which are characteristic for each chromosome pair. In G-11 banding the Giemsa stain is used at pH 11 instead of pH 6.8, which causes differential staining of chromosomes of different species. Thus G-11 banding may be useful when examining cell hybrids such as human–rodent.

The Q banding method uses quinacrine to produce fluorescent bands within chromosomes which mostly correspond to G-bands. This method is particularly useful for analysis of the Y chromosome in man which is more distinctive using this method. Thus cytogenetic analysis provides a number of methods with useful applications in cell culture. It is also a useful technique for confirmation of species, but lacks the speed and simplicity of isoenzyme analysis.

5.3 DNA fingerprinting

The discovery of hypervariable regions of repetitive DNA within the genomes of many organisms led to the development by Jeffreys *et al.* in 1985 of DNA fingerprinting, which can specifically identify human individuals. A range of probes and methods have been developed to exploit the repetitive DNA sequences found throughout the animal kingdom (22). Multilocus probes identify many loci in a given genome. Using the multilocus probes developed by Jeffreys, 33.15 or 33.6, the chance of finding two unrelated human individuals with identical fingerprints is reported to be less than 1 in 5×10^{19} which indicates the resolving power of this technique. Using this method it is possible to detect cross-contamination from a wide range of species. This is because, under the correct hybridization conditions, the multilocus probes are able to cross-hybridize with a wide range of repetitive DNA families which occur in many species (see *Figure* 4). Thus, the multilocus 'fingerprint' simultaneously identifies the culture and screens for intra-species as well as inter-species cross-contamination. The multilocus probes 33.15 and 33.6 used under licence by the ECACC are available from Cellmark Diagnostics.

Briefly, the Southern blot procedure used in this method involves complete digestion of extracted genomic DNA with the restriction enzyme *Hin*fI followed by separation of the fragments by agarose gel electrophoresis. After depurination, the DNA is blotted onto a nylon membrane and hybridized to an alkaline phosphate-labelled DNA probe (33.6 or 33.15) which binds to specific DNA sequences on the membrane. The labelled probe converts a luminescent substrate and gives out light, enabling the fragments to be visualized using X-ray film producing a characteristic fingerprint as shown in *Figure* 4. The use of standard cell lines on each gel enables adjustments to be made for migration differences between sets, and data from many gels can be compared simultaneously. However, more work needs to be carried out before these systems are able reliably and specifically to identify unknowns by screening such a database. These systems are also adaptable to storing isoenzyme patterns. For cytogenetic data, specialist equipment is available which can digitize karyotypes and store them. However, such equipment is very expensive and exceeds the means of most cell culture laboratories.

Very recently the method of multiplex PCR has been introduced by culture collections to replace the traditional profile method above for the authentication of human cell lines.

5.3.1 Multiplex PCR

Simple tandem repeats (STRs), which were originally developed for forensic applications, are now routinely used as a fast, effective, and cheap method to identify DNA from human material. A number of polymorphic STR loci are

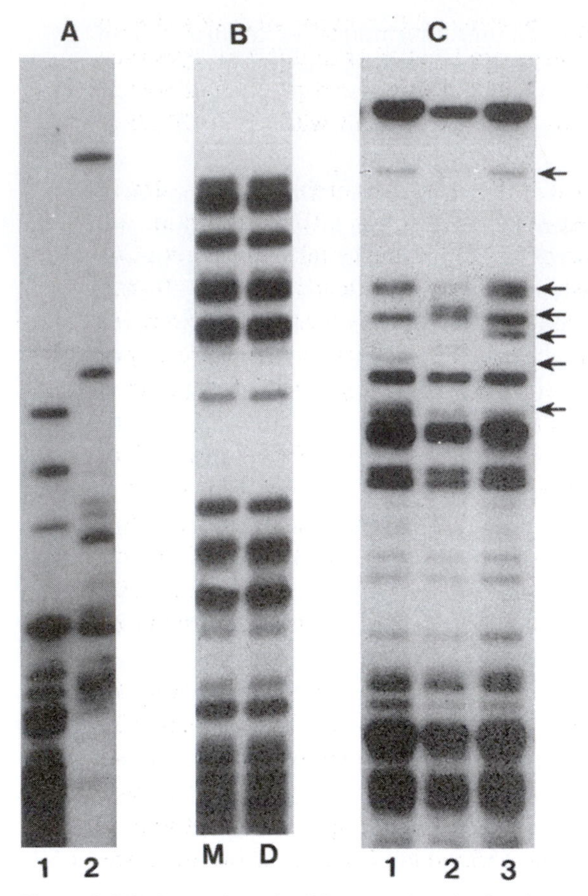

Figure 4 DNA fingerprints of cell lines produced using the enzyme *Hin*fI and probes 33.6 (A) and 33.15 (B and C). (A) Iguana (IgH2, lane 1) and bat (Tb1Lu, lane 2) cell lines. (B) Samples of a murine cell culture from master (M) and distribution (D) cell banks. (C) Three different murine hybridomas all produced from the NS1 myeloma and Balb/c spleen donors (1, DA6 231; 2, AFRC IAH CC14; 3, AFRC IAH CC13): differences in patterns are marked by arrows.

amplified using commercially available sets of primers. The PCR products are analysed simultaneously with size standards using automated fluorescent detection techniques. The result gives a simple numerical code corresponding to the lengths of the PCR products amplified at each locus. Indeed, this method has been shown to provide the basis for an international reference standard for human cell lines (23). The technique can be used to identify and differentiate members of the same family. This is a vital tool for ECACC where it is used to authenticate DNA produced from immortalized cell lines produced from the blood of over 30 000 individuals within its collections of over 650 disease types.

ECACC routinely uses ten primers (D11S902, D12S83, D16S404, D1S234, D2S165, D3S1292, D4S424, D5S436, D7S516, D8S260). Multiplex PCR reactions are carried out using ten different fluorescent dye-linked primers. Labelled products are detected by electrophoretic size fractionation on a PE-ABI Prism 377

Genetic Analyser and analysed using Genescan and Genotyper analysis software (Perkin Elmer). As non-overlapping alleles are labelled with different dyes they can be detected in a single lane on the ABI PRISM automated genotyping system. The end-result for each cell line is an electropherogram with each STR allele represented as one or more peaks.

The data is further analysed to categorize peaks according to their size in relation to an internal standard run in every lane in the gel. This analysis enables every peak to be allocated a size corresponding to the number of repeat units present. An algorithm was developed to compare the allelic profiles, with each profile (questioned profile) being checked against every other profile (reference profiles). For each comparison the number of alleles present in both reference and questioned profiles are scored and expressed as a percentage of the total number of alleles in the questioned profile.

For normal human DNA, STR profiling will show two alleles at most loci, as expected for highly polymorphic loci in a diploid genotype. Additional alleles at a locus are rarely seen in normal profiles, but theoretically can occur by trisomy, gene duplication, or mixed populations of cells or hybrids. If most of the loci show more than two peaks, this is an indication of an hybrid or mixture of cells, whereas single loci with more than two peaks are more likely explained by trisomy or gene duplication events.

DNA profiling could be applied to DNA samples from all species. However, the appropriate primers would have to be validated and available. This would require considerable work, and, whereas it might be feasible for popular cell line species such as mouse and rat, commercial sense would suggest that such primers will not be developed in the immediate future for the likes of iguana and potoroo.

In conclusion, all three authentication techniques described above have useful qualities, as outlined in *Table 5*.

Isoenzyme analysis can provide rapid determination of species which is a useful initial test. Cytogenetics will also confirm species and identify chromosome markers specific to an individual or cell line. However, this technique is labour-intensive and requires highly specialized skills to cover a wide range of species. The development of new techniques for chromosome analysis such as automated

Table 5 Comparison of the different cell authentication methods available

Application	Multilocus DNA fingerprinting	Multiplex PCR DNA profiling	Cytogenetic analysis	Isoenzyme analysis
Species determination		√	√	√
Individual identification	√	√ [b]		
Detection of cell line variation	√	√ [b]	√ [a]	
Regulatory authority recognition	*		√	√

√ Useful applications.

* Under consideration.

[a] Only if unique chromosome markers are present.

[b] Applicable to human DNA only.

karyotype interpretation and fluorescence-activated sorting of chromosomes may facilitate the routine use of direct chromosome analysis.

Multilocus DNA fingerprinting provides data representative of all cells in a sample, confirming identity between consecutive cell stocks and the absence of switched cultures or, within limits, cross-contamination. DNA profiling using single locus probes is a useful technique for identity testing but to date has been primarily validated in work with human cell lines alone.

There is no simple choice to be made from the methods available (24) and even use of all three techniques described above will not, in every case, uniquely identify a cell line. In such circumstances other tests such as surface antigen expression may be important. Nevertheless, in a general tissue culture laboratory the three techniques outlined above can provide a good basis for authentication, although the significance and value of molecular genetic techniques is increasingly recognized. For further detailed discussion on this topic see ref. 25.

6 Regulatory aspects

From the development of the first virus vaccines using cells in culture by Salk in 1954, the production of biologicals for human or veterinary use has meant the imposition of compliance with safety regulations laid down by the appropriate regulatory agency. It is increasingly important for all scientists engaged in the development of cell lines that they are fully aware of regulatory implications from the outset if an eventual product from cells is to gain regulatory approval (26) (see also Chapter 10).

Unreliability of primary cell culture led to the development of human diploid cell lines, e.g. WI 38 (24), which were fully characterized to a rigid standard which also dictated the range of population doublings within which the cells were considered stable enough to provide a substrate for vaccine production. In recent years the increasing confidence in the safety of these lines has permitted the 'usable' lifespan to be increased, e.g. from population doubling 30 to 40 with MRC5 (27).

One unfortunate aspect of the historical precedent set by the use of human diploid cells was that ground rules were set which were then applied to less suitable systems such as tumour-derived human and rodent cell lines. In some cases the tests applied were at best meaningless and at worst not only irrelevant and expensive but also controversial (e.g. *in vivo* tumorigenicity testing).

The present list of requirements is now more rational (*Table 6*) and several aspects relate to quality control and maintenance of cell stocks. This involves not only the quality control of master and working cell banks but also important details on early characterization data and the derivation of the cell line, and implications for late in-process testing. As has been mentioned already, several tests would seem redundant in the context of transformed cells, but guidelines are set by the US FDA Office of Biological Research & Review (now designated the Center for Biologics Evaluation & Research) in their *Points to consider* 1984 and 1993 (28–30), the CEC Committee for Proprietary Medicinal Products (14, 31),

Table 6 Information required for regulatory approval

Information
History and genealogy of the cell line
Records and storage information on master and working cell banks
Culture requirements
Growth characteristics
Sampling and testing procedures
Production and testing facilities
Quality control tests:
karyology*
isoenzyme analysis
DNA fingerprinting*
virus testing
retrovirus status
tests for contaminating DNA (final product)
purification procedures/validation data
characterization of product

* Only required in some applications.

World Health Organization regulations (32), and more recently the International Committee for Harmonization (33). Fundamental to this process is the existence of fully characterized master and working cell banks. Increasingly, as tests based on new technology (e.g. DNA fingerprinting) have become available these have been seen as supplementary tests rather than replacements for outmoded existing methodology. In practice, the biological product is evaluated closely in a series of meetings with the experts from the regulatory agencies to discuss results and to build up a picture of the cell and the product. Virus testing procedures are considered to be of particular importance and as a result the tests have to be exhaustive; if at first negative, further testing with induction studies may be required.

If there are adventitious agents present, what are the consequences? This depends upon the value of the product to the patient (benefit/risk analysis) and the efficiency of the post-production purification procedures. Validation of these procedures may also require 'spiking' experiments with high concentrations of viruses to mimic a possible contamination event and thus prove general inactivation or clearance potential within the system. As a consequence, the process leading to acceptance of a cell line-derived product by regulatory agencies is lengthy and extremely expensive.

In summary, quality control of cell lines is a requirement from the very outset of a research project leading to an eventual product or the development of a production process. Failure to do so could cost dearly both in financial terms and in scientific reputation. Many research projects these days lead to commercialization of products, whether these are diagnostic reagents or therapeutics. Therefore, it is vitally important to carry out necessary quality control measures to satisfy the various regulatory bodies and not leave the issue unresolved until the product is ready for the market. This also requires appropriate record keeping which may be overlooked. It is our view that scientific journals should take a more active role in ensuring that basic quality control is carried out (34). This could be done, for example, by insisting that articles involving the use of cell lines state that the cell lines have been tested for microbial contamination at the very least. These issues have been emphasized most recently by the formulation of new guidelines for the use of cell lines in cancer research (35) and a timely reminder on cross-contamination events (36).

References

1. Hay, R. J. (1988). *Anal. Biochem.*, **171**, 225.
2. Stulberg, C. S., Cornell, L. L., Krizaeff, A. I., and Sharron, J. E. (1970). *In Vitro*, **5**, 1.
3. Nelson-Rees, W., Daniels, D. W., and Flandermeyer, R. R. (1981). *Science*, **212**, 446.
4. McGarrity, G. J. (1982). *Adv. Cell Culture*, **2**, 99.
5. Doyle, A. and Griffiths, J. B. (ed.) (1998). In *Cell and tissue culture: laboratory procedures for biotechnology*, Appendix 3, p. 325. John Wiley & Sons, Chichester, UK.
6. Advisory Committee on Dangerous Pathogens. (1995). *Categorisation of biological agents according to hazard and categories of containment* (4th edn). HSE Books, Sudbury, Suffolk, UK.
7. Stacey, G. N., Doyle, A., and Hambleton, P. (ed.) (1998). *Safety in cell and tissue culture*. Kluwer, Netherlands.
8. Cowan, S. T. and Steel, K. J. (1979). *Manual for identification of medical bacteria*, p. 163. Cambridge University Press, UK.
9. *Detection of mycoplasma contamination*. (1986). C.F.R. 113.28, p. 379. Office of the Federal Register, National Archives and Records Administration, Washington.
10. *European pharmacopoeia, biological tests* (2nd edn). (1980). Part I, p. V2. 1.3. Maisonneave, Sainte-Ruffine.
11. Mowles, J. M. (1990). In *Methods in molecular biology*, Vol. 5, *Animal cell culture* (ed. J. W. Pollard and J. M. Walker), pp. 65–74. The Humana Press, Clifton, NJ.
12. Zuckerman, A. J. and Harrison, T. J. (1990). In *Principles and practice of clinical virology* (ed. A. J. Zuckerman, J. E. Banatavala, and J. R. Pattison) (2nd edn), p. 153. John Wiley & Sons, Chichester, UK.
13. Biggar, R. J., Schmidt, T. J., and Woodall, J. P. (1977). *J. Am. Vet. Med. Assoc.*, **171**, 829.
14. CEC Committee for Proprietary Medicinal Products. (1988). *Trends Biotechnol.*, **6**, G5.
15. Peebles, P. T. (1975). *Virology*, **67**, 288.
16. Rowe, W. P., Pugh, W. F., and Harley, J. W. (1970). *Virology*, **42**, 1136.
17. Hertig, C., Pauli, U., Zanoni, R., and Peterhans, F. (1991). *Vet. Microbiol.*, **25**, 65.
18. Transmissable spongiform encephalopathy agents: safe working and the prevention of infection. (1998). The Stationery Office, London. (ISBN 0 11 3221665)
19. Dawson, M. (1993). In *Cell and tissue culture: laboratory procedures* (ed. A. Doyle, J. B. Griffiths, and D. G. Newell), Section 7B.7. John Wiley & Sons, Chichester, UK.
20. Rooney, D. F. and Czepulkowski, B. H. (1986). *Human cytogenetics*. IRL Press, Oxford.
21. Macgregor, H. and Varley, J. (1988). *Working with animal chromosomes* (2nd edn). John Wiley & Sons, Chichester, UK.
22. Burke, T., Dolt, G., Jeffreys, A. J., and Wolff, R. (1991). *DNA fingerprinting: approaches and applications*. Birkhäuser Verlag, Basel.
23. Masters, J. R. W., Thomson, J. A., Daly-Burns, B., Reid, Y. A., Dirks, W., Packer, P., *et al.* (2001). *Proc. Natl. Acad. Sci. USA*, **98**, 8012–17.

24. Stacey, G. N., Hoelzl, H., Stephenson, J. R., and Doyle, A. (1997). *Biologicals*, **25**, 75.
25. Stacey, G. N., Bolton, B. J., and Doyle, A. (1992). *Nature*, **357**, 261.
26. Doyle, A. and Griffiths, J. B. (ed.) (1998). In *Cell and tissue culture: laboratory procedures in biotechnology*, p. 295. John Wiley & Sons, Chichester, UK.
27. Wood, D. J. and Minor, P. (1990). *Biologicals*, **18**, 143.
28. US Dept of Health and Human Services, Center for Biologics Evaluation and Research, Food and Drug Administration. (1993). *Points to consider in the characterisation of cell lines to be used to produce biologicals*. CBER, Rockville, MD, USA.
29. US Dept of Health and Human Services, Food and Drug Administration, Office of Biologics Research and Review. (1984). *Points to consider in the manufacture and testing of monoclonal antibody products for human use*. OBRR, Rockville, MD, USA.
30. Food and Drug Administration. (1992). *Federal Register. 21 Code of federal regulations, part 630.35*. Office of the Federal Register National Archives and Records, Washington DC.
31. CEC Committee for Proprietary Medicinal Products. (1988). *Trends Biotechnol.*, **6**, G1.
32. World Health Organization. (1987). *Technical Report Series*, Vol. 747. WHO, Geneva.
33. ICH Guidelines for Biotechnological Quality, Q5D: Cell substrates. (1997). Published in Federal Register, **62**, 24311.
34. Mowles, J. M. and Doyle, A. (1990). *Cytotechnology*, **3**, 107.
35. Masters, J. R. W., Twentyman, P., Arlett, C., Daley, R., Davis, J., Doyle, A., *et al.* (1999). *UKCCCR guidelines for the use of cell lines in cancer research*. United Kingdom Coordinating Committee on Cancer Research, London WC2A 3PX, UK.
36. Stacey, G. N. (2000). Nature, **403**, 356.

Chapter 10
Good Laboratory Practice in the cell culture laboratory

Barbara Clitherow
Quality Assurance Unit, Bio Products Laboratory, Dagger Lane, Elstree, Hertfordshire WD6 3BX, UK.

Stephen J. Froud
Lonza Biologics plc, 228 Bath Road, Slough, Berkshire SL1 4DY, UK.

Jan Luker
Eli Lilly & Co. Ltd., Erl Wood Manor, Sunninghill Road, Windlesham, Surrey GU20 6PH, UK.

1 What happens if you fall under a bus?

Suppose that you are about to subculture a cell line: an activity common to all cell culture scientists, so familiar as to be completed almost without thinking. You clean your laminar flow cabinet, warm your bottle of fresh medium, locate your culture, and ensure that all your other equipment and materials are to hand. Then you disinfect your gloves, transfer one-fifth of the culture volume into a new flask, aseptically add four volumes of fresh medium, similarly add any other separate but essential medium components, label the new culture, and return this to the incubator. Finally you clean your cabinet, discard your contaminated materials, and return the medium and components to the fridge or freezer. Comfortable?

Unfortunately, the next day you fall ill. At first nobody worries, but after you've been off for nearly a week one of your colleagues realizes that it would be helpful to subculture your cells. Your benefactor cleans the cabinet and proceeds to collect the cells. By asking every other member of the laboratory to identify their flasks, he is able to select the right culture. He has less success with the medium. Because it was a simple subculture you have not yet written it into your lab book (there would be more time in the morning), but by hunting along all the potential bottles of medium on your shelf in the fridge, he chooses the right one. So far so good. He's not sure how much you normally dilute the cells on subculture and the culture looks very thick, so he decides on a one in ten split. By now the job has taken longer that he had planned, so the medium is 'warmed' quickly in a nearby water-bath, and the culture is split and returned to the incubator. Your helper retires, satisfied.

Upon your return you find just dead cells. Your benefactor hadn't realized that this line would not tolerate anything over a one in eight dilution, or that it was sensitive to temperature shock when adding fresh medium (did he know that the water-bath he used was set at 30 °C for another experiment?). Finally, he had no idea that you were now adding interleukin-6 to your cultures and that this was stored in the freezer, to be added to the medium at each subculture.

Try to apply the 'fallen under a bus' principle to your own work. Could the experiment proceed if you fell under a bus? Given that cell culture studies are usually expensive and lengthy, this is a useful, if not cheerful, concept. By the application of the principles of Good Laboratory Practice (GLP) the above scenario would not occur, and your work would outlive you. But what is GLP? Our objective in this chapter is to illustrate GLP, in practical terms, for the cell culture scientist. We have included some examples from our own experience. We also provide some ideas on the most useful elements of GLP that every cell culture scientist may wish to consider, even in the absence of a full GLP framework.

2 GLP, an overview

Let us examine how the application of the principles of GLP would have led to a different outcome in the story given above. Each GLP study is the responsibility of a **Study Director**, an individual experienced in the field, and absolutely central to the performance of GLP. When our sick scientist failed to arrive, the Study Director would be informed, and he would ensure that someone was available to subculture the cells. The Study Director may not have the supervisory authority to tell someone to do this, but he must be able to communicate directly with someone who can. This latter person, the supervisor or **management**, has already ensured that a suitably trained scientist would be available for such an emergency. The Study Director will be close enough to the work to know roughly what needs to be done because he has written an overall **protocol** listing the key activities for this particular study before it began. In particular, the range of acceptable subculture ratios for this study was given in the protocol. After reading the scientist's (up-to-date) laboratory notebook or **study records**, he would be able to describe what was required of the surrogate scientist. To ensure easy and foolproof recognition of the correct culture these records would include details of how the cultures were labelled (e.g. cell line name, date, flask number, and scientist's name). They would also include a complete description of the materials, such as medium components, that were to be used. To provide further assistance, the experimental protocol would refer to written methods or **standard operating procedures** (SOPs) that in this case would require the scientist to record the temperature of the water-bath and to check if it lay in the range specified. Thus the incorrect temperature would have been identified and the medium correctly warmed.

To expand this scenario further, consider other key factors that could seriously influence the growth of the culture or the outcome of a cell culture study. Should the temperature of the incubator used for the cultures be checked regularly and the temperature recorded in a log book? Should the correct operation of the incubator, and all other critical equipment, be inspected routinely according to a predetermined **equipment maintenance schedule**? Should a senior scientist approve planned **amendments** to the protocol during the study (e.g. introduction of a new method of calculating results)? Similarly, should this Study Director assess the impact of any unexpected events or deviations (e.g. the transient temperature shock to a culture during an incubator failure, before the culture was moved to another incubator)? Should the Study Director check the raw data, calculations, and conclusions in the report prepared by the scientist? Before release should the report be submitted for an independent review for final checking? Finally, should a copy of the report and all raw data be archived in a safe place to prevent loss? In a GLP study all of these questions would be answered in the affirmative.

GLP then is all about ensuring that consistent results are obtained for important studies, that the conclusions are valid, and the data is safely stored for future reference. These are, of course, standard scientific principles. GLP merely provides the framework to ensure that all of the key areas have been considered. We have introduced most of the general terms in this section and many of these key words have been given in bold print. The flow of a GLP study is presented in *Figure 1*. The key words are explained in more detail in the sections below. But first let us consider when GLP should be used for cell culture and, if it is not used, which aspects could be useful in the general cell culture laboratory.

3 What is GLP and when should it be used?

Any study concerned with the testing of substances to obtain data on their properties or safety with respect to human health or the environment, must conform to certain guidelines in order to be accepted by regulatory agencies. Examples in the field of cell culture include *in vitro* toxicity studies, viral assays on biological products, the testing of cell lines used for the production of pharmaceuticals, and in some laboratories, even the preparation and cloning of cell lines. By adhering to the principles of GLP, laboratories ensure that the results are accurate, reproducible, and traceable. Data from such studies can be used, for example, in the product licence application that must be submitted before a pharmaceutical can be marketed.

A system for GLP was first defined in 1976 by the Food and Drug Administration in the United States (1) in response to studies it had received in support of such product safety submissions in which negligence and carelessness, if not fraud, were apparent. In 1982 the UK Health and Safety Commission published a Code of Practice, later modified in accordance with the 1982 GLP Principles produced by the Organisation for Economic Co-operation and Development (OECD). European Community Directives of 1987 and 1988 required member countries to implement legislation to enforce GLP standards for the conduct of safety studies, and in the UK The GLP Regulations 1997 were introduced. Subsequently the OECD GLP Principles were revised in 1997 (2), and a new UK Statutory Instrument has now been issued in accordance with these latest guidelines. All laboratories conducting safety studies in the UK are therefore inspected for compliance with The GLP Regulations 1999 (3), by the UK GLP Monitoring Authority. The regulations from the Food and Drug Administration have also been amended (4).

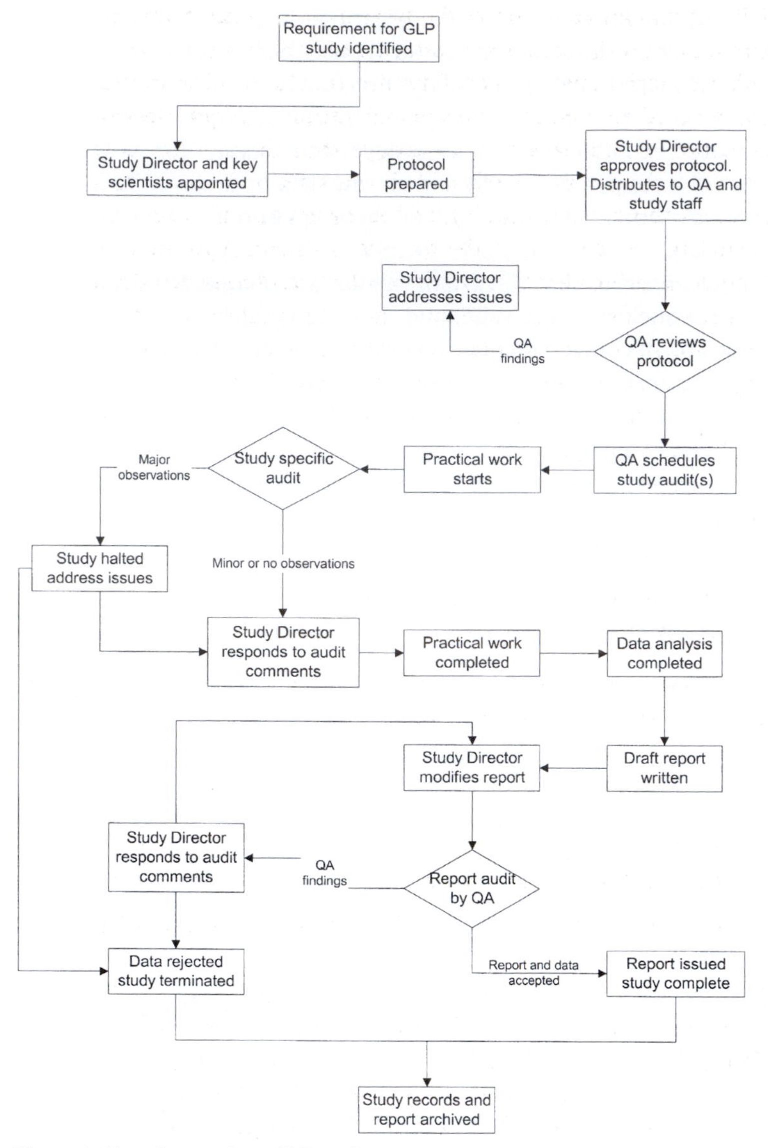

Figure 1 Flow diagram for a GLP study.

4 Planning the study

When the need for a GLP study is identified, the laboratory management must ensure that suitable facilities and equipment are available, provide the required materials, and agree to commit sufficient time and personnel to carry out the study. In considering available personnel for the new study, management can refer to the master schedule, an information system maintained in GLP laboratories. This includes key details of *all* studies performed within the laboratory (including the name of the study director, the nature of the study, and whether the study claims GLP compliance), and shows the progress of studies from protocol approval through the experimental stage to report issue.

No GLP study can begin until a study protocol has been written and approved. The protocol is prepared by the scientists involved and authorized by the Study Director, who is appointed by management to ensure that the study is conducted in accordance with the principles of GLP. All the scientific staff have a responsibility to perform the study according to the protocol, and maintain appropriate records. An independent body, the quality assurance (QA) unit, reviews the protocol, and also undertakes to monitor the study methods, raw data, and audit the final report.

The study protocol should list the following:

- unique study reference
- title and objective of the study
- location of the testing facility
- name of the Study Director
- identity of other senior scientists and their responsibilities within the study
- proposed experimental starting and completion dates
- identity of all reagents
- source of reference materials and controls
- key materials and storage facilities (e.g. for samples)
- technical methods to be used
- specific equipment
- methods for calculations
- documents to be archived at the end of the study

Many of the experimental details may be covered by reference to appropriate SOPs.

Once a protocol has been issued, any alteration or amendment can only be carried out with the agreement of the Study Director, and such alterations must be documented and retained with the protocol. It is the responsibility of the Study Director to ensure that all relevant personnel are then informed of the changes. In addition, any accidental 'deviation' from the protocol or any methodology or system it describes should likewise be recorded, and assessed for its likely impact on the validity of the results.

5 Performance of the study

5.1 Personnel

It is essential that all personnel involved in the study are suitably trained in the principles of GLP and in the relevant techniques and use of any equipment (including safety aspects), and, if appropriate, interpretation of the data. Furthermore, written evidence of this training should exist. All members of staff, therefore, should have their own training records, which are regularly reviewed and updated as new equipment or techniques are introduced or modifications are made to existing ones. No one is too junior or too senior to be trained or to have a reminder. One approach is to have a one hour training session each week, for laboratory scientists, supervisors, and the head of department. This ensures that new and revised methods are discussed, and that established procedures are re-visited at least annually. At the end of each session the relevant section of individuals' loose-leaf training files are signed.

All personnel involved in GLP studies must be aware of their responsibilities and level of authority within a study and to this end each must have a job description detailing the work he or she will be expected to perform. To demonstrate that the scientists' qualifications and experience are appropriate for their role in the study, each should also have a current curriculum vitae, detailing qualifications and work experience. It may be useful for both the job description and curriculum vitae to be included as one-page summaries at the front of each individual's training file.

The most important individual in any study is the Study Director. It is he or she who ensures that the study conforms to the principles of GLP. The Study Director does this by ensuring that the study protocol is followed, that only the appropriate equipment and methods are used, and that the technical standard of the work performed is reviewed. It is also the Study Director who is responsible for ensuring that the recording, interpretation, and archiving of data are performed correctly, and for assessing and reporting any deviations. The Study Director must be able to appraise the validity of any conclusions drawn from the results. Thus it is more important that the Study Director is experienced in the work performed and is thoroughly trained in GLP than that he or she is the most senior member of the team. The Study Director must also respond to audit reports from QA and ensure that any necessary actions are completed if any problems are noted. Ultimately it is also the Study Director who is responsible for scrutinizing the draft final report and submitting it to QA for audit, together with all of the raw data and an assessment of any discrepancies that occurred.

5.2 Methods

Wherever possible, Standard Operating Procedures should be utilized. These are clear detailed instructions for the techniques to be used. They should also include criteria that can be used to assess if a study is valid, usually by comparing the results of control cultures against the results expected. These expected results will have been determined during the development of the study method. This latter process is called validation. An SOP should cover all aspects of the tasks involved, including analytical methods and techniques, use of the relevant equipment and/or software, reagents used (with details of how they are to be prepared), and the calculation and interpretation of results. As with study protocols, SOPs should be individually issued and uniquely numbered documents so that their location can be tracked by QA. This ensures that all scientists have the most recent version. SOPs should be reviewed and revised at frequent intervals to ensure that they are current, relevant, and easy to understand. An incomprehensible or out-dated SOP is worse than none at all. One solution is to ensure that all SOPs are reviewed at least once every two years, although in practice many will be revised sooner as equipment or procedures are updated.

In the rare instances when it is not possible to use an SOP (e.g. when the operator has developed a new technique or process during the course of the study) all practical details must be recorded in the laboratory notebook. These notes should be in sufficient depth for another operator to repeat the work precisely and should be checked and verified by the scientist concerned.

When a deviation from the SOP does occur, this must be thoroughly and promptly documented, including an assessment by the Study Director of the potential impact of the deviation on the results that will be generated during the study.

5.3 Equipment

It is the responsibility of management to provide all necessary equipment, and to ensure that written instructions for its use, calibration, and maintenance are in place before the start of the study. It should be remembered that 'equipment' includes not only large, sophisticated items such as HPLC analysers, but also small or 'simple' items such as precision pipettes, refrigerators, and balances. All are equally important and must be treated with the same care and respect.

It is essential that all items of equipment are kept clean and in good working order. Chemicals left on a balance pan from a previous operator could affect the results of the next experiment. Dirty and neglected equipment will not perform as efficiently or as accurately as a well maintained instrument. Hence, written instructions (SOPs) must be immediately available for the routine use, cleaning, checking, and maintenance of all equipment. There must also be a documented programme of calibration and servicing of all equipment. The frequency and method of calibration and/or service will depend on the complexity and frequency of use of the instrument concerned. Some items such as pH meters may need calibration every time they are used, whereas for a vertical laminar flow cabinet, a six-monthly calibration and service may be sufficient. All equipment should also be tested for electrical safety and have documents to show that this is done. Some facilities prefer to contract out the calibration and servicing of equipment. This is acceptable, but as with internally maintained items there must be a documented schedule of work.

It is expected that major equipment items display their calibration or service status (as appropriate) detailing when work was last done and when it is next due. Adhesive or tied labels may be used to show such information. It is the responsibility of the scientists to be aware of the status of their equipment, to use only calibrated 'in-date' instruments, and to ensure that any necessary work is carried out as scheduled. If there is no evidence that a piece of equipment is in calibration or has been regularly maintained or serviced, it should not be used. If a fault occurs or a machine does not perform within the specified acceptance criteria, it should be taken out of use at once (e.g. by applying an adhesive label) and the engineering department or external contractor called immediately. The equipment should not be used again until the defect has been rectified and the repair has been documented.

An equipment usage log is very useful, including the name of the scientist who used the equipment and its performance characteristics at that time. This assures that only trained and competent staff have used the equipment and that it was operating correctly. It also helps to identify the studies that may have been affected if a piece of equipment is subsequently found not to be working correctly.

New equipment should be checked before being used in a GLP study. This includes relevant computer hardware and software. Dry runs will need to be performed to ensure that expected results are obtained. The validation records concerned need to be retained, e.g. the plan documenting the required tests and acceptable results, records of the conduct of the tests and formal assessment and acceptance.

5.4 Materials

We include with 'materials' all reagents and chemicals used in the study, and any reference, standard, or control substances. Chemicals should be described in the study protocol or SOPs. All purchased reagents should also be of the grade specified and the supplier and grade should be documented. Written instructions must be available for the make-up of all reagents. Pre-prepared blank forms ensure that the required details for reagents are recorded during a study. All chemicals and reagents must be correctly labelled with identity, composition, storage conditions (e.g. temperature and light), and expiry or re-test date. No reagent which has exceeded its expiry or re-test date should be used in a GLP study. For example, growth of some cell lines in media containing glutamine can be seriously impaired after storage of such media for a few weeks. For obvious reasons, refrigerators, freezers, and cold rooms must be regularly maintained, with appropriate temperature monitoring to ensure that storage conditions are as required.

The written procedures for a study should include the receipt or transfer, handling, sampling, and appropriate storage of all test and reference substances. Methods for ensuring the correct identity of all reference and test samples should be in place. Correct and thorough labelling is essential. Again, adhesive labels are commonly used to ensure that all of the relevant details are completed on the container. Information on the source, date of receipt and expiry date, as well as quantities received and used in studies should be recorded for each batch of test substance. In addition, the purity, stability, and composition should be known for all test substances. For all studies except short-term studies, samples of all batches of test substances should be retained in the archives. In the case of any unexpected observations, or queries from the regulatory authorities, they may be used for investigative purposes.

6 Monitoring

The GLP Regulations require the management of a company to appoint one or more independent personnel 'to conduct inspections to determine if all studies are conducted in accordance with the principles of GLP' (3).

Monitoring is carried out to maintain an 'external' objective view of studies and facilities and this is usually performed by the QA function within a company. Findings from all types of audit are reported to relevant Study Directors and to management. Written responses are required to any QA observations and the inspection records must also be archived. Auditing activities fall into the three main areas described below.

6.1 Facilities and systems

Facilities-type inspections according to a schedule are often used in laboratories where GLP studies are conducted. These would involve checking that the facility is of the correct size and location, that all necessary standards of cleanliness, hygiene, and safety are met, and that all equipment is calibrated, validated, and correctly maintained. The QA auditor should also inspect computer systems to check that both hardware and software are appropriate and have been validated for use in GLP studies. Support services (e.g. engineering calibration schedules and documentation systems) may be investigated. Audits of training schemes and training records should be carried out on a regular basis to ensure that staff are receiving continuous ongoing training in all aspects of their work, including adequate knowledge of GLP. Documentation systems and tracking and archiving of GLP documents, e.g. written procedures, protocols, and records, should also be audited on a regular basis.

6.2 Study-specific audits

At the initiation of the study, QA should plan a schedule of study audits to be performed at critical points in the study. The number and frequency of the audits will depend on the length and nature of the work. For example, cell line stability studies performed in one of our laboratories take approximately 12 weeks. At the start the principal scientists and QA personnel get together to perform a pre-study review of the protocol and training records to ensure that the procedure is fully understood by all and that the scientists have the appropriate level of training for their agreed tasks. Then, on one day during the study the QA auditor will watch the scientists at work and look at the available data to ensure that the protocol is being followed and that the written records are of the required quality. The auditor will check topics such as storage and maintenance of samples, standards, and reagents to ensure that they are being prepared, stored, and labelled correctly and that expired reagents are not being used. In other laboratories, where many identical short studies are performed (e.g. testing cell lines for viruses or mycoplasma) QA may review only a proportion of the studies in such detail.

6.3 Final report audit

In many laboratories the principal scientist(s) who has performed the laboratory work will write the report using a template report on a word processor. This template may include sections for the names of the scientists, a description of the facilities, equipment and methods, table and figure layouts, standard text, and the alternative conclusions that could be drawn. The author can then insert raw data into the template and delete the conclusions that do not apply. This draft report will be reviewed by the Study Director and the conclusions checked against the study records. Often the Study Director will then make minor amendments to the report before forwarding it to QA.

The QA auditor will review the report and study records, assess if the protocol has been adhered to, and check the validity of the study against the criteria for acceptance. He or she will also ensure that all deviations and amendments have been assessed for their potential impact on the study, and that they are attached to the raw data. QA auditors then report their findings to the Study Director, often using a standard form. The Study Director appraises these comments, makes any necessary amendments to the report or adds notes to the raw data, and prepares a written summary of his or her actions. The final report is then issued with a statement signed by the Study Director confirming that the study was carried out in accordance with the principles of GLP, and a signed QA statement listing the inspections relevant to the study. Laboratory management may also review and approve the study report.

QA may check each and every calculation and transcribed number, but where final study reports contain extensive tables of data it is normal for QA procedures to allow for the verification of a sample of the values quoted, providing that the basis for sampling is documented. The proportion of data points checked may vary depending on how critical to the study objective the figures are. The purpose of the QA report audit is to confirm that the report reflects the conduct and results of the study. The responsibility for the accuracy of the report remains with the Study Director, who should not assume that QA will find any and all errors.

7 Records

All experimental details and results should be recorded as they are generated in identified and traceable laboratory notebooks or study-specific files. All entries should be clear and in ink, dated, and signed or initialled by the scientist responsible. This will provide evidence of when the work was carried out and when it was recorded, and that this was done by a suitably trained person. The Study Director or a suitably qualified nominee may also check and countersign each page or critical entry to verify that the experiment was carried out correctly according to the study protocol, using agreed procedures, and that all calculations and conclusions drawn are correct and valid. The scientist must record enough detail of methods and calculations to enable another scientist to reproduce the work exactly. Where SOPs or other written procedures are used, these should be fully cross-referenced in the notebook with sufficient information to show that the procedures were followed, or where deviations occurred. All records whether in a notebook, on a form or instrument printout must be identified to the study. Any graphs, figures, photographs, or other documents should be signed and dated by the scientist and attached to the notebook or file. If a study generates large computer print-outs or other documents, these should be similarly annotated and stored in a study-specific file. Only approved, validated software programs should be used for processing data and any data disks should be archived with the hard copy in the study file at the end of the study. In addition, software versions also need to be archived when superseded. Rough notebooks from which data is transcribed at a later date should be avoided as they generate a great deal of work when checking for correct transcription.

Many laboratories use standard forms to provide memory prompts to ensure that all the relevant details are recorded. *Figure 2* shows an example of a form used to record the subculturing of a cell line. Similar forms are used to record details for the recovery of cells from cryopreserved stocks, for the preparation of media, for calculations, and to record conclusions. In fact, some cell line studies can be conducted without notebooks, recording everything on such forms. These forms are stored in a study-specific file. Use of study files must ensure that forms cannot be lost or 'disappear', for example by maintaining an up to date file index of all the study records.

All written material must remain legible at all times. If a mistake is made, the original erroneous entry should not be obliterated by overwriting or by use of labels or correcting fluid. The error should be simply crossed through, so that it remains legible, and the reason for the amendment explained briefly, signed, and dated by the scientist concerned.

At the end of the study, all notebooks, data files, computer data, and any other raw data should be archived. Typically a period of ten years is used. Material is archived because it may be required to be submitted to the regulatory authorities, for example for investigations into adverse reactions or other such enquiries. For studies performed on cell banks which may be used in production for many years, it may be necessary to reduce the raw data to microfiche and to store this indefinitely.

8 Advice for researchers creating new cell lines

GLP is essential in laboratories which generate data which will contribute to the assessment of pharmaceutical safety, efficacy, and consistency. The cost of maintaining a GLP system for cell culture laboratories has been estimated at 30% of the total laboratory expenditure. On the grounds of cost and flexibility, it is not normally appropriate to operate basic research laboratories to GLP. Such laboratories, however, may be the source of novel cell lines which are ultimately used for commercial purposes. There are key issues that will need to be addressed if the cell line is to be transported overseas or to be used in the production of a biopharmaceutical.

Cell Line: .... ID code: .... Study No. ....

Sheet No.: .... Date: .... Time: ....

Cell count details							Sub culture details								
		Viable count		Total count						Calculated					
Flask ID	Diln.	Cytometer count	Viable cells ($x10^5$/ml)	Cytometer count	Total cells ($x10^5$/ml)	Viability (%)	Initial volume (ml)	Transfer volume (ml)	Final volume (ml)	Discard volume (ml)	New cell conc. ($x10^5$/ml)	New flask ID	Gen. No.	Media Lot No.	Scientist

Comments:

Form No.: CCD298(3) Issued: ....

Page 1 of 1

Supervisor: ....

Date: ....

Figure 2 Example of a document for recording cell culture details.

Concerns over agents that potentially may be transmitted in biological materials, e.g. bovine spongiform encephalopathy (BSE), have led to an increase in the data required before cell lines can be transported to another country or used in the preparation of a pharmaceutical. The supplier, country of origin, and testing status of biological additives to media are often unknown. Yet an industrial organization will be asked such questions before the cell line can be transported overseas or be used in the production of a biopharmaceutical. This is the most common problem that is encountered when an industrial organization begins to apply excellent work from research institutions. It would be advisable for each researcher to keep a log of biological additives used, e.g. serum, trypsin, and even purified chemicals (albumin, insulin, transferrin, cholesterol, etc.). This will allow a suitable testing strategy for potential adventitious agents to be defined. Some non-GLP cell culture development laboratories use a simple photocopied form which prompts for all the relevant details to be recorded during medium preparation. This memory aid is stuck into the relevant laboratory notebook.

Similarly, before a product derived from cell culture can be administered to a human, details on the method of construction and clonality of a cell line will need to be provided to the regulatory authority. For recombinant cell lines the data required from all laboratories by the regulations governing the construction and storage of such cell lines (see for example, *A guide to the genetically modified organisms (contained use) regulations 2000*) (5) will mean that a suitable detailed description must be available. For a hybridoma cell line the species, sex, and source tissue of the fusion partners and the method of isolation and fusion should be recorded. When the donor has been immunized, a description of the immunogen and the methods for immunization must be provided. For other cell lines, the species, sex, and health status of the donor will be needed, in addition to details on the source tissue and method of transformation.

One area that is frequently neglected is the necessity to demonstrate the clonality of a cell line that is to be used in the production of a pharmaceutical. This is best described as the probability that the cell line is derived from a single cell (see Chapter 8). In order to avoid unnecessary recloning of a cell line on transfer from an academic to an industrial laboratory, a complete description of the methods used and the results obtained should be available. Typically, in order to make a clear assessment, data on the seeding concentration, the number of cloning plates used, the number of clones obtained and screened, the method of inspection, and the number of clones showing the appropriate activity will be needed.

Given the time-scales involved, it is essential that all of the above details are written in a robust notebook, suitable for long-term storage.

Acknowledgements

The authors would like to thank Dr Gillian Lees (Q-One Biotech Ltd.) and Dr Chris Morris (Wellcome Trust for Human Genetics) for helpful discussions during the preparation of the original manuscript.

References

1. Food and Drug Administration. (1976). *Federal Register. 21 Code of federal regulations*, **part 58**. Office of the Federal Register National Archives and Records Administration, Washington DC.
2. The Organisation for Economic Co-operation and Development. (1998). *OECD principles of good laboratory practice (as revised in 1997)*. OECD Publications Service, Paris.
3. Statutory Instruments 1999 No. 3106 Health and Safety. *The Good Laboratory Practice Regulations 1999*. HMSO, London.
4. Food and Drug Administration. (1998). *Federal Register. 21 Code of federal regulations*, **part 58**. Office of the Federal Register National Archives and Records Administration, Washington DC.
5. The Health and Safety Executive (2000). *A guide to the genetically modified organisms (contained use) regulations 2000*. HSE Books, Sudbury, Suffolk, UK.

Appendix 1
List of suppliers

3M Health Care, Morley Street, Loughborough, Leicestershire LE11 1EP, UK.

Abbott Laboratories, North Road, Queenborough, Kent ME11 5EL, UK.
Abbott Laboratories, 1 Abbott Park Road, Abbott Park, IL 60064-3500, USA.

Alpha Laboratories Ltd., 40 Parham Drive, Eastleigh, Hampshire SO50 4NU, UK.

Amersham Pharmacia Biotech UK Ltd, Amersham Place, Little Chalfont, Buckinghamshire HP7 9NA, UK (see also Nycomed Amersham Imaging UK; Pharmacia)
Tel: 0800 515313
Fax: 0800 616927
URL: http//www.apbiotech.com/

Amersham, see Nycomed Amersham.

Anderman and Co. Ltd, 145 London Road, Kingston-upon-Thames, Surrey KT2 6NH, UK
Tel: 0181 5410035 Fax: 0181 5410623

Antec International Ltd., Chilton Industrial Estate, Sudbury, Suffolk CO10 6XD, UK.

Applied Precision Inc., 121 High Street, Marlborough, SN8 1LZ, UK.

Asahi Chemical Industry Co. Ltd., 54 Grosvenor Street, London, W1X 9FH, UK.
Asahi Chemical Industry Co. Ltd., Room 7412, Empire State Building, 350 Fifth Avenue, New York, NY 10118, USA.

Ashby Scientific Ltd., 11 Atlas Court, Coalville, Leicestershire, UK.

Astell Scientific, Powerscroft Road, Sidcup, Kent DA14 5EF, UK.

ATCC, see Appendix 2.

AutoQuant Imaging Inc., 877 25th St, Watervliet, NY 12189, USA.

Barnstead–Thermolyne, 2555 Kerper Boulevard, PO Box 797, Dubuque, Iowa 52004-0797, USA; supplied in the UK via Philip Harris Scientific.

BBL: from Becton Dickinson.

BDH Laboratory Supplies, Poole, Dorset BH15 1TD, UK.

Beckman Coulter (UK) Ltd, Oakley Court, Kingsmead Business Park, London Road, High Wycombe, Buckinghamshire HP11 1JU, UK
Tel: 01494 441181 Fax: 01494 447558
URL: http://www.beckman.com/
Beckman Coulter Inc., 4300 N. Harbor Boulevard, PO Box 3100, Fullerton, CA 92834–3100, USA
Tel: 001 714 8714848
Fax: 001 714 7738283
URL: http://www.beckman.com/

Becton Dickinson and Co., 21 Between Towns Road, Cowley, Oxford OX4 3LY, UK
Tel: 01865 748844
Fax: 01865 781627
URL: http://www.bd.com/
Becton Dickinson and Co., 1 Becton Drive, Franklin Lakes, NJ 07417-1883, USA
Tel: 001 201 8476800
URL: http://www.bd.com/

Bibby Sterilin Ltd., Tilling Drive, Stone, Staffordshire ST15 0SA, UK.
Bibby Sterilin Ltd., Corning, New York, NY, USA.

Bio 101 Inc., c/o Anachem Ltd, Anachem House, 20 Charles Street, Luton, Bedfordshire LU2 0EB, UK
Tel: 01582 456666 Fax: 01582 391768
URL: http://www.anachem.co.uk/
Bio 101 Inc., PO Box 2284, La Jolla, CA 92038-2284, USA
Tel: 001 760 5987299
Fax: 001 760 5980116
URL: http://www.bio101.com/

Bionique Laboratories, Saranac Lake, NY 12983, USA.

Bioptechs Inc., 3560 Beck Road, Butler, PA 16001, USA.

Bio-Rad Laboratories Ltd, Bio-Rad House, Maylands Avenue, Hemel Hempstead, Hertfordshire HP2 7TD, UK
Tel: 0181 3282000 Fax: 0181 3282550
URL: http://www.bio-rad.com/
Bio-Rad Laboratories Ltd, Division Headquarters, 1000 Alfred Noble Drive, Hercules, CA 94547, USA
Tel: 001 510 7247000
Fax: 001 510 7415817
URL: http://www.bio-rad.com/

BioReliance, Innovation Park, Hillfoots Road, Stirling, FK9 4NF, UK.
BioReliance, 14920 Broschart Road, Rockville, MD 20850-3349, USA.

Biovation Ltd., Auris Business Centre, 23 St. Machar Drive, Aberdeen AB24 3RY, UK.

Biovest International Inc., (formerly Cellex Biosciences Inc.), 8500 Evergreen Boulevard, Minneapolis, MN 55433-6000, USA.
Biovest International Inc., 4 Ince Avenue, Great Crosby, Merseyside L23 7FX, UK.

BioWhittaker, BioWhittaker House, 1 Ashville Way, Wokingham, Berkshire RG41 2PL, UK.
BioWhittaker, 8830 Biggs Ford Road, Walkersville, MD 21793-0127, USA.

Bocknek Ltd., 165 Bethridge Road, Rexdale, Toronto, Ontario, Canada, M9W 1N4.

Boehringer Mannheim, see Roche Diagnostics.

Browne Healthcare Ltd., Hamilton Industrial Park, Leicester LE5 1QZ, UK.

Calbiochem, 3 Heathcote Building, Highfields Science Park, University Boulevard, Nottingham NG7 2QJ, UK.
Calbiochem, 10933 N. Torrey Pines Road, La Jolla, CA 92037, USA.

Cambio, 34 Newnham Road, Cambridge CB3 9EY, UK.

Cambridge Bioscience, 25 Signet Court, Newmarket Road, Cambridge CB5 8LA, UK.

Cansera International Inc., 165a Bethridge Road, Rexdale, Ontario, Canada, M9W 1N4.

Cascade Biologics Inc., 4475 SW Scholls Ferry Road, Portland, OR 97225, USA.

Cellmark Diagnostics, Blacklands Ways, Abingdon Business Park, Abingdon, Oxon OX14 1DY, UK.

Cellmark Diagnostics, 20271 Goldenrod Lane, Suite 101, Germantown, Maryland 20876, USA.

Chemicon International, 28835 Single Oak Drive, Temecula, CA 92590, USA.

Chroma Technology Corp., 72 Cotton Mill Hill, Brattleboro, VT 05301, USA.

Clonetics Corp., 9620 Chesapeake Drive, San Diego, CA 92123, USA; products available via BioWhittaker.

Cohu Inc., Electronics Division, PO Box 85623, San Diego, CA 92186, USA.

Collaborative Biomedical Products BD Discovery Labware, see Becton Dickinson.

Corning Costar Scientific Products, 45 Nagog Park, Acton, MA 01720 USA.
Corning Costar Scientific Products, 1 The Valley Centre, Gordon Road, High Wycombe, Bucks HP13 6EQ, UK.

Coulter, see Beckman Coulter.

CP Instrument Co. Ltd, PO Box 22, Bishop Stortford, Hertfordshire CM23 3DX, UK
Tel: 01279 757711 Fax: 01279 755785
URL: http//:www.cpinstrument.co.uk/

Cryo-Med, 51529 Birch Street, New Baltimore, MI 48047, USA.
Cryo-Med, via Saxon Micro, PO Box 28, Newmarket, Suffolk CB8 8NY, UK.

CSL-JRH, see JRH Biosciences.

DataCell, Falcon Business Park, 40 Ivanhoe Road, Finchampstead, Berkshire RG40 4QQ, UK.

Decon Laboratories Ltd., Conway Street, Hove, Sussex BN3 3LY, UK.

Difco Laboratories, PO Box 14B, Central Avenue, West Molesey, Surrey KT8 2SE, UK.
Difco Laboratories, PO Box 1058A, Detroit, MI 48322, USA.

Dow Corning Corp./Dow Chemical Co., PO Box 0994, Midland, NI 48686-0994, USA.

Dupont (UK) Ltd, Industrial Products Division, Wedgwood Way, Stevenage, Hertfordshire SG1 4QN, UK
Tel: 01438 734000 Fax: 01438 734382
URL: http://www.dupont.com/
Dupont Co. (Biotechnology Systems Division), PO Box 80024, Wilmington, DE 19880–002, USA
Tel: 001 302 7741000
Fax: 001 302 7747321
URL: http://www.dupont.com/

Eastman Chemical Co., 100 North Eastman Road, PO Box 511, Kingsport, TN 37662–5075, USA
Tel: 001 423 2292000
URL: http//:www.eastman.com/

ECACC, see Appendix 2.

Expression Systems LLC, 1242-D Commercial Avenue, Woodland, CA 95776, USA.

Falcon Plastics: from Becton Dickinson.

Fisher Scientific UK Ltd, Bishop Meadow Road, Loughborough, Leicestershire LE11 5RG, UK
Tel: 01509 231166 Fax: 01509 231893
URL: http://www.fisher.co.uk/
Fisher Scientific, Fisher Research, 2761 Walnut Avenue, Tustin, CA 92780, USA
Tel: 001 714 6694600
Fax: 001 714 6691613
URL: http://www.fishersci.com/

Flow: see ICN Biomedicals.

Fluka, Indistriestrasse 35, CH-9470 Buchs, Switzerland.
Fluka, The Old Brickyard, New Road, Gillingham, Dorset SP8 4XT, UK.
Fluka, PO Box 2060, Milwaukee, WI 53201, USA
Tel: 001 414 2735013
Fax: 001 414 2734979
URL: http://www.sigma-aldrich.com/

Gelman Sciences, see Pall Gelman Laboratory.

Genentech Inc., 1 DNA Way, South San Francisco, CA 94080-4990, USA.

Genesis Service Ltd, Unit 10, Kings Park Industrial Estate, Kings Langley, Hertfordshire WD4 8ST, UK.

Gen-Probe Inc., 9880 Campus Point Drive, San Diego, CA 91212, USA.

Gibco-BRL, see Invitrogen.

Greiner Labortechnik, Maybachstrasse, D-7443 Frickenhausen, Germany.
Greiner Labortechnik, Brunel Way, Stroudwater Business Park, Stonehouse, Gloucestershire GL10 3SX, UK.

G.R.I., Gene House, Queenborough Lane, Rayne, Braintree, Essex, CM7 8TF, UK
Tel: 01376 332900
URL: http://www.gri.co.uk

Guest Medical Ltd., Wimborne House, 136 High Street, Sevenoaks, Kent TN13 1XA, UK.

Hamamatsu Photonics UK Ltd., Lough Point, 2 Gladbeck Way, Windmill Hill, Enfield, Middlesex EN2 7JA, UK.

Hays Chemical Distribution Ltd., 215 Tunnel Avenue, East Greenwich, London SE10 0QE, UK.

Henry Schein Rexodent, 25–27 Merrick Road, Southall, Middlesex UB2 4BR, UK.

Heraeus Equipment, see Kendro.

Hitachi cameras are available from DataCell.

Hi-Tech Ltd., Brunel Road, Salisbury, Wiltshire SP2 7PU, UK.

Hybaid Ltd, Action Court, Ashford Road, Ashford, Middlesex TW15 1XB, UK
Tel: 01784 425000
Fax: 01784 248085
URL: http://www.hybaid.com/
Hybaid US, 8 East Forge Parkway, Franklin, MA 02038, USA
Tel: 001 508 5416918
Fax: 001 508 5413041
URL: http://www.hybaid.com/

HyClone Laboratories, 1725 South HyClone Road, Logan, UT 84321, USA
Tel: 001 435 7534584
Fax: 001 435 7534589
URL: http://www.hyclone.com/
Hyclone Laboratories, Nelson Industrial Estate, Cramlington, Northumberland NE23 9BL, UK.

ICN Biomedicals, 1 Elmwood, Chineham Business Park, Basingstoke, Hampshire, UK.
ICN Biomedicals, 3300 Hyland Avenue, Costa Mesa, CA 92626, USA.

In Vitro Scientific Procucts, 2686 Johnson Drive, Ventura, CA 93003, USA.

Innovatis GmbH, Meisenstrasse 96, D-33607 Bielefeld, Germany; supplied in the UK via G.R.I.

Innovative Cell Technologies Inc., PO Box 13717, La Jolla, CA 92039, USA.

Innovative Chemistry Inc., PO Box 90, Marshfield, MA 02050, USA.

Intergen Company, The Center at Purchase, 2 Manhattanville Road, Purchase, NY 10577, USA.

Intergen Company, The Magdalen Centre, Oxford Science Park, Oxford OX4 4GA, Uk.

International Products Corporation, 1 Church Row, Chislehurst, Kent BR7 5PG, UK.
International Products Corporation, PO Box 70, Burlington, NJ 080106, USA.

Invitrogen Corp., 1600 Faraday Avenue, Carlsbad, CA 92008, USA
Tel: 001 760 6037200
Fax: 001 760 6037201
URL: http://www.invitrogen.com/
Invitrogen BV, PO Box 2312, 9704 CH Groningen, The Netherlands
Tel: 00800 53455345
Fax: 00800 78907890
URL: http://www.invitrogen.com/

Invitrogen, 3 Fountain Drive, Inchinnan Business Park, Paisley PA4 9RF, UK.
Invitrogen, 8400 Helgerman Court, PO Box 6009, Gaithersburg, MD 20884-9980, USA.

Irvine Scientific, 2511 Daimler Street, Santa Ana, CA 92705-5588, USA.

Jencons (Scientific) Ltd., Cherrycourt Way Industrial Estate, Stanbridge Road, Leighton Buzzard LU7 8UA, UK.

Johnson & Johnson Medical Ltd., Coronation Road, Ascot, Berkshire, UK.

JRH Biosciences, Smeaton Road, West Portway, Andover, Hampshire SP10 3LF, UK.
JRH Biosciences, 13804 Wet 107th Street, Lenexa, KS 66215, USA.

Kendro Laboratory Products; 9 Wates Way, Brentwood, Essex CM15 9TB, UK.
Kendro Laboratory Products, Postfach 1563, D-6450 Hanau 1, Germany.
Kendro Laboratory Products, 111A Corporate Boulevard, South Plainfield, NJ 07080, USA.

Kinetic Imaging Ltd., 2 Brunel Road, Croft Business Park, Bromborough, Wirral, Merseyside CH62 3NY, UK.

Krüss GmbH, Borsteler Chaussee 85-99a, D-22453 Hamburg, Germany.

Kyokuto Pharmaceutical Industrial Co., Honmachi 3-1-1, Nihonbachi, Chuoh-ku, Tokyo 103, Japan.

Lab Impex, 111-113 Waldegrave Road, Teddington, Misslesex TW11 811, UK.

Laboratories Phagogene, Etablissement Pharmaceutique no. F85/56, Z. I. de Carros, BP 128, F-06513 Carros cedex, France.

Lancer UK Ltd., 1 Pembroke Avenue, Waterbeach, Cambridge CB5 9QR, UK.

LEEC Ltd., Private Road 7, Colwick Industrial Estate, Nottingham NG4 2AJ, UK.

Leica Microsystems UK Ltd., Davy Avenue, Knowlhill, Milton Keynes MK5 8LB, UK.

Life Technologies, see Invitrogen.

Linde, see Union Carbide.

MatTek, 200 Homer Ave., Ashland, MA 01721, USA.

Medical Systems, 1 Plaza Road, Greenvale, NY 11548, USA.

Merck Ltd., Hunter Boulevard, Magna Park, Lutterworth, Leicestershire LE17 4XN, UK.

Merck Sharp & Dohme Research Laboratories, Neuroscience Research Centre, Terlings Park, Harlow, Essex CM20 2QR, UK
URL: http://www.msd-nrc.co.uk/
MSD Sharp and Dohme GmbH, Lindenplatz 1, D-85540, Haar, Germany
URL: http://www.msd-deutschland.com/

Metrex Research Corp., Parker, CO 80134, USA.

Millipore (UK) Ltd, The Boulevard, Blackmoor Lane, Watford, Hertfordshire WD1 8YW, UK
Tel: 01923 816375 Fax: 01923 818297
URL: http://www.millipore.com/local/UKhtm/
Millipore Corp., 80 Ashby Road, Bedford, MA 01730, USA
Tel: 001 800 6455476
Fax: 001 800 6455439
URL: http://www.millipore.com/

Minerva Biolabs, Kopenickerstrasse 325, D-12555, Berlin, Germany.

Minnesota Valley Engineering, 407 7th Street NW, New Prague, MN 56071, USA.

Molecular Probes Inc., 4849 Pitchford Ave., Eugene, OR 97402-9165, USA.

Nalge Nunc International (NNI), Foxwood Court, Rotherwas, Hereford HR2 6JQ, UK.
Nalge Nunc International (NNI), 75 Panorama Creek Drive, Box 20365, Rochester, NY 14602-0365, USA.

National Collection of Type Cultures, PHLS Central Public Health Laboratory, 61 Colindale Avenue, London NW9 5HT, UK.

New England Biolabs, 32 Tozer Road, Beverley, MA 01915–5510, USA
Tel: 001 978 9275054

Nikon Inc., 1300 Walt Whitman Road, Melville, NY 11747–3064, USA
Tel: 001 516 5474200
Fax: 001 516 5470299
URL: http://www.nikonusa.com/
Nikon Corp., Fuji Building, 2–3, 3-chome, Marunouchi, Chiyoda-ku, Tokyo 100, Japan
Tel: 00813 32145311
Fax: 00813 32015856
URL: http://www.nikon.co.jp/main/index_e.htm/
Nikon UK Ltd., Nikon House, 380 Richmond Road, Kingston-upon-Thames, Surrey KT2 5PR, UK.

Nucleopore, Pleasanton, CA 94588-8008, USA; in UK via Corning Costar.

Nunc, see *Nalge Nunc International*; in the UK and numerous other countries, Nunc cultureware is also available via Life Technologies.

Nycomed Amersham Imaging, Amersham Labs, White Lion Rd, Amersham, Buckinghamshire HP7 9LL, UK
Tel: 0800 558822 (or 01494 544000)
Fax: 0800 669933 (or 01494 542266)
URL: http//:www.amersham.co.uk/
Nycomed Amersham, 101 Carnegie Center, Princeton, NJ 08540, USA
Tel: 001 609 5146000
URL: http://www.amersham.co.uk/

Olympus Optical Company (UK) Ltd., 2-8 Honduras Street, London EC1Y 0TX, UK.

Omega Optical Inc., 3 Grove St., Brattleboro, VT 05301, USA.

Oncor, see Intergen.

Oxoid, Wade Road, Basingstoke, Hampshire RG24 8PW, UK.

P.A.A. Laboratories, Wienerstrasse 131, A-4020 Linz, Austria.
P.A.A. Laboratories, 1 Technine, Guard Avenue, Houndstone Business Park, Yeovil, Somerset BA22 8YE, UK.
P.A.A. Laboratories, 2570 Route 724, PO Box 435, Parker Ford, PA 19457, USA.

Pall Process Filtration/Pall Gelman Laboratory, Europa House, Havant Street,

Portsmouth PO1 3PD, UK.
Pall Process Filtration/Pall Gelman Laboratory, 600 S. Wagner Road, Ann Arbor, MI 48103-9019, USA.

Panvera Corporation, 545 Science Drive, Madison, WI 53711, USA.

Perkin Elmer Ltd, Post Office Lane, Beaconsfield, Buckinghamshire HP9 1QA, UK
Tel: 01494 676161
URL: http//:www.perkin-elmer.com/

Pharmacia, Davy Avenue, Knowlhill, Milton Keynes, Buckinghamshire MK5 8PH, UK (also see Amersham Pharmacia Biotech)
Tel: 01908 661101 Fax: 01908 690091
URL: http//www.eu.pnu.com/
Pharmacia, 800 Centennial Avenue, Piscataway, NJ 08854, USA.

Philip Harris Scientific, Novara House, Excelsior Road, Ashby Park, Ashby-de-la-Zouche, Leicestershire LE65 1NG, UK.

Polaroid Corp., 784 Memorial Drive, Cambridge, MA 02139, USA.
Polaroid Corp., Polaroid Europe, Wheathampstead House, Codicote Road, Wheathampstead, Hertfordshire AL4 8SF, UK.

Princeton Instruments Ltd., (Roper Scientific), PO Box 1192, 43 High Street, Marlow, Buckinghamshire SL7 1GBB, UK.

Promega UK Ltd, Delta House, Chilworth Research Centre, Southampton SO16 7NS, UK
Tel: 0800 378994 Fax: 0800 181037
URL: http://www.promega.com/
Promega Corp., 2800 Woods Hollow Road, Madison, WI 53711–5399, USA
Tel: 001 608 2744330
Fax: 001 608 2772516
URL: http://www.promega.com/

PromoCell GmbH, Handschuhsheimer Landstrasse 12, D-69120 Heidelberg, Germany.

Protein Polymer Technologies, 10655 Sorrento Valley Road, San Diego, CA 92121, USA.

Pulnix Europe Ltd., Pulnix House, Aviary Court, Wade Road, Basingstoke, Hampshire RG24 8PE, UK.

Qiagen UK Ltd, Boundary Court, Gatwick Road, Crawley, West Sussex RH10 2AX, UK
Tel: 01293 422911 Fax: 01293 422922
URL: http://www.qiagen.com/
Qiagen Inc., 28159 Avenue Stanford, Valencia, CA 91355, USA
Tel: 001 800 4268157
Fax: 001 800 7182056
URL: http://www.qiagen.com/

Q-One Biotech Ltd., Todd Campus, West of Scotland Science Park, Glasgow G20 0XA, UK.

R&D Systems, 4-10 The Quadrant, Barton Lane, Abingdon, Oxfordshire OX14 3VS, UK.
R&D Systems Inc., 614 McKinley Place NE, Minneapolis, MN 55413, USA.

Research Instruments, Kernick Road, Penryn, Cornwall TR10 9DQ, UK.

Roche Diagnostics Ltd, Bell Lane, Lewes, East Sussex BN7 1LG, UK
Tel: 0808 1009998 (or 01273 480044)
Fax: 0808 1001920 (01273 480266)
URL: http://www.roche.com/
Roche Diagnostics Corp., 9115 Hague Road, PO Box 50457, Indianapolis, IN 46256, USA
Tel: 001 317 8452358
Fax: 001 317 5762126
URL: http://www.roche.com/
Roche Diagnostics GmbH, Sandhoferstrasse 116, 68305 Mannheim, Germany
Tel: 0049 621 7594747
Fax: 0049 621 7594002
URL: http://www.roche.com/

Roche Molecular Biochemicals, see Roche Diagnostics.

RS, PO Box 99 Corby, Northamptonshire NN17 9RS, UK.

Rubbermaid, 1147 Akron Road, Wooster, Ohio 44691-6000, USA.

Salzman Corp., 308 East River Drive, Davenport, Iowa 52801, USA.

Sartorius, Longmead Business Centre, Blenheim Road, Epsom, Surrey KT19 9QN, UK.
Sartorius North America Ind., 140 Wilbur Place, Bohemia, Long Island, NY 11716, USA.

Schleicher and Schuell Inc., Keene, NH 03431A, USA
Tel: 001 603 3572398

Schott Glasswerke, Geschäftsbereich Chemie, Produktgruppe Laborglas, Postfach 2480, D-6500 Mainz, Germany.
Schott Glass Ltd, Drummond Road, Aston Fields Industrial Estate, Stafford ST16 3EL, UK; Schott glassware is also sold by distributors, e.g. Fisher or Merck (see above).

Sera-Lab Ltd., Hophurst Lane, Crawley Down, Sussex RH10 4FF, UK.

Serologicals Proteins Inc., 195 West Birch Street, Kankakee, IL 60901, USA.
Serologicals Proteins Inc., Waterbeach, Cambridge CB5 9PS, UK.

SGI, 1530 Arlington Business Park, Theale, Reading, Berkshire RG7 4SB, UK.

Shandon Scientific Ltd, 93–96 Chadwick Road, Astmoor, Runcorn, Cheshire WA7 1PR, UK
Tel: 01928 566611
URL: http//www.shandon.com/

Shawcity Ltd., Units 12/13, Pioneer Road, Faringdon, Oxfordshire SN7 7BU, UK.

Sigma–Aldrich Co. Ltd, The Old Brickyard, New Road, Gillingham, Dorset SP8 4XT, UK
Tel: 0800 717181 (or 01747 822211)
Fax: 0800 378538 (or 01747 823779)
URL: http://www.sigma-aldrich.com/
Sigma Chemical Co., PO Box 14508, St Louis, MO 63178, USA
Tel: 001 314 7715765
Fax: 001 314 7715757
URL: http://www.sigma-aldrich.com/

Solmedia Laboratory Suppliers, 6 The Parade, Colchester Road, Romford, Essex RM3 0AQ, UK.

Sony cameras are available from DataCell.

South Pacific Sera Ltd., Blue Cliffs, RD 24, Timaru, New Zealand.

Sterling Medicare, Onslow Street, Guildford, Surrey GU1 4YS, UK.

Stratagene Inc., 11011 North Torrey Pines Road, La Jolla, CA 92037, USA
Tel: 001 858 5355400
URL: http://www.stratagene.com/
Stratagene Europe, Gebouw California, Hogehilweg 15, 1101 CB Amsterdam Zuidoost, The Netherlands
URL: http://www.stratagene.com/

Stratagene, Cambridge Innovation Centre, Cambridge Science Park, Milton Road, Cambridge CB4 4GF, UK.
Stratagene, 11099 North Torrey Pines Road, La Jolla, CA 92037, USA.

SVI, Scientific Volume Imaging B.V., Alexanderlaan 14, 1213 XS Hilversum. The Netherlands.

Takara Biomedical; in Europe – Europarc des Barbanniers 6, Place du Village, F-92230 Gennevilliers, France; in UK via BioWhittaker; In USA via Intergen.

Taylor-Wharton, PO Box 5685, Theodore, AL 36590, USA; in UK via Jencons.

TCS CellWorks Ltd., Botolph Claydon, Buckingham MK18 2LR, UK.
Tel: 01296 713120

Technical Video Ltd., PO Box 693, Woods Holes, MA 02543, USA.

Till Photonics GmbH, Lena-Christ-Str. 44, D-82152 Martinsried, Germany.

UBI, 199 Saranac Avenue, Lake Placid, NY 12946, USA.

Union Carbide, Somerset, NJ 08873, USA.

United States Biochemical (USB), PO Box 22400, Cleveland, OH 44122, USA
Tel: 001 216 4649277

Universal Imaging Corporation, 502 Brandywine Parkway, West Chester, PA 19380, USA.

Upjohn Co., Kalamazoo, MI 49001, USA.

V.A. Howe & Co Ltd.., Beaumont Close, Banbury, Oxon OS16 7RG, UK.

Ventrex, from JRH Biosciences.

Weber Scientific International, Marlborough Road, Lancing Business Park, West Sussex BN15 8TN, UK.

Whatman, St Leonard's Road, 20/20 Maidstone, Kent ME16 0LS, UK.
Whatman, 5285 NE Elan Young Parkway, Suite A-400, Hillboro, Oregon 97124, USA.

William Pearson Ltd., Clough Road, Hull HU6 7QA, UK.

Worthington Biochemical Corp., Freehold, NJ 07728, USA.

Zeiss, Carl Zeiss Ltd., PO Box 78, Woodfield Road, Welwyn Garden City, Hertfordshire AL7 1LU, UK.

Appendix 2
Culture collections and other resource centres

American Type Collection (ATCC), 10801 University Bl., Manassas, VA 20110, USA.
Telephone: +1 703 365 2700
Fax: +1 703 365 2701
Website: www.atcc.org

The ATCC has a large and comprehensive stock of material built up over a considerable time. A certain number of deposits carry 'certified cell line' status and extensive testing programmes have been carried out on this material. The whole of the collection comprises over 4000 cell lines and hybridomas from approximately 150 different species.

DSMZ — Deutsche Sammlung Von Mikro-organismen und Zellkulturen GmbH, Mascheroder Weg 1B, D-3300 Braunschweig, Germany.
Telephone: +49 531 618760
Fax: +49 531 618750
Website: www.dsmz.de

Description of holdings: over 400 cell lines and hybridomas.

European Collection of Cell Cultures (ECACC), Centre for Applied Microbiology and Research, Porton Down, Salisbury SP4 0JG, UK.
Telephone: +44 1980 612512
Fax: +44 980 611315
Website: www.ecacc.org.uk

Description of holdings: 1500 cell lines and hybridomas, 360 HLA-defined cell lines, and over 25000 human genetic and chromosome abnormality cell lines.

Interlab Cell Line Collection, Servizio Biotechnologie, Instituto Nazionale per la Ricerca sul Cancro, Largo Rosanna Benzi 10, 16132 Genova, Italy.
Telephone: +39 105737 474
Fax: +39 105737 295
Website: www.biotech.ist.unige.it

Description of holdings: 200 cell lines and hybridomas.

National Institute of General Medical Sciences (NIGMS), Human Genetic Mutant Cell Repository and National Institute on Aging Cell Culture Repository (NIA), Coriell Institute for Medical Research, 401 Haddon Avenue, Camden, NJ 08103, USA.
Telephone: +1 609 757 4848
Fax: +1 609 964 0254
Website: locus.umdnj.edu.ngms

Description of holdings: 5270 cell cultures and 275 DNA samples mainly derived from patients with genetic and chromosome abnormalities.

Riken Gene Bank, 3-1-1 Yatabe, Koyadai, Tsukuba Science City 305, Japan.
Telephone: +81 2983 63611
Fax: +81 2983 69130

Description of holdings: 300 cell lines and hybridomas.

Appendix 3
Online resources for cell biology

Julian A. T. Dow
IBLS Division of Molecular Genetics, Anderson College, University of Glasgow, Glasgow G11 6NU, UK.

Introduction

Cell biology is a diverse field that embodies a challenging range of approaches and techniques. Yet surely the most universal of those techniques is now the use of Internet resources. This author has visited the subject in the past (1); but it is fair to assume that, by now, all cell biologists have either adapted to the Internet or left the field. Accordingly, I will go easy on the tutorial aspects, and concentrate on categorizing, listing, and describing a range of resources that are likely to prove useful to practicing cell biologists. In this way, the reader can get quickly to areas of interest. There is also a glossary of Internet terms. The links are available at the site below, so that you don't need to type the URLs individually. There are 171 URLs in this article (about 15/page), so if at least one or two of them turn out to be of value to you, I hope you will consider the article to have been worth reading!

Discussion groups

I will begin with discussion groups, as they are often overlooked. Although less glossy than web sites, the text-only discussion groups are a very useful source of information and help. Accessible both through dedicated software and through the major Internet browsers, there are several thousand Internet discussion groups. Of these, the most important are the bionet newsgroups. Subscribing to newsgroups isn't as straightforward as surfing the web, and so many readers may not have taken the plunge. Accordingly, I reproduce here the full list of bionet newsgroups (*Table 1*). Although most readers will be aware of the existence of bionet, the full range of newsgroups available is remarkable.

Before using the newsgroups, you should read the documentation and mini-FAQ (http://www.bio.net/), which describes permissible postings, etiquette, and access to the archives. In essence, you should:

- search the archives before posting a request for information or help, as many questions crop up repeatedly

(The links given as underlined URLs in this article are available in http://www.mblab.gla.ac.uk/tubules/basiccellculture.html)

- not try to advertise products on the sly, pretending it is a posting
- not insult people or say anything libelous
- keep your postings short and clear
- not cross-post to multiple newsgroups, but rather only to those central to your topic

As I alluded, to subscribe to newsgroups requires that you either have specialist newsreader software (like Newsreader on the Mac); or that you have correctly configured your browser with the IP name of your local news (or NNTP) server. You may need to contact your computing service for this. You will then need to visit the newsgroups regularly, because many servers keep only a week's worth of postings. A far simpler procedure – and one preferred by the bionet organizers – is to visit the archive online at http://www.bio.net. In this way, you can stick with the familiar web-browser interface.

Table 1 Full listing and description of Bionet newsgroups (from http://www.bio.net/archives.html)

Newsgroup	Description
ACEDB/bionet.software.acedb	Discussions by users of genome DBs using ACEDB.
ADDRESSES/bionet.users.addresses	Who's who in Biology
AFCR/bionet.prof-society.afcr	American Federation for Clinical Research announcements (Moderated)
AGEING/bionet.molbio.ageing	Research into cellular and organismal ageing.
AGROFORESTRY/bionet.agroforestry	Agroforestry research.
AIBS/bionet.prof-society.aibs	American Institute of Biological Sciences announcements. (Moderated)
AMYLOID/bionet.neuroscience.amyloid	Research on Alzheimer's disease and related disorders (Moderated)
ANNELIDA/Prototype	Discussions of the scientific study of Phylum *Annelida*
ARABIDOPSIS/bionet.genome.arabidopsis	Information about the *Arabidopsis* project (Moderated)
AUDIOLOGY/bionet.audiology	Research on audiology and hearing science (Moderated)
AUTOMATED-SEQUENCING/bionet.genome.autosequencing	Research and support on automated DNA sequencing (Moderated)
BIGBLUE/Prototype	Discussions between researchers who use transgenic animal systems for mutation assays.
BIO-MATRIX/bionet.molbio.bio-matrix	Computer applications to biological databases.
BIO-SOFTWARE/bionet.software	Information about software for biology.
BIO-SRS/bionet.software.sts	Discussions about Sequence Retrieval System (SRS) software
BIO-WWW/bionet.software.www	WWW resources for biologists. (Moderated)
BIOCAN/bionet.prof-society.cfbs	Newsgroup for the Canadian Federation of Biological Societies (Moderated)
BIOFILMS/bionet.microbiology.biofilms	Research on microbial biofilms (Moderated)
BIOFORUM/bionet.general	General BIOSCI discussion.
BIOHERV/Prototype	Discussions about human endogenous retroviral elements.
BIOLOGICAL-INFORMATION-THEORY/bionet.info-theory	Biological information theory.
BIONEWS/bionet.announce	Announcements of widespread interest to biologists. (Moderated)
BIOPHYSICAL-SOCIETY/bionet.prof-society.biophysics	Biophysical Society announcements. (Moderated)
BIOPHYSICS/bionet.biophysics	The science and profession of biophysics.
BIOTECHNIQUES/bionet.journals.letters.biotechniques	Discussion of articles from the journal Biotechniques. (Moderated)
BTK-MCA/bionet.metabolic-reg	Kinetics and thermodynamics at the cellular level.
CARDIOVASCULAR-RESEARCH/bionet.biology.cardiovascular	Research discussions between scientists engaged in cardiovascular research
CELEGANS/bionet.celegans	Research discussions on *Caenorhabditis elegans* and related nematodes (Moderated)

Table 1 Continued

CELL-BIOLOGY/bionet.cellbiol
Cell biology research.
CHLAMYDOMONAS/bionet.chlamydomonas
Research on *Chlamydomonas* and other green algae (Moderated)
CHROMOSOMES/bionet.genome.chromosomes
Mapping/sequencing of eukaryote chromosomes.
COMPUTATIONAL-BIOLOGY/bionet.biology.computational
Computer and mathematical applications. (Moderated)
CSM/bionet.prof-society.csm
Canadian Society of Microbiologists announcements. (Moderated)
CYTONET/bionet.cellbiol.cytonet
Cytology research.
DEEPSEA/bionet.biology.deepsea
Research in deep-sea marine biology, oceanography, and geology (Moderated)
DIAGNOSTICS/bionet.diagnostics
Problems and techniques in all fields of diagnostics (Moderated)
DROS/bionet.drosophila
Research into the biology of fruit flies.
ECOPHYSIOLOGY/bionet.ecology.physiology
Research and education in physiological ecology (Moderated)
EMBL-DATABANK/bionet.molbio.embldatabank
Info about the EMBL, nucleic acid database.
EMF-BIO/bionet.emf-bio
Interactions of EM fields with biological systems. (Moderated)
EMPLOYMENT-WANTED/bionet.jobs.wanted
Requests for employment in the biological sciences.
EMPLOYMENT/bionet.jobs.offered
Job openings in the biological sciences. (Moderated)
FLUORESCENT-PROTEINS/bionet.molbio.proteins.fluorescent
Research on fluorescent proteins and bioluminescence
FREE-RADICALS/bionet.molecules.free-radicals
Research on free radicals in biology and medicine. (Moderated)
G-PROTEIN-COUPLED-RECEPTOR/bionet.molbio.proteins.7tms_r
Research on G-protein coupled receptor systems (Moderated)
GDB/bionet.molbio.gdb
Messages to and from the GDB database staff.
GENBANK-BB/bionet.molbio.genbank
Info about the GenBank nucleic acid database.
GENETIC-LINKAGE/bionet.molbio.gene-linkage
Research into genetic linkage analysis.
GENSTRUCTURE/bionet.genome.gene-structure
Genome and chromatin structure and function (Moderated)
GLYCOSCI/bionet.glycosci
Research issues re carbohydrate and glycoconjugate molecules.
GRASSES-SCIENCE/bionet.biology.grasses
Research into the biology of grasses.
HIV-BIOL/bionet.molbio.hiv
Research into the molecular biology of HIV.
HUMAN-GENOME-PROGRAM/bionet.molbio.genome-program
Discussions regarding the international Human Genome Project (Moderated)
IMMUNOLOGY/bionet.immunology
Research in immunology.
INFO-GCG/bionet.software.gcg
Discussions about using the GCG software.
INSULIN-ACTION/bionet.cellbiol.insulin
Biology and chemistry of insulin and related receptors (Moderated)
JRNLNOTE/bionet.journals.note
Advice on dealing with journals in biology.
MAIZE/bionet.maize
Research on maize (Moderated)
METHDS-REAGNTS/bionet.molbio.methds-reagnts
Requests for information and lab reagents.
MICROBIOLOGY/bionet.microbiology
The science and profession of microbiology.
MOLECULAR-EVOLUTION/bionet.molbio.evolution
Discussions about research in molecular evolution (Moderated)
MOLECULAR-MODELLING/bionet.molec-model
Physical and chemical aspects of molecular modelling.
MOLECULAR-REPERTOIRES/bionet.molecules.repertoires
Generation and use of libraries of molecules. (Moderated)
MOLLUSC-MOLECULAR-NEWS/bionet.molbio.molluscs
Research on mollusc DNA. (Moderated)
MUTATION/Prototype
Discussions about mutation research.
MYCOLOGY/bionet.mycology
Research on mycology (Moderated)
N2-FIXATION/bionet.biology.n2-fixation
Biological nitrogen fixation research.
NAVBO/bionet.prof-society.navbo
Forum for the North American Vascular Biology Organization. (Moderated)
NEUROSCIENCE/bionet.neuroscience
Research issues in the neurosciences.
P450/bionet.molecules.p450
Research on cytochrome P450
PARASITOLOGY/bionet.parasitology
Research into parasitology.
PEPTIDES/bionet.molecules.peptides
Research involving peptides (Moderated)
PHARMACEUTICAL-BIOTECHNOLOGY/Prototype
Discussion about research in pharmaceutical biotechnology.
PHOTOSYNTHESIS/bionet.photosynthesis
Research into photosynthesis. (Moderated)

Table 1 Continued

Newsgroup	Description
PLANT-BIOLOGY/bionet.plants	Research into plant biology.
PLANT-EDUCATION/bionet.plants.education	Education issues in plant biology
PLANT-SIGNAL-TRANSDUCTION/bionet.plants.signaltransduc	Research on plant signal transduction (Moderated)
POPULATION-BIOLOGY/bionet.population-bio	Population biology research.
PRENATAL-DIAGNOSTICS/bionet.diagnostics.prenatal	Research in prenatal diagnostics
PROTEIN-ANALYSIS/bionet.molbio.proteins	Research on proteins and protein databases.
PROTEIN-CRYSTALLOGRAPHY/bionet.xtallography	Research into protein crystallography,
PROTISTA/bionet.protista	Discussion on protists (protozoa, algae, zoosporic fungi) (Moderated)
PSEUDOMONADS/bionet.organisms.pseudomonas	Research on the genus *Pseudomonas*.
RADIATION-ONCOLOGY/Prototype	Radiation Oncology journal club (moderated list)
RAPD/bionet.molbio.rapd	Research on Randomly Amplified Polymorphic DNA.
RECOMBINATION/bionet.molbio.recombination	Research on the recombination of DNA or RNA (Moderated)
RNA/Prototype	Discussions about RNA editing, RNA splicing, and ribozyme activities of RNA.
RUST-MILDEW/Prototype	Research about the biotrophic foliar fungal diseases of cereals, including rust, powdery mildews and downy mildews.
SCHISTOSOMA/bionet.organisms.schistosoma	Discussions about *Schistosoma* research (Moderated)
SCIENCE-RESOURCES/bionet.sci-resources	Information about funding agencies, etc. (Moderated)
STADEN/bionet.software.staden	The Staden molecular sequence analysis software.
STRUCTURAL-NMR/bionet.structural-nmr	Structural NMR of macromolecules (Moderated)
SYMBIOSIS-RESEARCH/bionet.biology.symbiosis	Research in symbiosis (Moderated)
TIBS/bionet.journals.letters.tibs	Letters to the Editor of Trends in Biochemical Science (Moderated)
TOXICOLOGY/bionet.toxicology	Research in toxicology
TROPICAL-BIOLOGY/bionet.biology.tropical	Research in tropical biology.
TWADDLE/bionet.twaddle	
URODELES/bionet.organisms.urodeles	Research on urodele amphibians. (Moderated)
VECTOR-BIOLOGY/bionet.biology.vectors	Research and control of arthropods which transmit disease. (Moderated)
VIROLOGY/bionet.virology	Research into virology. (Moderated)
WOMENINBIOLOGY/bionet.women-in-bio	Discussion of issues related to women in biology (Moderated)
X-PLOR/bionet.software.x-plor	X-PLOR software for 3D macromolecular structure determination
YEAST/bionet.molbio.yeast	Molecular Biology and Genetics of Yeast (Moderated)
ZBRAFISH/bionet.organisms.zebrafish	Discussion group for Zebrafish (Daniorerio) researchers (Moderated)

It is fair to point out that the newsgroups attract widely varying levels of activity. In addition, not all are moderated, leaving them open to junk postings. By far the most active, and useful to a cell biologist seeking help, is **bionet.molbio.methds-reagnts**. On here, one can expect to see plaintive pleas for help with ligations, detailed discussion about genetic backgrounds of cell lines for transfection, or information about GFP tagging proteins. The general cell biology discussion group, **bionet.cellbiol**, is considerably less active, though potentially more relevant. Unfortunately, this newsgroup is not moderated, so there are more get-rich-quick cross-postings than biological. More active is the neuroscience board (**bionet.neurosci**), with very wide coverage. Bionet is also a useful place for model organisms – for example, all the official *Drosophila* genome-related messages and meeting announcements are posted at **bionet. drosophila**. Conferences, calls for papers and similar are posted at

bionet.announce, and the jobs newsgroups (**bionet.jobs.offered** and **bionet.jobs.wanted**) do a lively and useful trade.

More general laboratory support is available from the **comp/sys** newsgroups. These are perhaps the best way to seek help and advice related to computing. Viruses and bugs are often spotted and dealt with in hours, and advice on computer systems and peripherals is freely given.

Meta-indices and search engines

The function of a meta-index is to compile a list in a cognate area. To some extent, this article and its website (http://www.mblab.gla.ac.uk/tubules/basiccellculture.html) are meta-indices for cell biology, but are necessarily frozen in time by the need to publish hardcopy. As online resources can be updated regularly, they do not suffer from this limitation. Judicious use of a couple of meta-indices can help greatly in answering relevant questions, as an alternative to the whole-Internet search engines. In this section, I list a few of my favourites; you are free to browse them and bookmark those that seem most useful to you.

The **WWW virtual library of cell biology** (http://vl.bwh.harvard.edu/) is the central index for this area. Analogous indices are available for **biosciences** in general (http://vlib.org/Biosciences.html), **genetics** (http://www.ornl.gov/TechResources/Human_Genome/genetics.html), **developmental biology** (http://sdb.bio.purdue.edu/_Other/_VL_DB.html), and **neurobiology** (http://neuro.med.cornell.edu/VL/).

Cell and Molecular Biology Online (http://www.cellbio.com/) is a site aimed specifically at cell biologists. It is divided into four sections: *research* covers protocols and overviews of current research, *communication* covers professional societies and jobs, *education* covers general educational links, and *random*, background information and new sites.

The **Bio-Web** (http://cellbiol.com/ – notice the single letter difference from the preceding URL!) serves a very similar purpose, with links to meetings, protocols and companies, downloadable software, sequence searches, and electronic journals.

Biotechnology Information Directory (http://www.cato.com/biotech) is hosted by Cato Research, and although providing a useful meta-index, it has an understandable emphasis on clinical drug trials.

The **Biotool kit** (http://www.biosupplynet.com) is another portal to supply companies, with the support of Cold Spring Harbor Laboratories.

BioMedNet (http://www.biomednet.com/) tries to make their resource more of a club. You need to sign in (although registration is free) and then have access to a collection of over 170 journals. You need to provide your credit card details, and pay real money for any articles you download, but the convenience may outweigh the cost. The site also includes job pages, annotated links to other resources, evaluated Medline, and BioSupplyNet, an attempt to automate supplier searching and ordering.

A crude but effective strategy to find something - or someone - on the Internet is to use a search engine (such as Lycos, Excite, Hotbot, or Alta Vista). These are surprisingly effective at turning up hits even for specialized scientific topics, and I'd also recommend this strategy to find where pages have moved to, should any of the links in this article go 'dead' with time. However, each engine has coverage of at best 10% of all Internet pages, and they are not equally up to date. So it is usually necessary to repeat your search with several engines. This topic is covered in depth at http://dir.yahoo.com/Computers_and_Internet/Internet/World_Wide_Web/Searching the_Web/Comparing_Search_Engines/.

An alternative is to use a meta-search engine, which sends your search to multiple search engines, and curates and presents the results to you, in about the same time that any individual search engine would have taken. I highly recommend this approach!

Google (http://www.google.com) is my favourite, but you can also experiment with:

Metacrawler http://www.metacrawler.com

SavvySearch http://www.savvysearch.com/

Dogpile http://www.dogpile.com/

Inference Find http://www.infind.com/

ProFusion http://www.profusion.com/

Mamma http://www.mamma.com/

The Big Hub http://www.thebighub.com/

For the ultimate abstraction, you can visit

http://www.searchenginewatch.com/links/Metacrawlers/, which is a meta-index of available meta-indices!

Protocols

In cell culture, reliable protocols are invaluable. Most labs keep a dog-eared file of protocols, and many have hit on the idea of mounting it on the group's web server, so avoiding stains and wear. Many of these protocols are thus available on the web, and have often been collated into meta-indices. Of course. there is no guarantee that these protocols are effective, but if you know of the lab and respect their output, it seems reasonable to trust their protocols too.

Protocol-online (http://www.protocol-online.net/) has a very useful collection of protocols in cell and molecular biology, and immunology. The cell biology section contains protocols for cell culture, cell adhesion and motility, microscopy, cell cycle, organelles and cell parts, cytogenetics, transgenics, and extracellular matrix.

Cell Biology Laboratory Manual of Gustavus Adolphus College http://www.gac.edu/cgi-bin/user/~cellab/phpl?index-1.html is a useful listing of miscellaneous cell biology techniques.

Cell and Molecular Protocols (http://www.cellbio.com/protocols.html) is a useful curation of protocols from the Cell and Molecular Biology Online website, discussed later.

On a more narrow basis, there is an excellent collection of protocols for baculovirus work at the eponymous website (http://www.baculovirus.com).

There is a useful collection of sequencing and DNA protocols curated by the Roe lab at http://www.genome.ou.edu/proto.html.

Biological Procedures Online (http://sciborg.uwaterloo.ca/bpo/) from the University of Waterloo in Canada, provides a new twist to online protocols. This is a peer-reviewed journal that is available free-of-charge on the Internet. The coverage is obviously broader than cell culture *per se*, but it makes for interesting reading nonetheless.

Similarly, **Molecular Biology Today** provides a list of molecular biology protocols online (http://www.horizonpress.com/gateway/protocols.html).

Another journal that is familiar in its free or library incarnations is **BioTechniques** (http://www.biotechniques.com/). It provides a useful source of wizard techniques for making PCR subtraction libraries, and similar molecular virtuosity. Searching the full-text online is free, but you need to be a subscriber to the US commercial edition to view the articles.

The University of Michigan's **transgenic animal web resource** hosts a list of laboratory protocols (http://www.med.umich.edu/tamc/protocols.html) for mouse work. Here for example, you can find protocols for DNA purification for micro-injection, lacZ tissue staining, and preparation of DNA from tail biopsies. Perhaps most usefully for the neophyte, there are sensible timelines for transgenic and gene-targeted mouse production and analysis.

Microscopy and image analysis

Microscopy is intrinsic to cell culture, if only to check on health and confluence. The area is reasonably well served by web sites. The **WWW virtual library of microscopy** hosts a techniques page (http://www.ou.edu/research/electron/www-vl/), although this covers more than just biology. The **Microscopy Society of America** (http://www.msa.microscopy.com/) is a learned society that offers a useful page with links.

Douglas Kline (http://www.kent.edu/projects/cell/INDEX.HTM) runs a site, '**views of the cell**' with some nice fluorescence images of cultured cells, together with resources for educators. One of the most potent image analysis packages available is actually free. **NIH Image** (http://rsb.info.nih.gov/nih-image/) is a magnificent, multi-talented beast that bears comparison with very expensive commercial packages. Image input can be through scanners or a select list of cameras that the authors consider to be up to the job. There is an extensive macro language for customization, and there is a range of third-party macros for purposes like gel densitometry and particle counting. The package runs on both the Mac and Windows platforms.

The *supply companies* section (later) lists companies that supply microscopes and microscopy materials.

Genome projects

Cell culture provides a simple model system that frees the experimenter from the vagaries of whole organism biology. However, the future of biology is now post-genomic, and the intoxicating mixture of genomics and genetics offers really exciting prospects for the re-integration of cell and molecular approaches into whole organism biology - the so-called 'integrative' biology. This section provides pointers to some useful meta-indices of genome project resources, together with some tutorial sites if needed.

If your interest in genome projects extends beyond mouse and human, then the **WWW virtual library of genetics** (http://www.ornl.gov/TechResources/Human_Genome/genetics.html) is an ideal starting point. It contains curated resources for Bacteria, *C. elegans*, Cattle, Dog, *Drosophila*, *E. coli*, Fish, Frog, Fungus, Horse, Human, Marsupial, Microbial, Mosquito, Nematode, Parasite, Plant, Poultry, Rodent, *Schistosoma*, Sheep, Slime Mold, Swine, and Yeast genomes. A closely related page in the virtual library (http://ceolas.org/VL/mo/) covers **model organisms**, so you should see which suits you best.

If your interest in cell culture is based on transgenic organisms, then there are other databases you should know about. The **transgenic animal web** (http://www.med.umich.edu/tamc/), hosted by the University of Michigan, is an ideal starting place. The **Transgenic/Targeted Mutation Database** (http://tbase.jax.org/), hosted by The Jackson Laboratory, Bar Harbor, Maine, provides an alternative view, emphasizing reverse genetic analysis of gene function. The **Medical Research Council's Mammalian Genetics Unit** at Harwell, UK, has more than 200 mouse strains available for distribution. These include both genetically engineered mice and spontaneous mutants. A searchable database is available at http://www.mgu.har.mrc.ac.uk/stocklist/stocklink.html.

Post-genomics

As the genome projects run to completion, two major technologies will come to the fore: microarrays and proteomics. These are well-served with portal and link sites. An excellent starting point for microarrays is Leming Shi's Microarray page (http://www.gene-chips.com/). Another set of links can be found on http://linkage.rockefeller.edu/wli/microarray/, and more on protocols and useful information is at http://www.microarrays.org/. It is also nice to pay homage at Pat Brown's site (http://cmgm.stanford.edu/pbrown/); by publishing instructions on how to build your own array equipment (the 'M guide'), and by releasing free software, he has helped to drive down prices of software and equipment in the commercial sector. Proteomics is well-served by both equipment supply companies and the large post-genomic companies like Incyte (www.incyte.com). A useful portal can be found at http://www.highveld.com/genomics.html#others. A rare public domain offering in this area is the Danish Centre for Human Genome Research (http://www.biobase.dk/cgi-bin/celis), which features a database of 2D gels.

Given the multiplication in –omes, there's a whimsical table for links to all the known variants, including the metabolome, physiome, and biome, at http://arep.med.harvard.edu/.

Bioinformatics

Bioinformatics is a horribly vague term that implies merely a science of disseminating scientific information. To some extent, everything here can be considered to be bioinformatics; however, I'll dedicate this section to the more popular concept of organizing and annotating sequence data with the corresponding publications and functional data. One site is pre-eminent in this area: the **National Center for Biotechnology Information** (NCBI), at http://www.ncbi.nlm.nih.gov/. Although the centre of the Genbank database, it has added value by annotating it and linking it to other resources. It is chastening for the newcomer to discover this site; it is clear that some bright minds have travelled along the same mental paths and produced resources that anticipate most likely needs. It is even more remarkable that the NIH has provided it free. In doing so, it has undoubtedly fulfilled its remit to accelerate biomedical research, and in spades.

Why the hyperbola? It is because one of the most common threads in bioinformatics (from the researcher's viewpoint) is to go from a new DNA sequence to a putative function; or conversely, to go from an interesting paper to work up a new project all the way to PCR primers, perhaps to seek the gene in a different species. Whichever line of approach, there are tools available. The core set of the NCBI resources is made of:

(a) **Pubmed** – an extensive superset of the Medline (MEDlars onLINE) database.

(b) **Entrez** – a database that explicitly matches DNA and protein sequences to the papers that describe them.

(c) **BLAST** (Basic Local Alignment Sequence Tool) – given a chunk of DNA or peptide sequence, a quick and dirty way of looking for similarities in both the DNA and protein databanks.

(d) **OMIM** (Online Mendelian Inheritance in Man) – curated reviews on all genetic disease loci in humans, and of genes that are candidates for genetic diseases, with links across to Entrez and Pubmed.

These tools are augmented by other, more specialist resources. For example, the **Unigene** set is a curation of all known human ESTs, sorted into unique sets that correspond, if not to all genes, certainly to all abundant human transcripts. This is a slightly more frustrating resource to use, because most ESTs are not in the public domain, but are being offered to drug companies for huge fees, in exchange for any intellectual property rights. So it can actually be quite hard to identify a clone of interest and order it. However, within a couple of years, the position should have regularized itself.

How do these utilities fit together? For example, imagine that I have just cloned a novel cDNA that, by virtue of its expression pattern, is likely to be of interest to me. I use my first sequencing results for a BLASTN (nucleotide to nucleotide) similarity search. This will tell me in under a minute whether the gene is novel, or if it has been sequenced before. If the latter, I can cover my disappointment and link directly to the citations that describe it (in Entrez), and then to related citations in Pubmed, to see where my research fits. If the gene is genuinely novel, I then perform a BLASTX search, which translates my sequence in all six possible reading frames and aligns the translations to the protein database. This is a much more sensitive search, that will show up any conserved coding domains anywhere in the sequence. Based on what I find, I can then branch out into Entrez and Pubmed as before. If there is a human homologue described, there's a good chance that there will be an OMIM article on it, perhaps linking it to a known disease. And, given that the most interesting papers always seem to be at the binders when I go to the library, it is a relief that most journals now have institutional arrangements that let me print out the full paper (as a pdf file) at my desktop. (Eerily, Pubmed seems able to relay you directly to the relevant publisher's website, which will automatically offer you a pdf file if your institution has an e-journal subscription.) So, if there is any informative sequence similarity for my novel gene, I can hope to have identified it and read into the research field in one afternoon, without leaving my desk. In this way, an awful lot of thrashing around with different databases and packages has been elegantly circumvented.

Highwire Press, the Stanford University publishing arm responsible for many online journals, claims to have the second largest collection of free articles in the world at http://highwire.stanford.edu/lists.freeart.dtl. These are largely trial offers, or reflect a new publishing ethic that articles beyond a certain age (say one year) can safely be put in the public domain without damaging paid access to the latest articles. In doing so, of course, the journals increase their exposure and thus inevitably their impact factors. This can only feed back positively on their subscription revenue.

Specialized informatic resources

The Internet is famously a place where everyone who has ever curated anything has mounted their life's work for all to see. This can often be surprisingly useful, and there are world-class resources curated by anything from a large group to a single individual. An example is the Interactive Fly (http://flybase.bio.indiana.edu/allied-data/lk/interactive-fly/aimain/1aahome.htm), a valuable resource for *Drosophila* developmental biologists, originally written by Thomas Brody as a one-man show, but so useful that it has been rolled into the official Flybase website. For cell signalling, the Protein Kinase Resource page (http://www.sdsc.edu/kinases) is an excellent all-round source, but particularly strong on the classification and structural biology of the protein kinase families. Another excellent resource for cell signalling (albeit protected by 'free' registration) is the signal transduction knowledge environment (http://www.stke.org), run by *Science* magazine. Cancernet (http://cancernet.nci.nih.gov) has the weight of NIH behind it, and provides a valuable resource for clinicians, researchers, patients, and carers alike.

Online analysis tools

Molecular biology software is excellent but prohibitively expensive. Many labs try to economize on their workstation software by using online tools. In many cases, for example gene prediction and promoter analysis, the online resources are actually better than you can buy, and are those used by genome projects themselves.

There are two indices to start: the excellent **Baylor College of Medicine search launcher** (http://www.hgsc.bcm.tmc.edu/SearchLauncher/); and the older, but still useful, **Pedro's biomolecular search tools** (http://www.public.iastate.edu/~pedro/research_tools.html). Pedro's tools haven't been updated since 1996, so you can expect a few broken links, but it is a very comprehensive listing, split into three sections: Molecular Biology Search and Analysis; Bibliographic, Text, and WWW Searches; and Guides, Tutorials, and Help Tools. Here, drawn from both indices, are the ones that I find most useful, grouped by function:

Protein–protein: The simple BLASTP search from NCBI (http://www.ncbi.nlm.nih.gov/blast/blast.cgi) is a good first resort, but in addition, BCM offers automatic BEAUTY post-processing of gapped BLASTP searches (http://dot.imgen.bcm.tmc.edu:9331/seq-search/protein-search/html). This means that the alignments are annotated with known protein motifs and domains, providing valuable pointers to possible function in a novel peptide sequence. The FASTA algorithm (also available at the BCM site above) is much more sensitive (and thus slower). From the search launcher, it's easy to try both.

DNA–protein: A similar argument exists. The simple BLASTX search from NCBI (http://www.ncbi.nlm.nih.gov/blast/blast.cgi) is fast and good, but BCM also offer BEAUTY post-processing that can help to make sense of the results (http://dot.imgen.bcm.tmc.edu:9331/seq-search/nucleic acid-search.html).

DNA–DNA: For this, I've always found the BLASTN search from NCBI (http://www.ncbi.nlm.nih.gov/blast/blast/cgi) to be sufficient.

Multiple sequence alignment: You can download ClustalX to your own machine at http://www-igbmc.u-strasbg.fr/BioInfo/ClustalX/Top.html, or run the older ClustalW online at BCM (http://dot.imgen.bcm.tmc.edu:9331/multi-align/multi-align.html) or at EBI (http://www2.ebi.ac.uk/clustalw/). If needed, you can colour your multiple alignments with Boxshade (http://www.ch.embnet.org/software/BOX form.html). If you're more interested about the phylogenetic implications of your alignment, you should try downloading Phylip – PHYLogeny Inference Package – at http://evolution.genetics.washington.edu/phylip.html.

Protein analysis: To colour amino acids by hydrophobicity, etc. try the protein colourer (http://www2.ebi.ac.uk/cgi-bin/translate/visprot.pl). To identify protein sorting signal sequences, use PSORT (http://psort.nibb.ac.jp:8800/). To identify known protein motifs in your sequence, use Scan Prosite (http://www.expasy.ch/

tools/scnpsite.html). There are several tools that perform a battery of analyses of single protein sequences. Try Predict Protein at http://dodo.cpmc.columbia.edu/predictprotein/.

To seek putative genes in a stretch of genomic sequence: Try BCM's gene finder (http://dot.imgen.bcm.tmc.edu:9331/gene-finder/gf.html), or Genie http://www.fruitfly.org/seq tools/genie.html.

To identify binding sites of known transcription factors: Use TRANSFAC (http://www.biobase.de/).

To design PCR primers for a particular DNA sequence: Try xprimer at the Virtual Genome Center (http://alces.med.umn.edu/VGC.html).

To calculate the Tm of an oligo: http://alces.med.umn.edu/rawtm.html

Educational resources

In this area, I include material useful in teaching both at undergraduate and postgraduate level. This section thus includes both eye-catching visual material and primers for workers new to the field. In the former category, **Cells alive!** (http://www.cellsalive.com/) is a fun place to spend an afternoon, with a range of still and moving pictures that act as promotion material for the host (Quill Graphics), and their commercial CD. Where else could you find a cancer cell webcam? You can use up to three images from the Cells alive! site on your own academic website, provided you seek permission.

For stunning colourized scanning electron micrographs, try **Dennis Kunkel's microscopy** (http://www.pbrc.hawaii.edu/~kunkel/). This even includes a virtual scanning E.M., written in Javascript.

The **Access Excellence** program curates a useful collection of graphics that not only includes cell biology, but the whole molecular biology/biotech area (http://www.accessexcellence.org/AB/GG/).

Microbiology Video Library (http://www-micro.msb.le.ac.uk/Video/Video.html) has a collection of videos, particularly of microbes and viruses.

There are a few useful meta-indices for educational resources in cell biology that can be recommended:

Cell and Molecular Biology Online (http://www.cellbio.com/education.html). This is a useful general meta-index.

Ken's Bio-web references (http://www.hotlink.com/~house/Cell%20Structure.html). This is particularly good on classification of cell types and a source of useful images for teaching.

The **Dictionary of Cell Biology** (http://www.mblab.gla.ac.uk/dictionary/) is one of the very first online dictionaries. This author was attracted to the idea, because of the possibility of actually seeing how the book was used, and using feedback from unsuccessful searches to improve it. There are around 7000 entries, searchable by keyword, and no graphics. The site has accumulated well in excess of a half a million hits in four years. At present, the Dictionary is resting, while access issues for the Third Edition are negotiated with the publishers.

At a simple level, there is a nice, Alberts-style cutaway diagram of an **animal cell**, together with hotlinks to descriptions of the major structural components of the cell, at http://www.ultranet.com/~jkimball/BiologyPages/A/AnimalCells.html. A **virtual plant cell** http://ampere.scale.uiuc.edu/~m-lexa/cell/cell.html provides a similar model for a plant cell.

Virtual Reality Modelling Language (VRML) is a hypertext protocol that allows three-dimensional web browsing for browsers with appropriate plug-ins. This technology has been applied to the **VRML tour of a cell** http://library.advanced.org/11771/english/hi/biology/vrml.shtml. This is a lovely idea, and in principle permits a virtual 'fly-through' of the inside of a eukaryotic cell. Unfortunately, as every surface must be mathematically specified in VRML, the resolution is pretty crude, and I struggled to recognize all the organelles. Another resource is the VRML Biology Page, on http://verbena.fe.uni-lj.si/~tomaz/VRML/. Although these are pilot projects, with time, VRML is likely to mature into an exciting teaching and research methodology.

Learned societies

Once you have your data, you need someone to tell about it. There are several learned societies in the area. Here is a list:

American Society for Biochemistry and Molecular Biology ASBMB)	http://www.faseb.org/asbmb/
American Society for Cell Biology (ASCB)	http://www.ascb.org/ascb/
American Society of Immunologists	http://12.17.12.70/aai/default.asp
Biochemical Society	http://www.biochemistry.org
British Society for Cell Biology (BSCB)	http://www.kcl.ac.uk/kis/schools/life_sciences/biomed/bscb/top.html
European Society for Animal Cell Technology	http://www.esact.org/
European Tissue Culture Society	http://www.uni-stuttgart.de/etcs
Federation of American Societies for Experimental Biology (FASEB)	http://www.faseb.org/
Federation of European Biochemical Societies	http://www.febs.unibe.ch/
International Union of Biochemistry and Molecular Biology	http://iubmb.unibe.ch/
Microscopy Society of America (MSA)	http://www.msa.microscopy.com/
Society for Developmental Biology	http://sdb.bio.purdue.edu/
The Royal Microscopical Society	http://www.rms.org.uk/
The Society for *in-vitro* biology	http://www.sivb.org/

In addition, there are good indices from the WWW virtual library of Cell Biology, at: http://vl.bwh.harvard.edu/organizations_meetings.shtml and the University of Waterloo at http://library.uwaterloo.ca/society/biol_soc.html.

Stock centres for cell lines, hybridomas, etc.

The major stock centre for cell lines is the **American Type Culture Collection** (ATCC) at http://www.atcc.org. It holds over 4000 cell lines from over 150 different species; over 950 cancer cell lines (including 700 human cancer cell lines); and over 1200 hybridomas for production of monoclonal antibodies. There is an analogous European Collection of Cell Cultures (**ECACC**) at Porton Down (http://www.ecacc.org.uk) that includes:

- General Cell Collection (1000+ lines)
- Hybridoma Collection (400+ lines)
- Human Genetic Disorder Collection (25 000+ lines)
- HLA Defined Human B lymphoblastoid Collection (350 lines)

There is a German centre: **DSMZ**—Deutsche Sammlung von Mikroorganismen und Zellkulturen GmbH (German Collection of Microorganisms and Cell Cultures) at http://www.dsmz.de/. Its coverage is phylogenetically broader than some, with 8700 bacteria and archaea, 100 bacteriophages, 2300 filamentous fungi, 500 yeasts, 1200 plant cell cultures, 700 plant viruses, and 420 human and animal cell lines.

The analogous Japanese Centre (JCRB: Japanese Collection of Research Bioresources) at http://cellbank.nihs.go.jp/, has a bilingual site, and 550 cell lines.

For human lines, the National Institute of Health (NIH) sponsored the **Coriell Cell Repositories** (CCR) (http://locus.umdnj.edu/ccr/). This includes:

- NIGMS Human Genetic Cell Repository
- NIA Ageing Cell Repository
- ADA Cell Repository Maturity Onset Diabetes Collection
- HBDI Cell Repository Juvenile Diabetes Collection

There is a searchable database, and it is possible to order online by credit card.

The NIGMS Human Genetic Cell Repository contains thousands of samples from humans with genetic diseases, and the database is searchable by diagnosis or gene, and cross-linked to OMIM entries. It couldn't get much easier! The NIA Ageing Cell Repository has human cell cultures from individuals with ageing-related conditions (e.g. progeria, Werner's syndrome, Cockayne syndrome, Rothmund-Thomson syndrome, and Down's syndrome) and cell cultures from familial Alzheimer's disease pedigrees.

Since 1960, yeast mutant strains have been curated, and these have recently been hugely augmented. The ***Saccharomyces* deletion project** (http://www-sequence.stanford.edu/group/yeast_deletion_project/deletions3.html) has generated deletion mutants of every identified reading frame, and these are being made available through **ATCC** (http://phage.atcc.org/searchengine/ygsc.html) or **Research Genetics** (http://www.resgen.com/products/YEASTD.php3).

The European Union has funded CABRI: Common Access to Biological Resources and Information (http://www.cabri.org). To some extent, this is a meta-index, because its 95 000 lines includes the ECACC and DSMZ collections, together with BCCM, CABI, CBS, and ICLC. The collections cover:

- animal and human cell lines
- bacteria and archea

- fungi and yeasts
- plasmids
- phages
- DNA probes
- plant cells and viruses

Antibodies

It's nice to find you don't need to make an antibody, because it's already available. Surprisingly, there are several sites that actively curate listings of antibodies from publications and supply company brochures, and these can be a huge timesaver. There is a useful resource page for antibody work at http://www.antibodyresource.com/. From here, you can find lists of antibody suppliers, custom antibody services, and searchable databases of known antibodies.

Perkin Elmer has a guide to 'antibodies; from design to assay' http://www.pebio.com/pa/340913/340913.html, emphasizing antibodies against synthetic peptides.

The **Developmental Studies Hybridoma Bank** at the University of Iowa (http://www.uiowa.edu/~dshbwww/) provides 300 monoclonal antibodies against some of the most famous developmental and cytoskeletal gene products, like cut, even-skipped, laminin, spectrin, wingless, and engrailed. Better yet, prices range from $10/ml crude supernatant to $200/vial for growing cells.

Linscott's directory contains over 67 000 entries, with heavy emphasis on antibodies, and is available both on paper and online at http://www.linscottsdirectory.com/, although registration is required.

Supply companies

Anderson's Timesaving Comparative Guides (ATCG - gettit?) provide a good meta-index for the area (http://www.atcg.com/atcg/atcghome.htm). Although access is free, it is necessary to register first, an unpleasant superfluity that is becoming more common. The reason, I suspect, is that Anderson are actually trying to set up as a supply company, and by imposing most of the formalities on all visitors, it makes the extra steps needed to register to order directly online that bit less onerous. Similar portal sites (with similar irksome formalities) are run by BioMedNet (http://www.bmn.com) and Bioresearchonline (http://www.bioresearchonline.com).

Individual companies are listed below. I have tried to limit this list to companies relevant to cell culture, with usable web sites, that I particularly find useful. Although most companies' URLs are guessable (www.companyname.com), a few cause surprises. For example, www.dynal.com takes you to a military equipment supplier. Similarly, many companies have regional pages, perhaps to protect readers from the discovery that pricing is also regional; so I've pointed you to the most international URLs available for each company where possible.

Company	Site	Main area
ABI	http://www.abionline.com/	Virology
Ambion	http://www.ambion.com/	Molecular biology, esp. RNA
Amersham Pharmacia	http://www.apbiotech.com/	Molecular and cell biology, radiochemicals
Beckman Coulter	http://www.beckmancoulter.com/	Flow cytometry, equipment
Becton Dickinson Biosciences	http://www.bdbiosciences.com/	Cell sorting, culture plasticware
Bio-Rad	http://www.bio-rad.com/	Electrophoresis, equipment
Calbiochem/Novabiochem	http://www.calbiochem.com/	Biochemicals
Carolina Biological	http://www.carolina.com/	Useful lab-related items for high-school and above
Chemicon	http://www.chemicon.com/	Antibodies and kits
Clontech	http://www.clontech.com/	Molecular biologicals, GFP etc.
Corning	http://www.scienceproducts.corning.com/	Tissue culture plastics, cell biology, biochemistry
Dako	http://www.dakousa.com/	Microscopy, immunocytochemistry, Dako wax pens
Dupont /NEN	http://www.nen.com/	Biochemicals, radiochemicals, equipment, sequencing, microarrays
Dynal	http://www.dynal.no/	Everything to do with cell/molecule separation by magnetic beads
Eppendorf 5 Prime	http://www.5Prime.com/	Molecular biologicals
Eppendorf Scientific	http://eppendorfsi.com/	Molecular and cell biologicals
Fisher	http://www.fishersci.com/	Everything-for-the-lab company
FMC Bioproducts	http://www.bioproducts.com/	Gel electrophoresis
Fuji	http://www.fujimed.com/	X-ray film, imaging
Gelman	http://www.pall.com/gelman/lifesci/	Filtration
Genosys	http://www.genosys.com/	Custom oligos, peptides, antisera
Hamilton	http://www.hamiltoncomp.com/	Liquid handling, esp. Hamilton syringes
HyClone	http://www.hyclone.com/	Culture media, sera
ICN Biomedicals	http://www.icnbiomed.com/	Cell culture, labware, biochemicals
Invitrogen	http://www.invitrogen.com/	Cloning, expression vectors and molecular biology
J.T. Baker	http://www.jtbaker.com/	Lab chemicals
Jencons	http://www.jencons.co.uk/	Labware and consumables
Kodak	http://www.kodak.com/	Medical imaging, digital photography
Labsystems	http://www.labsystems.fi/	Liquid handling (e.g. Finnpipettes)
LC Laboratories	http://www.lclabs.com/	Pharmacologicals for signal transduction
Leica	http://www.leics-microsystems.com/	Microscopy
Life Technologies (and Gibco BRL)	http://www.lifetech.com/	Cell and molecular biologicals
Millipore	http://www.millipore.com/	Filtration, membranes, labware
Molecular Probes	http://www.probes.com/	Fluorescent probes for cells, organelles, etc.
Nalge Nunc	http://www.nalgenunc.com/	Nalgene and Nunc cell culture plasticware
New England Biolabs	http://www.neb.com/	Molecular biology, esp. excellent restriction enzyme data resource
Nikon	http://www.nikon.co.jp/inst/Biomedical/index.html	Microscopes, photography
Ohaus	http://www.ohaus.com/	Lab balances
Olympus	http://www.olympus.com/	Microscopy
Omega Optical	http://www.omegafilters.com/	Leading supplier of specialized filters for fluorescence work
Oxford Glycosciences	http://www.ogs.com/	Proteomics
Peninsula Laboratories	http://www.penlabs.com/	Peptides, antibodies, immunoassay
Perkin Elmer Biosystems	http://www.pebio.com/	PCR, genomics, proteomics
Pierce	http://www.piercenet.com/	Protein biochemistry and modification, Western blots, immunochemicals

Company	Site	Main area
Polaroid Scientific	http://www.polaroid.com/work/scientific/	Photography and imaging
Promega	http://www.promega.com/	Molecular biologicals, enzymatics, cloning, PCR, neurochemicals
Qiagen	http://www.quiagen.com/	Molecular biologicals, esp. plasmid kits, transfection
R & D Systems	http://www.rndsystems.com/	Antibodies, esp. cytokines, cell separation
Rainin	http://www.rainin.com/	Pipettes and liquid handling
Research Biochemicals International (RBI)	http://www.callrbi.com/	Major supplier of pharmacologicals
Research Genetics	http://www.resgen.com/	Major supplier of genomics resources
Roche (and Boehringer-Mannheim) Biochemicals	http://biochem.roche.com/	Molecular biologicals/general biochemistry
Sartorius	http://www.sartorius,com/	Lab balances and filtration
Schleicher & Schuell	http://www.s-und-s.de/	Membranes for blotting and filtration
Sigma Aldrich	http://www.sigma-aldrich.com/	Leading supplier of lab, chemicals
Stratagene	http://www.stratagene.com/	Molecular equipment and consumables
Techne	http://www.techneuk. co.uk/	Cell and molecular biology equipment
US Biological	http://www.usbio.net/	Chemicals and biochemicals, molecular biology and immunochemicals
Vector Laboratories	http://www.vectorlabs.com/	Immunochemicals, molecular
VWR Scientific	http://www.vwrsp.com/	Everything-for-the-lab supplier
Whatman	http://www.whatman.com/	Separations and filtration
World Precision Instruments	http://www.wpiinc.com/	Microdissection, cell and tissue culture, neuroscience
Zeiss	http://www.zeiss.com/micro	Microscopy

Several of these companies now allow credit card, or other facilitated online ordering. This means that – if your institution allows it – you can place orders directly, get them faster, and have no-one but yourself to blame for any mistakes!

Patents

It is common for academic biologists to consider patenting an invention at some stage in their career. Most universities and institutions have some form of Enterprise or Industrial Office, and once your idea develops far enough, these should be your first port of call. The most important concepts in patenting are utility and novelty: it must be useful enough for your institution to be prepared to spend £2000–20 000 ($3000–30 000) protecting the idea with a patent, or to make it possible to sell on to another company to complete the process. For novelty, it is necessary to make a claim for some new process or invention that has not been described before. In this context, an exhaustive search of patent databases is a *sine qua non*, and it will be undertaken (for a large fee) by the patent agent appointed by your institution. However, forewarned is forearmed, and so you may well want to conduct a less exhaustive search yourself at the outset. There are several databases available online, and although a few of them will charge you for their services, you can usually make a pretty good search and view the results online for free. This latter step lets you study the structure of patents in your area in some detail. This can be very useful when sketching out a draft of your idea, before your patent agent's meter starts running.

The IBM Intellectual Property Network (http://www.delphion.com/ or http://www.patents.ibm.com/) is foremost in the field of online patent databases. It covers:

- United States patents
- European patents and patent applications
- PCT application data from the World Intellectual Property Office
- Patent Abstracts of Japan
- IBM Technical Disclosure Bulletins

These searches operate predominantly on keywords. If you are confident that you can identify the areas in which your proposed invention sits, then you can go to the **Index to U.S. Patent Classification** (http://metalab.unc.edu/patents/index.html), maintained by the University of North Carolina at Chapel Hill. So (for example) you can go directly from 'alanine' to the 285 associated patents. Be warned though, that they're pretty hardcore organic chemistry.

The **US Patent and Trademark Office** (PTO) (http://www.uspto.gov/) provides a more detailed interface to the US databases. It is interesting to view the full text of patents in your area, to see what people consider worth patenting.

The same site has a **guide for the independent inventor** (http://www.uspto.gov/web/offices/com/iip/index.htm) which answers most of the questions that might occur in the run-up to making a patent application. Although this is understandably geared to the American market, the fact remains that for inventors from any country, protection of the invention in the USA is by far and away the most important and profitable.

UK and European patents are searchable through Esp@ce http://gb.espacenet.com/. This returns 3852 hits for the search term 'cell culture'.

The **DNA Patent Database** (http://www.genomic.org/), provided by the Georgetown University's Kennedy Institute of Ethics and the Foundation for Genetic Medicine, allows free public access to the full text and analysis of all DNA patents issued by the United States Patent and Trademark Office (PTO). As it is based on patents that include the term 'DNA' or which fit into a relevant classification, this database forms a more focused master list from which to search in the biotechnology area.

Interestingly, an unpublished academic paper (if suitably written) is admissible as a patent application in the UK (http://www.patent.gov.uk/patent/indetail/academic.htm), and there is no charge for the initial filing.

Future directions

It's an exciting and worrying time to be in science. Exciting, because anyone who has been in science for at least a decade is aware that the Internet has made their job at least an order of magnitude more powerful. By driving your computer insightfully, you can read up around a new area, find new genes, design and order the relevant PCR primers, order cDNA and genomic clones, and identify and order mutant stocks, all in one afternoon and without leaving your chair. The basic work-up around a new project is then only a matter of a couple of days, rather than weeks. This is worrying, because the expectation is correspondingly higher. There is no doubt that there is a new selection pressure acting on life scientists, and only those who are utterly at home with Internet resources will remain in science for more than a few years. I hope this article will have improved your fitness slightly!

Glossary

Bionet Set of Usenet newsgroups that refer to biology. See http://www.bio.net.

BLAST Basic Local Alignment Search Tool. Method of comparing a sequence with one or more other sequences. Relies on previously assembled chunks of sequence in the database, so relatively fast, but not as sensitive as FASTA.

cgi Common Gateway Interface. A URL that ends .cgi implies that the data is being generated dynamically by a computer program on the server, rather than simply a static .html page. Most responses from Genome Project servers will be from cgis.

DNS Domain Name System. When you ask for a web service, the URL contains the IP address, usually in text form (e.g. www.mrc.ac.uk). Your browser sends a request to your local DNS server to translate this into dotted decimal form (e.g. 123.456.789.123) before going to look for the actual page. Reverse DNS lookup allows the opposite to take place, if you're curious as to a dotted decimal address. It follows that your computer must be set up correctly to know where its local DNS server is.

FAQ Frequently asked question(s). Bionet newsgroups, and many Internet sites, publish lists of FAQs, with answers, to try to head off multiple requests for the same information.

FASTA Method of comparing a sequence with one or more other sequences. More sensitive, but also much slower, than BLAST searches.

Firewall On all corporate and most university networks, data packets are routed through a firewall to protect the local network. These can be more or less invasive, but typically scan for obscenities, viruses, and attempts to run unauthorized services on local machines. On some networks, strict firewalls can interfere with normal operation: for example, they may prevent Java applets from loading.

ftp File Transfer Protocol. The most economical protocol for shifting files around on the internet, e.g. for downloading software packages. Although there are specialized ftp client programs, most web browsers can interact with ftp servers.

Gopher Obsolescent Internet protocol. Now largely superceded by http, but most web browsers can interact with any gopher site you might come across.

HTML Hypertext Markup Language. A system of encoding formatting (bold, italic, headings, etc.) in an http data stream.

http Hyper Text Transfer Protocol. The dominant means of sending information around the Internet. If you see it on a web browser, you probably got it via http.

Java A computer language widely touted as the successor to individual operating systems, because of its nominal platform independence. Web pages, e.g. from genome project resources, will sometimes load and run Java applets on your machine. If this starts to happen, you'll have plenty of time for coffee before you see anything.

Javascript A scripting language for Netscape browsers that bears a superficial resemblance to a subset of Java.

Meta-index Strictly, an index of indices, but frequently used for any page that has a collection of high level links. Meta-indices are useful starting points to search for resources in a particular area.

Moderated newsgroup A Newsgroup where postings are diverted to a human moderator, who decides whether they should go online. An effective way of reducing spam.

Newsgroup An Internet bulletin board dedicated to a particular subject, where email-like messages can be posted, read, and searched.

nntp Network News Transfer Protocol. The protocol used for Usenet newsgroups.

pdf Portable Document Format. A proprietary format from Adobe that is the *de facto* standard for downloading pre-formatted documents (like academic papers or manufacturers' data sheets). Your browser needs a free copy of Adobe Acrobat Reader (downloadable from http://www.adobe.com) to be able to handle this format.

Perl Practical extraction and report language. The language in which most cgis are written, although not confined to Internet applications. Very useful as a scientific tool! (http://www.perl.org).

Spam Junk messages (get rich schemes, advertising, etc.) that clog up the system.

URL Uniform Resource Locator. The address of a web page, usually in the format http://sitename/pagename.html

Usenet The whole Internet Newsgroup system, of several thousand groups.

XML Extensible Markup Language. A successor and superset of HTML, that permits more advanced, stylesheet-based, formatting of complex documents. Not yet in general use. For more details, see http://www.w3.org/XML/.

Acknowledgments

I am most grateful to the Wellcome Trust and the BBSRC for funding my research, so allowing this section to be written.

References

1. Dow, J. A. T. (1997). In *Cell biology: a laboratory handbook*, Vol. 3 (ed. J. E. Celis), p. 518. Academic Press, San Diego.

Index

Note: references to figures are indicated by 'f' and references to tables are indicated by 't' when they fall on a page not covered by the text reference.